DIAMOND FILMS

CHEMICAL VAPOR DEPOSITION FOR ORIENTED AND HETEROEPITAXIAL GROWTH

DIAMOND FILMS

CHEMICAL VAPOR DEPOSITION FOR ORIENTED AND HETEROEPITAXIAL GROWTH

Koji Kobashi

2005

ELSEVIER

AMSTERDAM • BOSTON • HEIDELBERG • LONDON • NEW YORK
OXFORD • PARIS • SAN DIEGO • SINGAPORE • SYDNEY • TOKYO

ELSEVIER B.V.
Radarweg 29
P.O. Box 211, 1000
AE Amsterdam
The Netherlands

ELSEVIER Inc.
525 B Street, Suite 1900
San Diego
CA 92101-4495
USA

ELSEVIER Ltd
The Boulevard
Langford
Kidlington
Oxford OX5 1GB, UK

ELSEVIER Ltd
84 Theobalds Road
London WC1X 8RR
UK

First edition 2005

ISBN: 0-08-044723-6

Printed in Great Britain.

Preface

The major purpose of this monograph is to review process technologies for oriented and heteroepitaxial growth of diamond by microwave plasma chemical vapor deposition (CVD). There are many CVD methods to synthesize diamond films using, for instance, hot filament, radio-frequency (rf) plasma, microwave plasma, DC plasma, arc-jet plasma, and combustion flame. Among these, this monograph mainly focuses on microwave plasma CVD. The microwave plasma is an electrode-free discharge so that the contamination of diamond films by electrode materials can be avoided. The microwave plasma is very stable over many days, and a subtle control of the plasma, and hence the properties of synthesized diamond films, is possible. This is particularly important to quantitatively investigate the diamond nucleation and the film morphology in oriented and heteroepitaxial growth. This monograph also includes research results that are related to oriented and heteroepitaxial growth.

Historically, a vapor growth of diamond was first demonstrated by W. G. Eversole, which was followed by an intensive research in B. V. Derjaguin's group in Russia and J. Angus' group in the USA. A revolutionary advancement in diamond CVD occurred in the early 1980s, when B. V. Spitsyn et al. demonstrated a growth of diamond particles by CVD (by the chemical transport method), and N. Setaka and co-workers disclosed three methods to synthesize diamond films: hot filament, rf plasma, and microwave plasma. This triggered intensive R&D in Japan, Europe, US and then all around the world. The progress of diamond film research has been very steady and consistent, and numerous findings have been made during the past two decades.

One of the most remarkable findings in the area of diamond CVD is the growth of azimuthally oriented diamond films with (100) faces aligned in the same direction at the film surfaces by B. Stoner and J. Glass in 1992. Such films are now called highly oriented diamond (HOD) films. A key to make HOD films is to use bias-enhanced nucleation (BEN) discovered by S. Yugo, and thus a number of works have been done on diamond nucleation using BEN. The standard substrates for HOD film synthesis are β-SiC(100) and Si(100). One of the best-coalesced films was successfully synthesized by H. Kawarada, P. Koidl and their co-workers, as well as X. Jiang and his co-workers. In the mid 1990s, it was discovered that (111) and (100)-oriented, spontaneously coalesced diamond films could be synthesized on Pt(111) and Ir(100) by Y. Shintani and A. Sawabe, respectively, showing the possibility of producing single crystal diamond films by CVD. Since the growth rate of diamond films has been remarkably increased in recent years from the past rate of ~0.2 μm/h to ~50 μm/h, and large CVD reactors of maximum 60–100 kW are commercialized, production of single crystal diamond plates is expected to begin in the not-too-distant future.

Under these circumstances, it would be of significance to review the articles on oriented and heteroepitaxial growth of diamond films by CVD, and particularly summarize the processing conditions for the readers to further develop and elaborate the science and technology of diamond CVD. It is expected that this monograph would be useful for such purposes.

In completing this monograph, I must acknowledge many people. First of all, I wish to thank late V. Chandrasekharan, D. Fabre, M. L. Klein, and R. D. Etters for my education in physics before I became an independent researcher. On diamond CVD, I am particularly indebted to N. Setaka, Y. Sato, and M. Kamo for kindly allowing me to learn diamond CVD in their group. I certainly would like to thank my colleagues in Kobe Steel, Ltd.: K. Nishimura, K. Miyata, T. Tachibana, Y. Yokota, K. Hayashi, and N. Kawakami. I wish to particularly thank T. Tachibana for supplying a number of papers related to heteroepitaxy. I would also like to thank my superiors in the company: T. Horiuchi, Y. Kawate, Y. Kawata, Y. Sugizaki, J. Miyazaki, H. Sato, and S. Suzuki for their consistent support of R&D of diamond films over two decades. A longtime friendship with J. Glass, R. Nemanich, and A. Gicquel was a strong encouragement for me to write this monograph. Finally, I would like to thank my wife Toshiko and my son Akira for creating a calm environment for me to work at home. This monograph would not have been published without the support of the people mentioned above.

Finally, I wish to acknowledge I. Craig, S. Foster, and N. Jones of Elsevier Ltd. for their very kind assistance in publishing this monograph.

Koji Kobashi
2005

Acknowledgment

In this monograph, a number of figures and tables are quoted from different sources, and the author would like to thank the following organizations and individuals for generously giving permission to use them in this monograph.

American Physical Society
Figures: 8.4, 8.5, 8.6, 8.7, 8.8, 10.7, 10.8, 10.9, 11.17, 11.18, 11.19, 11.21, 11.22, 11.24, 11.54, 12.1, 12.2, 12.3, 12.4, 12.5, 12.6, 12.7, 12.9, 12.10, 12.11

American Institute of Physics
Figures: 5.4, 5.5, 5.21, 5.22, 6.4, 7.7, 7.10, 9.1, 9.8, 9.9, 9.11, 9.12, 9.13, 9.16, 9.18, 9.21, 10.1, 10.23, 10.24, 10.25, 10.26, 11.1, 11.4, 11.14, 11.15, 11.16, 11.23, 11.34, 11.42, 11.44, 11.45, 11.46, 11.47, 11.48, 11.51, 11.52, 11.53, 11.55, 11.56, 13.4, 13.21, B.1
Tables: 11.2, 11.6, 11.11

ASM International
Figures: F.1, F.2, F.3, F.4, F.5

Electrochemical Society
Table: 11.7

Elsevier B. V.
Figures: 5.6, 5.7, 5.9, 5.10, 5.11, 5.12, 5.13, 5.14, 5.15, 5.20, 5.23, 6.1, 6.2, 6.3, 7.5, 7.6, 7.11, 7.12, 9.17, 10.2, 10.3, 10.4, 10.5, 10.6, 10.10, 10.18, 10.19, 10.20, 11.8, 11.9, 11.10, 11.11, 11.12, 11.20, 11.25, 11.26, 11.27, 11.40, 11.41, 11.57, 12.19, 12.20, 12.21, 12.22, 12.23, 12.24, 13.3, 13.5, 13.6, 13.7, 13.8, 13.12, 13.15, 13.17, 13.18, 13.19, 13.20, 13.23
Tables: 5.1, 11.3, 11.4, 11.5, 11.8, 11.9, 11.10, 12.2, 12.3, 12.4, 12.5, D.1, E.1, G.1

Japanese Journal of Applied Physics
Figures: 6.7, 6.8, 7.3, 7.4, 8.1, 8.2, 8.3, 8.9, 9.2, 9.3, 10.11, 11.2, 11.3, 11.5, 11.6, 11.31, 12.12, 12.13, 12.17, 12.18, 13.1, 13.2
Tables: 11.1, 12.1

Materials Research Society (permissions from the authors)
Figures: 5.16, 5.17, 5.18, 5.19, 6.5, 6.6, 6.10, 6.11, 6.12, 7.1, 7.2, 7.8, 7.9, 9.4, 9.10, 9.14, 9.22, 10.17, 10.21, 10.22, 11.7, 11.13, 11.28, 11.29, 11.30, 11.32, 11.33, 11.35, 11.36, 11.43, 12.14, 12.15, 12.16, A.1, A.2
Tables: 9.1, 10.1, 10.2

MYU, Inc.
Figures: 13.9, 13.10, 13.11, 13.13, 13.14

Nature
Figures: 9.19, 9.20

Taylor & Francis Group LLC
Table: 2.1

The Royal Society
Figures: 9.5, 9.6, 9.7

SPIE-The International Society for Optical Engineering
Figure: 5.8

Springer-Verlag GmbH
Figures: 11.37, 11.38, 11.39

Wiley InterScience
Figures: 10.12, 10.13, 10.14, 10.15, 10.16, 11.49, 11.50

Prof. A. Koma
Figure: 2.1

Dr. L. Plano
Figure: 9.15

Prof. A. Sawabe
Figure: 4.2

AIXTRON AG
Figure: 3.6

Kobe Steel, Ltd.
Figures: 1.1, 3.1, 5.1, 5.2, 6.9, 11.58, 12.8, 13.16, 13.22

Lambda Technologies, Inc.
Figure: 3.5

Seki Technotron, Corp.
Figures: 3.2, 3.3, 3.4

CONTENTS

Chapter 1

Overview of Oriented Growth

Chapter 1

Overview of Oriented Growth*

Science and technology of diamond film growth by chemical vapor deposition (CVD) have markedly advanced during the past decade. One of the most notable achievements is the growth of azimuthally oriented diamond particles on β-SiC by Stoner and Glass [1, 2] using Yugo's method of bias enhanced nucleation (BEN) [3], which then led to the growth of (100)-oriented diamond films on Si(100) that were later named as highly oriented diamond (HOD) films [4]. This technique was further elaborated by the groups of Glass, Koidl, Klages, and Kobe Steel [5] amongst others. Most recently, Kawarada et al. [6, 7] were successful in growing perfectly coalesced, (100)-oriented, 300-μm thick HOD films, where there was no grain boundary at the film surface. In the meanwhile, a new method of diamond heteroepitaxy was heuristically found by Shintani [8, 9], i.e. spontaneously coalesced, (111)-oriented diamond films can be grown on (111) surfaces of Pt that have been polished with diamond powder for diamond nucleation. This finding was followed by diamond heteroepitaxy on Ir(100) by Sawabe's group [10] using direct current (DC) plasma CVD, in which (100)-oriented, perfectly coalesced diamond films were grown. This work was reproduced by Morooka's group [11], and Schreck and Stritzker's group [12] using microwave plasma CVD (MPCVD), where the diamond nucleation was done by BEN.

Figures 1.1 (a)–(c) show a polycrystalline diamond film deposited on Si substrate, an HOD film grown on Si(111) surface, and a partially coalesced film on Pt(111) substrate, respectively, all synthesized by MPCVD. Diamond films grown on Ir(100) are so perfectly coalesced that there is no feature in the film surface image by scanning electron microscopy (SEM) (see Figure 12.18 [13]). At the present stage, the coalesced area is the largest for diamond films on Ir(100) [25 mm] followed by those on Pt(111) (~10 mm, see Figure 12.8 [14]), while the crystal facets are the smallest for polycrystalline diamond films (see Figure 1.1 (a)). In the latest technology, however, the edge lengths of crystal facets become extremely large even for polycrystalline diamond films. For instance, they are

*In this monograph, a number of notations, units, and abbreviations will be used, and they are summarized in Appendix A. It contains lists of notations for crystal orientations, process parameters for CVD, analytical techniques, CVD reactors, crystal growth, and carbon materials in addition to a description of standard diamond film characterizations, i.e. Raman spectroscopy and cathodoluminescence (CL). The readers are recommended to just quickly read through Appendix A at this point.

30–50 μm for few hundred-μm thick polycrystalline diamond films, and 50–100 μm for HOD films.

Judging from the recent progress of R&D on heteroepitaxial growth of diamond, it seems that the growth of single crystal diamond films over significantly large areas,

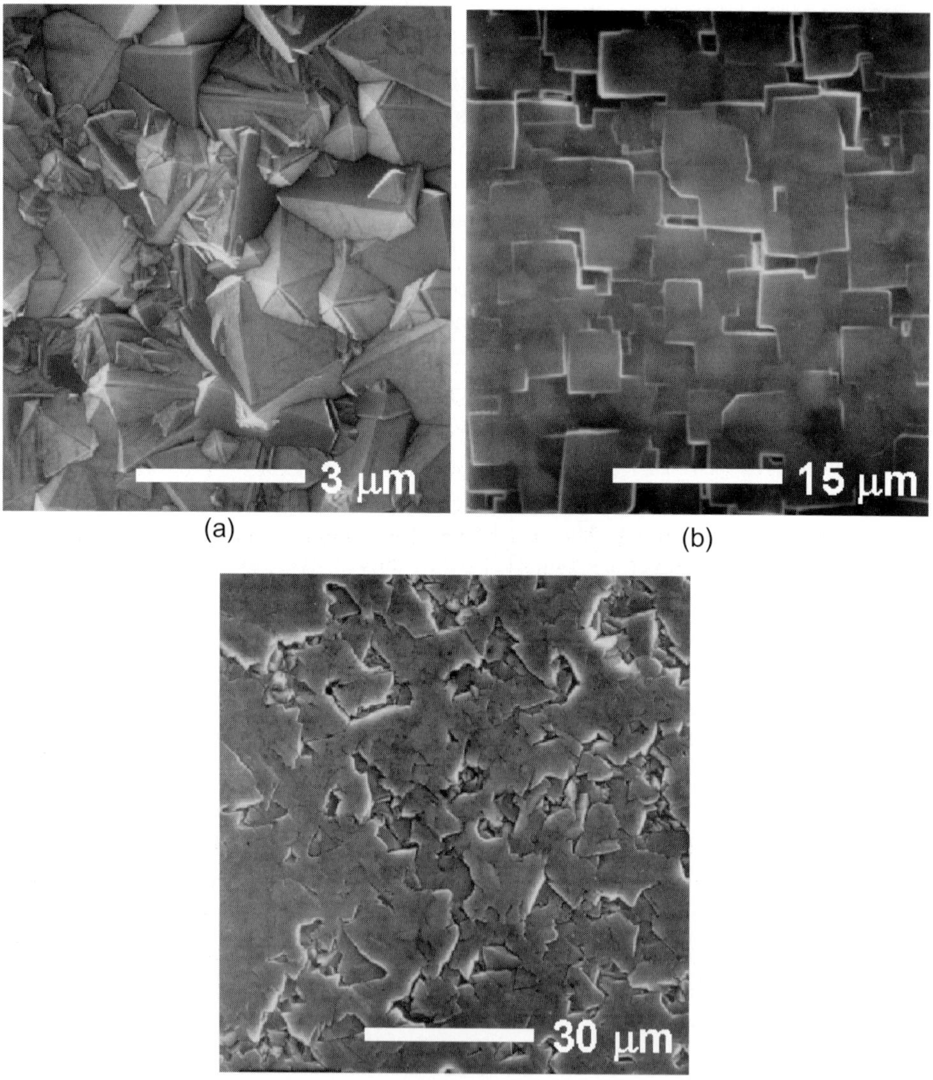

Figure 1.1. (a) Polycrystalline diamond film deposited on Si substrate, (b) HOD film grown on Si(100) substrate, and (c) partly coalesced diamond film on Pt(111) substrate.

such as one inch in diameter, will soon be achieved by CVD, and thence electronic applications of diamond films [15–17] will proceed much faster in the coming years than in the past. One of the purposes in the present monograph is to follow the trail of science and technology of oriented diamond film growth that has involved a number of researchers mostly in the past decade. This monograph was not able to cover all the papers related to oriented and heteroepitaxial growth of diamond films, and the readers are recommended to consult excellent reviews for heteroepitaxy [18–20], and nucleation and growth [21]. The readers can also consult Ref. [22] for defects, and Ref. [23] for a brief history of diamond research before 1988. Also, useful books on bulk diamond and diamond films are available [24–31].

Apart from MPCVD for diamond film synthesis, a number of works have been published on oriented growth using hot filament CVD, DC plasma CVD, and molecular beam epitaxy (MBE) [32], but most of them are left out because this monograph mainly concentrates on the MPCVD process technology. Among such papers that we missed, interesting and important papers are; homoepitaxial growth kinetics [33]; texture and morphology of diamond films [34]; HOD film growth by electron-assisted hot filament CVD (EACVD) [35]; diamond growth on highly oriented pyrolytic graphite [36]; TEM observation of the interface between the Si(100) substrate and the HOD layer [37]; heteroepitaxial nucleation of diamond particles on Si(100) using BEN [38]; heteroepitaxially (111)-oriented diamond films on Si(100) [39]; growth of both (100)- and (111)-oriented diamond films on Si(100) and Si(111) using BEN, respectively, where D[110] // Si[110] and D{100} // Si{100}, while D[$1\bar{1}0$] // Si[$1\bar{1}0$] and D{111} // Si{111} [40]. In the above expressions, D stands for diamond, and this abbreviation will be used throughout this monograph. Likewise, a notation G will also be used for graphite. Reference [40] stated that diamond could epitaxially grow directly on Si, and the existence of the native oxide on the Si wafer gave deleterious effects on hetero-epitaxy. An observation of interface structures by high-resolution transmission electron microscopy (HRTEM) [41,42] showed that there was a misorientation angle of 7.3° between D(100) and Si(100), but later a more detailed study showed that it was actually 9° [43]. Other papers left out in this monograph are; growth of (111)-oriented diamond films on 6H-SiC(0001), demonstrating that D[110] // 6H-SiC[$11\bar{1}0$] and D{111} // 6H-SiC{$000\bar{1}$} [44,45]; heteroepitaxial growth of diamond film on scratched Si(100), followed by β-SiC(111) formation at the Si surface, which demonstrated that D{111} // β-SiC{111} and D{771} // Si{111} [46]; effects of biasing time in the BEN step to synthesize HOD films [47]; study of BEN process by ellipsometry, showing that β-SiC is unnecessary for HOD film growth [48, 49]; bias current effects on nucleation density, showing that a nucleation density of 10^9/cm^2 at a nucleation rate of 10^6/cm$^2 \cdot$ s is most appropriate to grow HOD films [50]. In Ref. [51], different bias voltages were applied on two electrodes above the substrate to grow HOD films by HFCVD. Reference [52] showed that HOD films could be grown also by DC plasma CVD. Finally, it is of interest that a single crystal diamond film was grown on Si by thermal CVD [53, 54], but this work awaits a further development of technology.

Chapter 2

Diamond – Structure and CVD Growth

Chapter 2

Diamond – Structure and CVD Growth

2.1. STRUCTURE OF DIAMOND

Diamond has a well-known structure, as shown in Figure 2.1. It consists of two face-centered-cubic (fcc) lattices, one of which is displaced by (1/4 a, 1/4 a, 1/4 a), where a is the lattice constant of diamond, $a = 3.567$ Å (1 Å = 0.1 nm). The space group of diamond is $Fd3m$-O_h^7, and there are two atoms in a crystallographic unit cell. Each carbon atom has four sp^3 bonds with the neighboring four atoms where the bond length is 1.54 Å. Since there are two identical atoms per unit cell, there is no infrared (IR) absorption in the one-phonon region, and only a single band is observed in the Raman spectrum at 1333 cm^{-1}. The atomic density is 1.77×10^{23}/cm^3. The X-ray diffraction (XRD) intensities of diamond powder are listed in Table 2.1. The (111) diffraction is located at $2\theta = 41.9°$ and the strongest. From XRD patterns of diamond films, where the relative intensity of each line usually deviates from the data of Table 2.1, one can know the degree of uniaxial orientation for diamond grains in the film with respect to the direction normal to the substrate surface. For HOD films, for instance, the (400) diffraction line is dominant, indicating that the {100} crystal planes of diamond grains in the film are oriented normal to the film surface.

2.2. CVD GROWTH OF DIAMOND

In this section, basics of diamond growth process will be briefly described for the readers who are not very familiar with diamond CVD. One will find more detailed descriptions for each process factor later in this monograph. Note that the descriptions below are only for typical cases using MPCVD, and do not always apply to all cases.

Unlike other thin film deposition processes, conditions for diamond CVD have three unique features: (i) high substrate temperature T_s typically at 700–1200 °C, (ii) high gas pressure P at 20–150 Torr (1 Torr = 133.3 Pa), and (iii) low methane (CH_4) concentration of usually 1–5% with respect to the dilution gas, hydrogen (H_2). A standard temperature for diamond growth, monitored by an optical pyrometer without emissivity correction, is 800 °C. It is, however, considered that the surface temperature of the specimen exposed to the plasma is actually higher. Under these conditions, at least more than 95% of the deposited film can be crystalline diamond, if other process conditions and the system setup are appropriate. For a diamond film deposited on Si substrate, the film is polycrystalline just like the one shown in Figure 1.1 (a), which contains a high density of grain boundaries. The atomic structures along the grain boundaries are significantly disordered, hence they are the major origin of non-diamond carbons in the film. Since T_s for diamond CVD is so high that the substrate materials for diamond deposition must stand the temperature. Thus, the substrate materials that can be used are ceramics such as Si, SiC, Si_3N_4, cBN, and Al_2O_3, refractory metals such as W and Mo, metal carbides and nitrides, and other metals such as Cu, Ni, Pt, Ir, and Pd.

Regarding T_s, the deposited film includes more hydrogenated amorphous carbon (*a*-C:H) at lower temperatures. There is an indication that diamond still exists in the film deposited even at $T_s \sim 450\,^{\circ}C$, but the diamonds are micro- to nanometer-size,

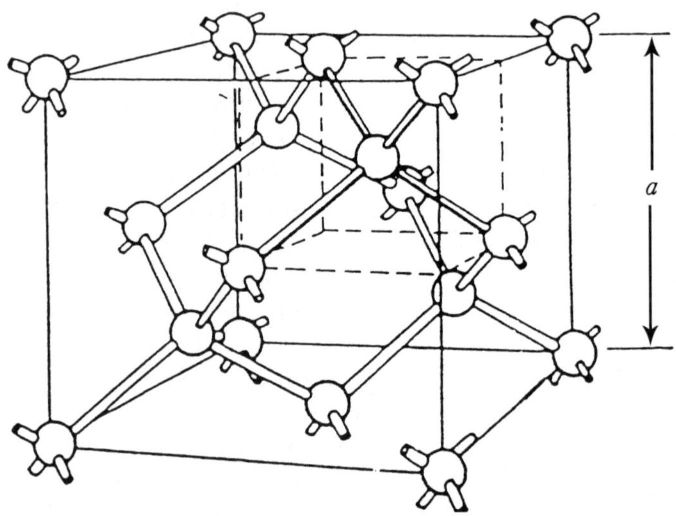

Figure 2.1. Diamond structure. The dotted cube indicates a crystallographic unit cell containing two carbon atoms. The lattice constant of diamond is $a = 3.567$ Å.

Table 2.1. Powder diffraction pattern of XRD of diamond [24].

hkl	2θ (°)	Relative intensity
111	41.9	100
220	75.3	2.5
311	91.3	16
400	119.5	8

X-ray source: Cu $K\alpha_1$ ($\lambda = 1.5405$ Å)

and its density is low. On the other hand, for T_s higher than 800 °C, there are more non-diamond components such as graphite and *a*-C in the film. The content of non-diamond components are usually studied by Raman spectroscopy (see Appendix A.5.).

The appropriate gas pressure P depends on the reactor used. For the NIRIM-type reactor (see Section 3.1), it is usually 30–50 Torr, while for 5-, 8-, and 100-kW reactors of Seki Technotron/ASTeX (see Section 3.2), it is around 100 Torr. These pressures are much higher than those used to deposit amorphous hydrogenated silicon (*a*-Si:H), in which case it is only a few milliTorr. At such high pressures used for diamond CVD, the plasma tends to be confined in a small region. For the case of the NIRIM-type reactor, for instance, the plasma has roughly a spherical shape with a diameter of 2–3 cm, when the input microwave power is $P_m = 300$–500 W, and hence the substrate size is usually only 1-cm square. Recently, high power reactors have been developed by Seki Technotron/ASTeX (Figure 3.4) and AIXTRON (Figure 3.6), and diamond film coatings on 6-inch wafers are now possible. The growth rate of diamond film depends on the gas pressure P and the input microwave power P_m as well as the source gas composition c. For the NIRIM-type reactor under conditions of $P = 30$ Torr and $P_m = 400$ W, it is 0.2 μm/h when 0.5%CH_4/H_2 is used as the source gas. On the other hand, a growth rate of approximately 10 μm/h can be achieved for the high power reactors.

As for the source gas for diamond CVD, CH_4 diluted to around 1% with H_2 is used to make well-crystallized diamond grains and films. This will be denoted as $c = 1\%CH_4/H_2$ hereafter. The possible range of c also depends on the reactor used and the film quality that one wishes to obtain. In this regard, it must be noted that Okushi's group used c of as low as 0.025%, and was successful in depositing an extremely high quality, atomically smooth diamond layer on a (100) surface of single crystal diamond [55–58]. One can use other hydrocarbons such as C_2H_2, alcohols, and acetone as the carbon source. Addition of small amount of oxygen in any form (e.g. O_2, CO, CO_2, alcohol, acetone) to the source gas tends to improve the quality of diamond, as oxygen reacts more readily with graphite and other forms of non-diamond components than with diamond, and remove them out of the diamond film. The shape of diamond crystals depends on the CVD conditions used, and is expressed by a parameter called "α-parameter," which will be described in Section 5.5. Normally, square (100) and triangular (111) faces of diamond appear on CVD diamond crystals and films, and it is very rare that (110) faces are seen. Define the growth rate of a diamond crystal face as its growth rate in the direction normal to the face, then it is a general rule that the fastest growth face does not appear in the crystals and film surfaces. This will be obviously understood if one imagines a crystal evolution in which a fastest growth face is present adjacent to a slower growth faces in the beginning.

Therefore, the fact that (110) faces are only rarely seen means that the growth rate of (110) face is significantly faster than those of (111) and (100).

Regarding the mechanism of diamond nucleation and growth, it is considered that CH_4 molecules are dissociated to be a variety of hydrocarbon ions and radicals in the plasma. They diffuse to be adsorbed on the substrate surface, forming carbon clusters with sp^2 and sp^3 bonds. Simultaneously, H_2 molecules also are dissociated to be molecular and atomic hydrogen ions and H radicals (i.e. atomic hydrogen) in the plasma. Carbon clusters consisting of mostly sp^3 bonds, i.e. diamond clusters, are more difficult to be etched by atomic hydrogen than clusters consisting of mostly sp^2 bonds, i.e. *a*-C and graphitic clusters. Consequently, only diamond clusters remain on the substrate to grow. Since oxygen also has an ability to preferentially etch sp^2 clusters, an addition of less than 1% of O_2 to the source gas has a similar effect as atomic hydrogen. The detailed mechanisms of diamond nucleation and growth processes are still inconclusive. There are two theories: in the first case, it is CH_3 radicals that play a major role in the diamond growth. This theory is based on the fact that radicals are very reactive. In the second case, it is C_2H_2 that contributes to the diamond growth. Indeed, theory shows that C_2H_2 is the dominant chemical species when CH_4 gas is heated up to more than 1000 °C.

Undoped diamond is a good insulator because of its large band gap (5.47 eV). To make a p-type semiconducting diamond film, usually diborane (B_2H_6) is added to the source gas. Boron atoms are substitutionally incorporated in diamond, and the B-dopants create acceptor levels at 0.37 eV above the valence band maximum. A presence of more than 10^{20} B-atoms/cm^3 makes the diamond metallic due to the Mott transition [59]. In terms of the crystal morphology, an addition of B increases the areas of (111) faces. On the other hand, to make an n-type semiconducting diamond film, PH_3 is usually added to the source gas. P-dopants create donor levels at 0.6 eV [60, 61] below the conduction band minimum, which is approximately 3/4 away from the zone center to the X-point in the Brillouin zone. Nitrogen atoms can be very easily taken into diamond, and also create donor levels that are as deep as 1.7 eV below the conduction band minimum. Thus, they do not give any contribution to generate free electrons in the conduction band at room temperature. As will be seen later, an addition of 10 ppm N_2 to the source gas increases the growth rate by 8 times higher than without N_2 addition, and tends to form (100) faces at the film surface.

The nucleation density of diamond on a pristine Si wafer surface is in the order of only 10^6/cm^2. Since the average distance between diamond nuclei is 10 μm, it is virtually impossible to make a continuous film on this substrate. To increase the nucleation density, the Si surface is mechanically polished with diamond powder or paste. Alternatively, the Si substrate is immersed in alcohol with diamond powder suspension, and ultrasonicated. As a result, the Si surface is subject to mechanical scratches, and the nucleation density increases to a level higher than $10^8/cm^2$. Since the average distance between diamond nuclei is now only 1 μm, a continuous film can be formed by CVD within a few hours. It is considered that atomic scale ridges on the Si substrate, created by scratching, can be the nucleation sites. However, if the Si surface is polished with SiC or Al_2O_3 powder, the nucleation density is lower than when it is done with diamond powder. It was later found that nanometer-size diamond fragments were embedded beneath the Si surface after mechanical polishing with diamond powder, and thus they also can be the nuclei of diamond [62]. At present, it is considered that both provide the nucleation sites. On the other hand, the BEN technique was found in 1991 [3]. It results in a nucleation density of more than $10^{10}/cm^2$, and becomes an important process to synthesize HOD films. These three methods are currently used to increase the diamond nucleation density.

Once a continuous diamond film is formed on the substrate, diamond grains grow in the direction normal to the substrate surface. The initial shape of diamond grains and the surface morphology of the film depend on the process conditions, which is represented by the α-parameter. After the initial phase of growth competition, each grain that has survived undergoes a columnar growth, if secondary nucleation is avoided, and the surface area of each column is almost unchanged. An enlargement of the surface area might occur, however, by changing the process parameters, but no detailed study has been done so far. Growth and properties of diamond films are closely related to those of *a*-C films, and the readers are recommended to read an excellent review by Silva et al. [63] that includes various properties and characterizations.

Chapter 3
Microwave Plasma CVD Reactors

Chapter 3

Microwave Plasma CVD Reactors

There are various types of CVD reactors for diamond film synthesis, and they are presented in this chapter. A recent advent of production-type CVD reactors is revolutionary changing diamond film from research to production phase. In reading the articles of CVD diamond, it should be noted that in some reactors, the substrate temperature T_s cannot be controlled independently of other parameters. The gas presssure P, the microwave power P_m and other parameters influence T_s, and thus the plasma condition is concurrently changed. Therefore, a meticulous care is necessary to see whether the results intrinsically arise from T_s, or from the plasma condition due to the change in other parameters, when one interprets experimental data.

3.1. NIRIM-TYPE REACTOR

The first microwave plasma CVD reactor was designed at NIRIM [64, 65] using a quartz tube of 40–55 mm in diameter that perpendicularly penetrates the waveguide for 2.45 GHz, as schematically shown in Figure 3.1. By this reactor, a diamond film coating is possible on a 1-inch Si wafer at the maximum, but in most cases, a piece of Si that is only less than 1-cm square is used as the substrate. Typical CVD conditions are as follows: gas pressure $P = 30\,\text{Torr}$, microwave power $P_m = 300–400\,\text{W}$, the source gas $c = 0.5\%CH_4/H_2$, and the gas flow rate is 100 standard cubic cm per minute (sccm). The substrate holder is usually either quartz or Mo, and placed in the center of the cross section between the quartz tube and the waveguide. The optimum substrate temperature T_s is at around 800 °C. Sometimes, approximately 0.1% O_2 or CO_2 is added to the source gas to efficiently remove non-diamond carbons, thus improving the film quality. Under these conditions, the growth rate of diamond film is approximately 0.2 μm/h. In the reactor shown in Figure 3.1, a bias voltage can be applied to the substrate for BEN.

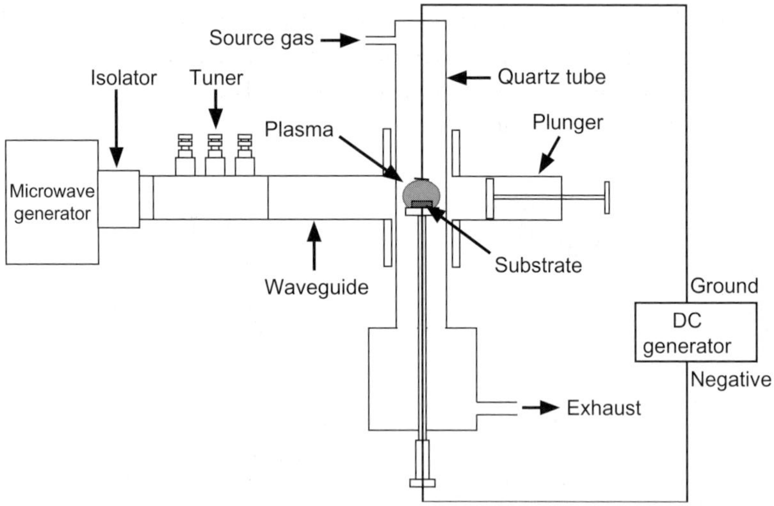

Figure 3.1. NIRIM-type reactor with biasing capability.

3.2. ASTeX-TYPE REACTORS [66]

Diamond film coating on larger area is possible using an ASTeX-type reactor, which is shown in Figure 3.2 (a). In this reactor, a uniform coating area is 2 inches in diameter, and the maximum microwave power is 1.5 kW. Standard CVD conditions are as follows: the gas pressure $P = 40$ Torr, the microwave power $P_m = 1000$–1200 W, and the source gas $c = 0.5$–$3\%CH_4/H_2$. The substrate is heated to maintain T_s at around 800 °C. The schematic figure of the reactor structure is shown in Figure 3.2 (b). Seki-Technotron/ASTeX also produces a similar reactor with the maximum microwave power of 5 kW. In this case, the standard CVD conditions are as follows: $P = 100$ Torr, $P_m = 1000$–1200 W, and $c = 0.5$–$5\%CH_4/H_2$. The substrate holder is water-cooled to maintain T_s at around 800 °C. The company recently produces semi-production type MPCVD reactors of 8 kW (2.45 GHz) and 100 kW (915 MHz), as shown in Figures 3.3 and 3.4, respectively. The film morphologies studied by the 5- and 100-kW reactors are presented in Refs. [67–70].

3.3. WAVEMAT REACTOR

Wavemat, Inc., USA, used to produce a 30-kW reactor, shown in Figure 3.5. This system also uses 915-MHz microwave, and can generate a 13-inch diameter plasma in the 18-inch inner-diameter chamber. It can accommodate an 8-inch substrate, and is operational under a wide range of gas pressure, from 1 to over 200 Torr. Recently, the technology was trasferred to Lambda Technologies, Inc., USA, and the system is going to be upgraded.

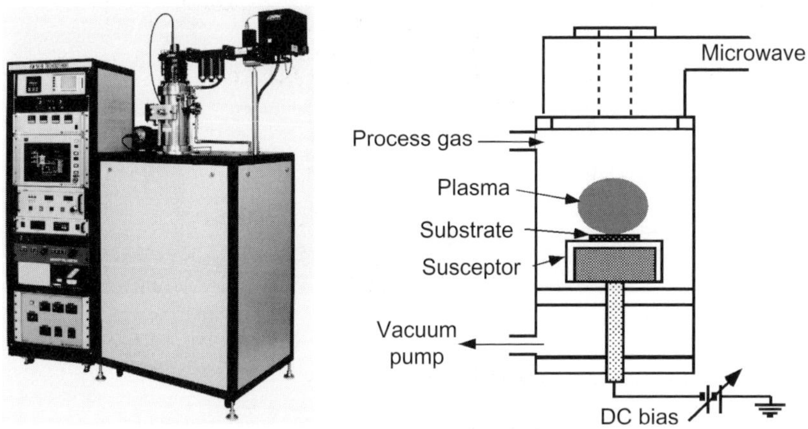

(a) (b)

Figure 3.2. (a) ASTeX-type reactor (1.5 kW, 5 kW) and (b) its schematic structure (http://www.sekitech.co.jp).

Figure 3.3. 8-kW, 2.45-GHz MPCVD reactor.

Figure 3.4. 100-kW, 915-MHz MPCVD reactor.

3.4. AIXTRON REACTORS

AIXTRON AG, Germany, produces 6-kW (2.45 GHz) and 60-kW (915 MHz) MPCVD systems that have been developed in Fraunhofer Institute. It has a very unique oval shape microwave cavity, as shown in Figure 3.6. They can accommodate 3- to 6-inch substrates, respectively.

3.5. SAIREM REACTOR

SAIREM, France, produces a 1.2-kW, 2.45-GHz MPCVD reactor with a quartz bell jar, as shown in Figure 3.7 [71].

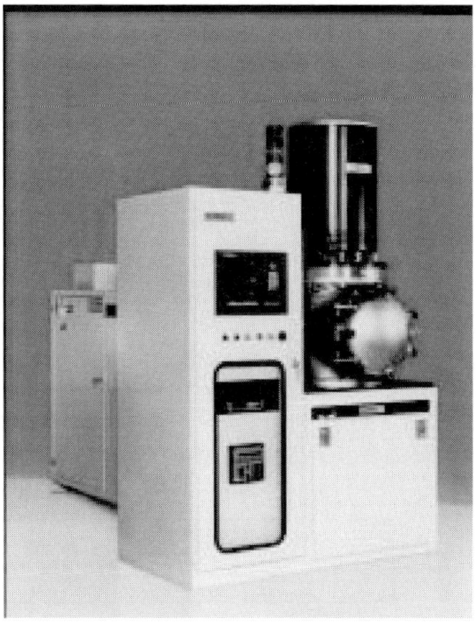

Figure 3.5. Wavemat reactor (http://www.microcure.com).

(a) (b)

Figure 3.6. AIXTRON reactors: (a) 2.45 GHz, 6 kW and (b) 915 MHz, 60 kW (http://www.aixtron.com).

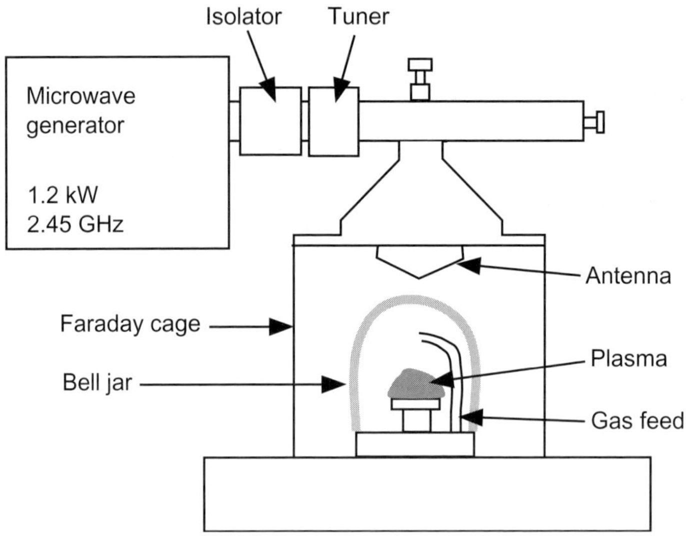

Figure 3.7. Schematic diagram of SAIREM 1.5-kW reactor (http://www.sairem.com).

Chapter 4
Other Reactors

Chapter 4

Other Reactors

Besides the MPCVD reactors, other CVD reactors are also used for diamond deposition. They are hot filament, DC plasma, radio-frequency (rf) plasma, thermal rf plasma, plasma jet, and combustion CVD reactors. In the following, hot filament and DC plasma CVD reactors will be described, because they have been used for oriented growth of diamond.

4.1. HOT FILAMENT CVD REACTOR

In the hot filament CVD (HFCVD) reactor, there are filaments of tungsten (W) or tantalum (Ta), which are heated up to 2000 to 2200 °C. They are placed at 5 to 10 mm above the substrate. A schematic structure of the reactor is shown in Figure 4.1. The gas pressure is usually 20–30 Torr. One can apply a bias voltage either to the filaments or the substrate. In this reactor, CH_4 and H_2 molecules that have touched the hot filament are thermally dissociated to be fragmented hydrocarbons and atomic hydrogen. They diffuse onto the substrate to create diamonds. Since the filament material is carburized during the diamond CVD, usually the filaments have been carburized in a CH_4/H_2 gas for more than 1 h before they are used for diamond growth. An advantage of hot filament CVD reactors over MPCVD reactors is the simplicity of the system and a uniform diamond deposition over such a large area as 40 cm × 60 cm. However, it is more difficult for HFCVD reactors to precisely control and reproduce the film morphology than for MPCVD reactors. Furthermore, degradation and bending of the filaments during CVD, incorporation of filament material in diamond films, generation of carbon flakes are the disadvantages. The growth rate of diamond, however, is more than 1 μm/h under the above mentioned conditions, better than the NIRIM-type MPCVD reactor.

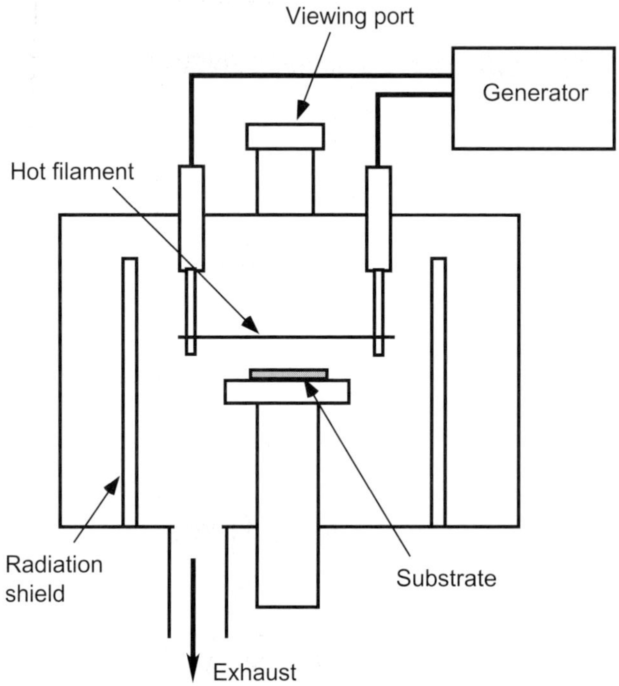

Figure 4.1. Schematic diagram of a hot filament CVD reactor.

4.2. DC PLASMA CVD REACTOR

In this reactor, a negative DC bias voltage is applied to the cathode that is about 10 cm away from the substrate. Standard growth conditions are: $P = 100$ Torr, the applied voltage is 1 kV, the current is 10 mA, and $c = 5\%CH_4/H_2$. Since the gas pressure and the energy density in the plasma are so high that a growth rate of 10 μm/h can be achieved. A schematic diagram of the reactor is shown in Figure 4.2 This is one of the most sophisticated reactors constructed by Sawabe's group [72] to make heteroepitaxial diamond films on Ir(100) substrates. Recently, a DC plasma CVD reactor with multiple cathodes was invented [73, 74], which is capable of depositing a diamond film on 4-inch area. The plasma in the reactor is shown in Figure 4.3.

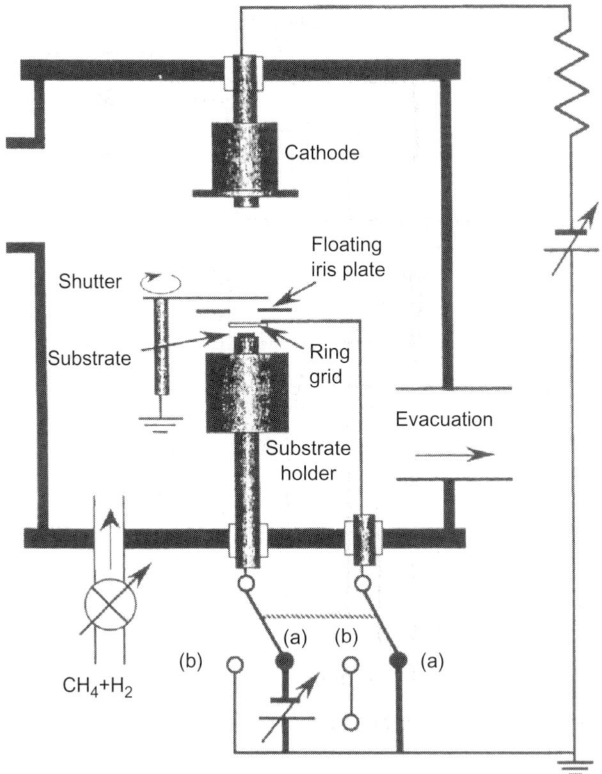

Figure 4.2. Schematic diagram of DC plasma CVD reactor with BEN capability [72].

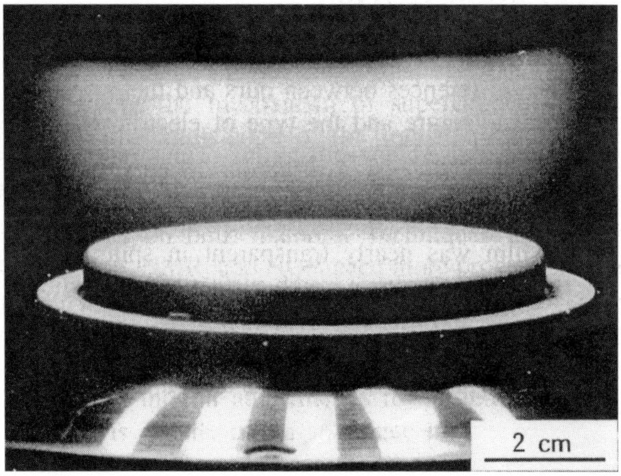

Figure 4.3. Plasma generated by the multiple cathode type DC plasma CVD reactor developed at KIST. The source gas was $c = 3\%CH_4/H_2$, and $P = 100$ Torr [73, 74].

Chapter 5

Crystal Orientations and Film Surface Morphology

Chapter 5
Crystal Orientations and Film Surface Morphology

Figures 5.1 (a) – (c) show a growth process of diamond crystals to be a continuous film by intermittently holding the diamond CVD and taking SEM micrographs of the same position on the specimen. It is seen that diamond crystals uniformly increase their sizes until they are in contact with each other, and then start to grow in the direction normal to the substrate surface. Once the vertical growth started, certain diamond grains grow faster than the others in both vertical and horizontal directions, and consequently, diamond grains with a selected orientation determine the film morphology as a result of growth competition. Under appropriate CVD conditions, these grains concurrently undergo a vertical growth, and form columns of diamond grains. Such a growth mode is called as a "columnar growth." Figure 5.2 shows a fractured edge of a 10-μm thick HOD film (see Figure 1.1 (b)), and the columnar growth is obviously seen. If the CVD conditions are not appropriate, randomly oriented secondary nucleation occurs very frequently and takes over the basal diamond grains, thus creating a randomly oriented polycrystalline diamond film (see Figures 1.1 (a) and 5.1(c)).

In polycrystalline diamond films, the orientations of diamond grains are random because of secondary nucleation and their growth so that the X-ray diffraction pattern is very similar to that of diamond powder (see Table 2.1). Namely, the relative intensity ratios are (111):(220):(311):(400) = 100:2.5:16:8 for Cu Kα_1 radiation ($\lambda = 1.5405$ Å) [24]. There are a number of studies in which the diamond film undergoes a columnar growth, and each column has a stack of {111}, {100}, or {110} layers parallel to the substrate surface. Such uniaxial growth states can be observed by X-ray diffraction, in which either (111), (400), or (220) diffraction intensity is significantly stronger than the relative intensity ratio of diamond powder given in Table 2.1. The deviation from the perfect uniaxial orientation can be quantified by the full width at half maximum (FWHM) of the X-ray diffraction lines or their rocking curves.

When a substrate material with a well-defined crystal orientation, or with a special treatment, is used for diamond CVD, it is possible to synthesize a diamond film in which either (100) or (111) diamond faces are parallel to the film surface, and make them align in the same direction. In other words, diamond faces can be azimuthally (in-plane) oriented in the same direction, as seen in Figures 1.1 (b) and (c). In such cases, it often happens that adjacent diamond faces coalesce with each other to form a larger face. If the coalescence develops over the entire surface, then the grain boundaries vanish and a single crystal diamond film is formed on the

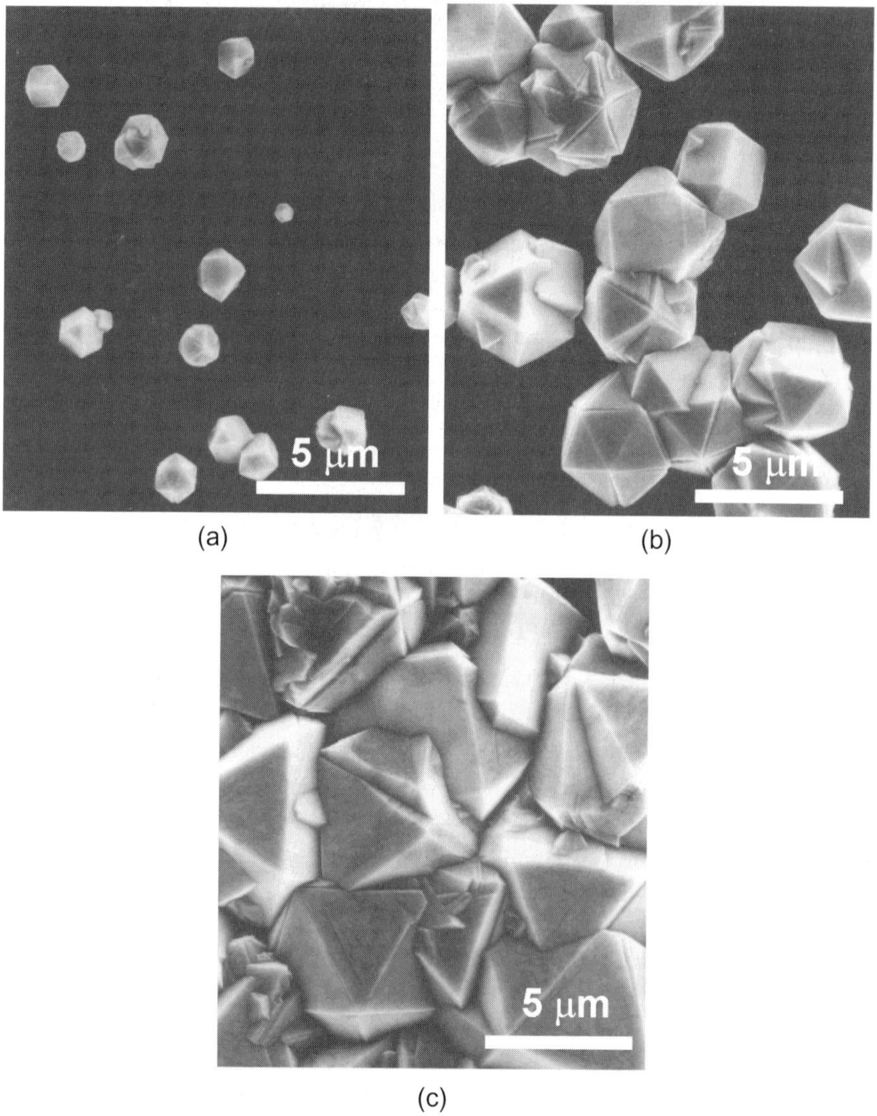

Figure 5.1. Growth process of diamond crystals: (a) ~5 h, (b) ~10 h, and (c) ~20 h.

substrate, as seen later in Figures 12.8 and 12.19. Since grain boundaries are the strongest barriers and traps for transport of electrons and holes, the coalesced films should have much better carrier (electron or hole) mobilities than polycrystalline films. Thus, even though the film surface is not fully coalesced, their electrical properties are better than those of randomly oriented polycrystalline films.

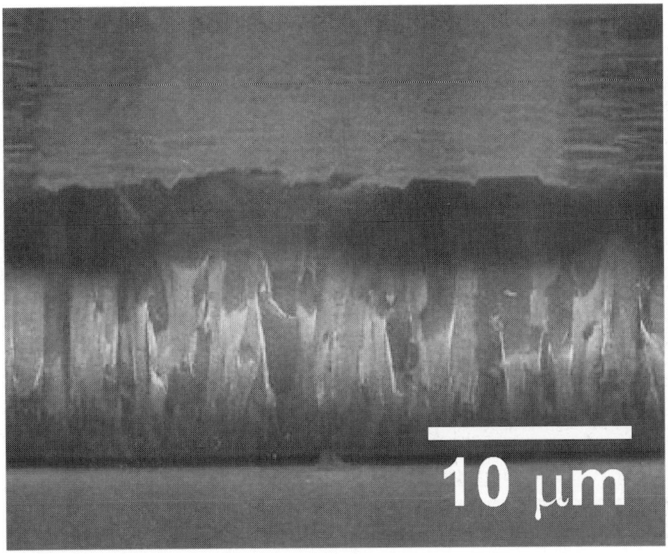

Figure 5.2. SEM image of a fractured edge of HOD film. A columnar growth of diamond is visible.

In studying the growth of CVD diamond films on non-diamond substrates such as Si, homoepitaxial growth studies of diamond are very useful. Indeed, Okushi and co-workers [55–58, 75] have synthesized extremely high quality diamond layers homoepitaxially using extremely low CH_4 concentrations. The same method is applied to oriented and heteroepitaxial growth of diamond films. Thus, homoepitaxial diamond growth is described whenever appropriate in this monograph.

5.1. X-RAY POLE FIGURE (XPF)

It was Koidl's group that has intensively carried out comprehensive studies of uniaxially and azimuthally oriented growth of diamond films in the early stage of diamond film research using XPF measurements. Thus, we will start with tracing their works to learn general results on oriented growth. Figure 5.3 schematically shows the X-ray scattering geometry in XPF measurements. The vector $\boldsymbol{k}$ is a scattering wave vector. The detector position is fixed at a specific angle 2θ of diffraction from $\{hkl\}$, and the specimen is rotated about χ and ϕ axes. Thus, only the diffraction from the lattice planes perpendicular to the $\boldsymbol{k}$-vector contributes to the observed X-ray intensity. In the following, the XPF obtained by this setup will be expressed as (hkl) XPF. Thus, it is $\{hkl\}$ diffraction peaks that are observed by (hkl) XPF.

In Ref. [76], diamond films of 150–400 μm thicknesses were deposited by a combination of HFCVD and MPCVD on Si substrates, and then removed from the substrates to obtain free-standing diamond films. The grain size at the film surface was in the order of tens of micrometers. Figure 5.4 (a) shows a (220) XPF. There is a strong (220) diffraction peak in the center of the XPF, indicating that the 400-μm thick film grew in the ⟨110⟩ direction perpendicularly to the Si substrate surface. On the other hand, Figure 5.4 (b) is a (111) XPF. It is seen that this diagram misses the central peak but exhibits a circular band. From the angular difference $\Delta\chi = 35.3°$ between the ⟨111⟩ direction and the circular peak, the circular band was identified to be {110} diffractions. The circular shape indicates that the azimuthal (in-plane)

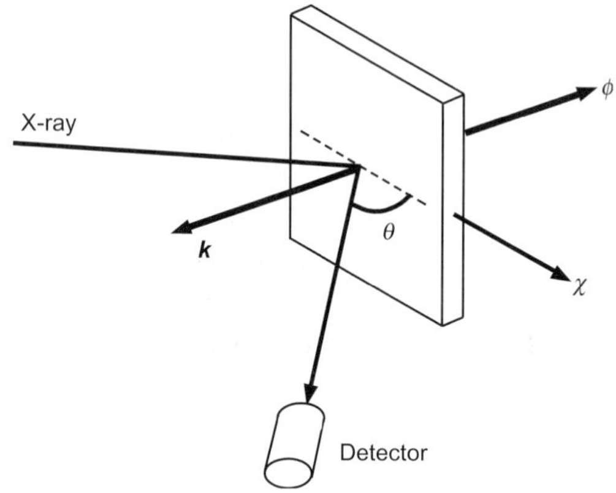

Figure 5.3. Scattering geometry of XPF.

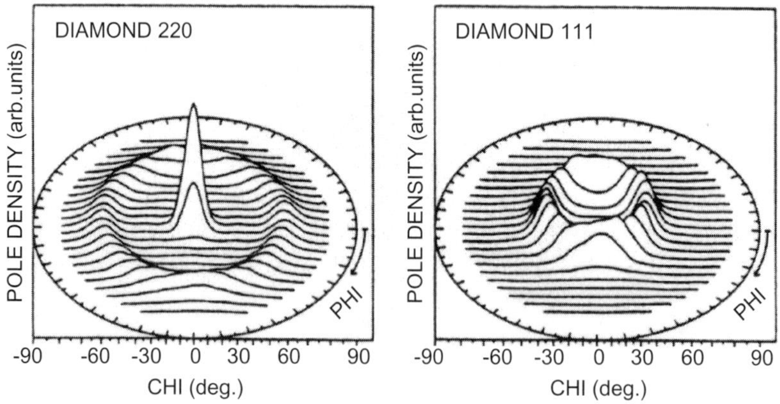

Figure 5.4. (220) and (111) XPFs of a 400-μm thick, free-standing diamond films. The X-ray source was Cu $K\alpha_1$ radiation [76].

orientations of diamond grains were random with respect to the growth direction ⟨110⟩. Note, however, that the crystal faces that appeared on the film surface were (111), which was 35.3° tilted from the growth direction ⟨110⟩. An existence of {111} twins also was detected by the XPF measurements.

5.2. ORIENTATIONAL EVOLUTION

In Ref. [76], X-ray diffraction patterns of diamond films, thinned to different thicknesses by oxygen plasma etching, indicated that the (220) diffraction intensity, relative to the (111) diffraction intensity, decreased as the film thickness was decreased. This implies that on average, the orientations of diamond grains are more random when the film thickness is thinner: namely, the diamond grains are better aligned along the ⟨110⟩ direction as the film thickness is increased. According to a two-dimensional computer simulation of the film growth based on the van der Drift model [77], the transition of the growth mode from random to ⟨110⟩ with time was a consequence of growth competition between different diamond grains, as shown in Figure 5.5. This figure also indicates that the columnar growth occurs with diamond faces of fastest growth at the film surface, and the average size of the faces gradually increases with the film thickness.

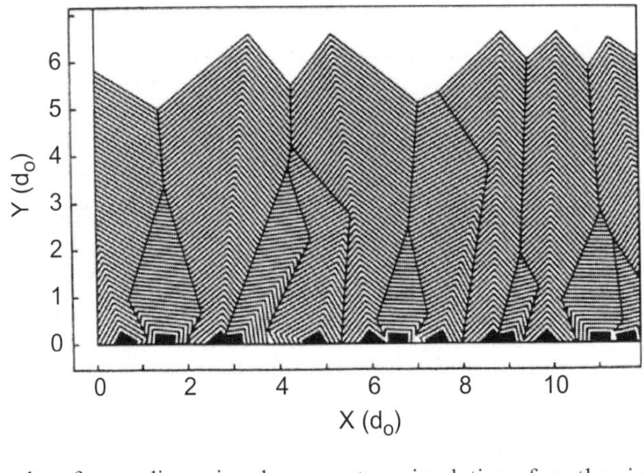

Figure 5.5. Result of two-dimensional computer simulation for the initial phase of polycrystalline diamond film growth. The X and Y-axes are normalized with respect to the mean nuclei distance (d_o) [76].

5.3. UNIAXIAL ⟨110⟩-GROWTH

An uniaxial ⟨110⟩-growth has also been achieved by MPCVD, where a microcrystalline diamond film was first deposited on Si substrate using a relatively high CH_4 concentration, and then a standard diamond CVD was undertaken using proper growth conditions [78, 79]. It was inferred that there was a tendency of ⟨110⟩ orientation in the diamond grains of the microcrystalline film, and this tendency was carried over to the overgrown layer, even though the second growth conditions were not perfectly suitable for the ⟨110⟩-growth. It seems from experience that ⟨110⟩-growth often takes place when the nucleation density is high. A strong ⟨110⟩-growth was also observed for the film grown by HFCVD [80]. In this case, however, the nucleation density was quite normal (in the order of $10^8/cm^2$) so that the ⟨110⟩-growth seems to occur because of proper process parameters. Uniaxial ⟨110⟩-growth is also described in Section 11.16.

5.4. UNIAXIAL ⟨100⟩-GROWTH

An uniaxially ⟨100⟩-oriented growth was first found at NIRIM in 1985 (see Ref. [23]). It was intensively studied by Wild et al. [81–83] using a NIRIM-type MPCVD reactor. A SEM image of a ⟨100⟩-oriented film surface is shown in Figure 5.6. It is seen that the (100) faces are co-planar (the angular spread was only 1°), and the

Figure 5.6. Uniaxially ⟨100⟩-oriented diamond film surface. The film was synthesized using $c = 2\% CH_4/H_2$ as the source gas and at $T_s = 900\,°C$ [82].

surface roughness is small, compared with randomly oriented polycrystalline diamond films (Figure 1.1 (a)). The degree of ⟨100⟩ orientation was examined by XPF, X-ray rocking curve, proton channeling, and light scattering. The substrate temperature gave a significant influence on both the surface morphology and the orientation of (100) faces. For 100-μm thick diamond films grown using $c = 2\% CH_4/H_2$ and $P = 38$ Torr, the best-aligned (100) faces were achieved when $T_s = 835\,°C$. For higher T_s, the (100) faces inclined in such a way that the corner of the (100) face was up, as is consistent with the theory on the growth of fastest direction, which is described later in Section 5.5. In order to make large flat faces, a strongly ⟨100⟩-textured film was first deposited, and then the growth conditions were switched so that (111) faces grow faster than (100) either by increasing T_s or by decreasing the CH_4 concentration. This causes a lateral expansion of the (100) faces. The procedure is now called as a "smoothing process" or a "lateral growth." Ion channeling experiments showed that the (100) faces tilted only within 0.5° from the substrate surface normal.

It should be noted here that the measurements of growth orientation is influenced by the bending of the Si surface due to diamond growth [81], and thus a proper correction is needed to obtain the "true" values.

In Ref. [84], it was investigated how the growth direction, or film texture, is related with the CH_4 concentration c and T_s, and the results for $P = 37.5$ Torr using a NIRIM-type reactor is shown in Figure 5.7. At low CH_4 concentrations and high temperatures, the film is ⟨110⟩-textured with (111) faces inclined by 35.3° from the substrate surface normal. At medium CH_4 concentrations, there is a transition in the uniaxial growth direction from ⟨110⟩ to ⟨100⟩. For higher CH_4 concentrations, the film becomes microcrystalline. Although it is not shown in Figure 5.7, there should be a region where ⟨111⟩ texture is dominant at low CH_4 concentrations and high T_s. Such a behavior is consistent with the results of Ref. [23] studied at a fixed temperature of $T_s = 800\,°C$, those of Ref. [24] studied for different P, T_s, CH_4 concentration and the gas flow rate, and those of Ref. [85], where both T_s and CH_4 concentrations were changed, as seen in Figure 5.8. The results of Ref. [23] is equivalent to the morphologies along $T_s = 990\,°C$ in the figure.

5.5. α-PARAMETER

In order to describe the crystal morphology, Wild et al. [82] defined a growth parameter α by:

$$\alpha = \sqrt{3}\frac{\nu_{100}}{\nu_{111}} \tag{5.1}$$

where v_{100} and v_{111} are the growth rates of (100) and (111), respectively. Assuming that a diamond crystal consists of (100) and (111) faces for the reason described in Section 2.2, the crystal shape is cubo-octahedral, and changes with the α-values, as shown in Figure 5.9. In this figure, the numbers are the α-values that correspond to the cubo-octahedrons, and the arrows indicate the directions of the fastest growth. These cubo-octahedrons correspond to the α-value between 1 and 3.

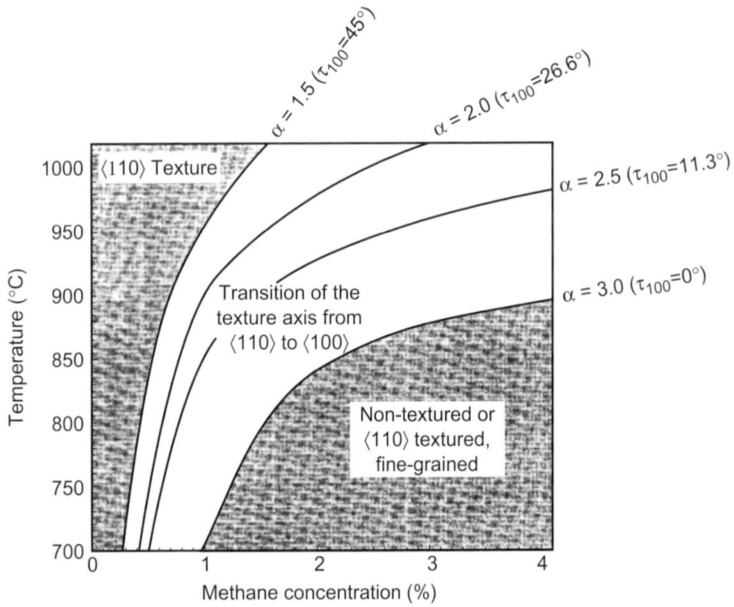

Figure 5.7. Growth orientation (film texture) and film morphology [84].

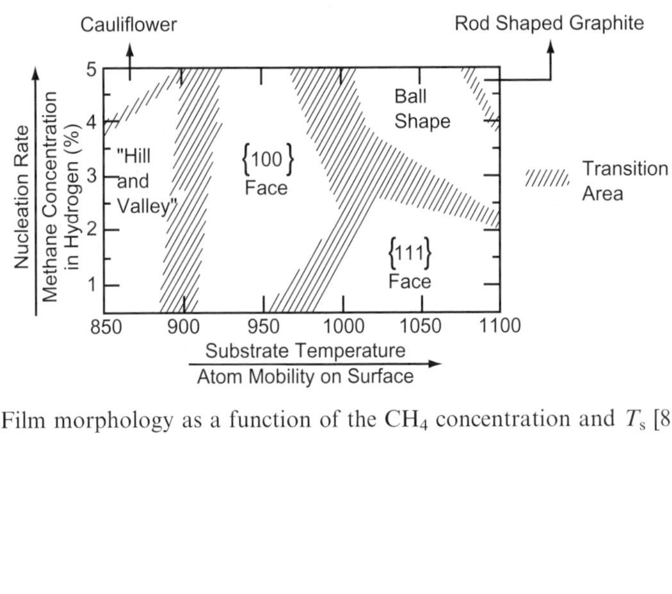

Figure 5.8. Film morphology as a function of the CH_4 concentration and T_s [85].

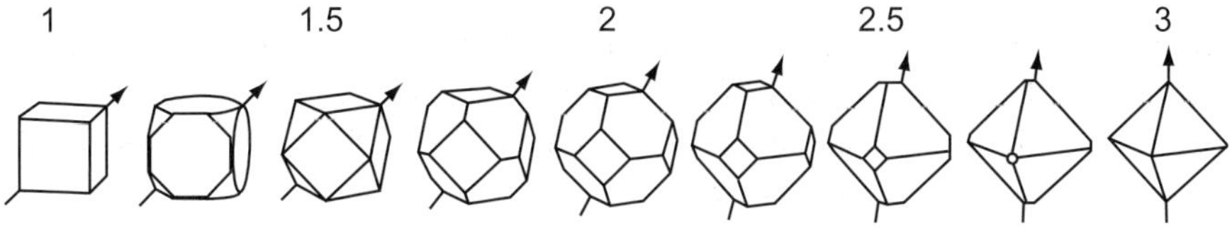

Figure 5.9. Crystal shapes that correspond to the values of the α-parameter. The arrows indicate the directions of the fastest growth [82].

The α-parameter can be $\alpha < 1$ and $\alpha > 3$, but the crystal shape is no longer cubo-octahedral.

The direction vector $\boldsymbol{v}$ for the fastest growth is expressed by [84]:

$$\boldsymbol{v} = \begin{bmatrix} 1 \\ 3/\alpha - 2 \\ 1 \end{bmatrix} \quad \text{for } 1 \leq \alpha \leq 1.5 \tag{5.2}$$

and

$$\boldsymbol{v} = \begin{bmatrix} 1 \\ 0 \\ 3/\alpha - 1 \end{bmatrix} \quad \text{for } \leq \alpha \leq 3 \tag{5.3}$$

The tilt angle of the (100) face, τ_{100}, from the direction of the fastest growth $\boldsymbol{v}$ is expressed by:

$$\cos(\tau_{100}) = 1/|\boldsymbol{v}| \tag{5.4}$$

In Figure 5.10, the tilt angles τ_{100} were measured as a function of the CH_4 concentration c and the substrate temperature T_s. It is seen that when c is low, the diamond film is $\langle 110 \rangle$-textured, while it is more $\langle 100 \rangle$-textured as c is increased. The $\langle 100 \rangle$ texture is achieved at lower c when $T_s = 800\,°C$ than when $T_s > 800\,°C$. Thus, this figure clearly shows that α is dependent on both c and T_s.

It is also of importance that the tilt angle τ_{100} depends on the oxygen atom concentration in the source gas. Figure 5.11 shows the effect of CO addition to the source gas (CH_4/H_2) on the tilt angle τ_{100} [84]. It is seen that a higher CO concentration results in the $\langle 100 \rangle$ texture when the CH_4 concentration is low. The results of Figure 5.11, and Figure 5.10 as well, were obtained using a NIRIM-type MPCVD reactor for $P = 37$ Torr. A similar effect was also observed by adding O_2 to the source gas using a 5-kW ASTeX MPCVD reactor under conditions of $c = 4\%CH_4/H_2$, $P = 100$ Torr, and $P_m = 4$ kW [67]. The XRD peak height ratios

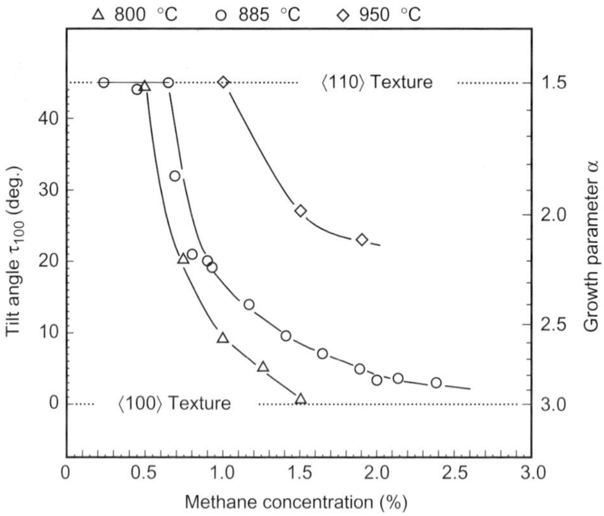

Figure 5.10. Tilt angle τ_{100} as a function of the CH_4 concentration and T_s [84].

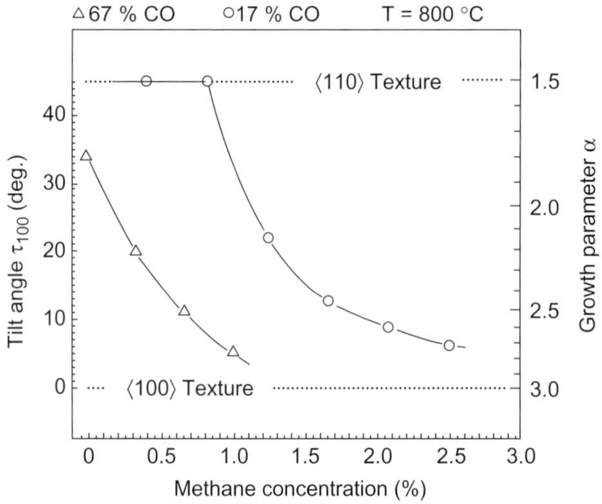

Figure 5.11. Effect of CO addition to CH_4/H_2 on the tilt angle τ_{100} [84].

of (400)/(111) and (111)/(400) as a function of O_2 concentration are shown in Figure 5.12. This and above results mean that the α-parameter can also be controlled just by adding O_2 or oxygen-containing gases to the source gas (CH_4/H_2) to achieve ⟨100⟩-textured diamond films. This knowledge is particularly useful in synthesizing HOD films.

It should be noted that the location of constant α curves in the c–T_s plane like Figure 5.7 is significantly different, if the gas pressure P and the input microwave power P_m are different. Indeed, the diagram of the α-parameter that was obtained using the 5-kW MPCVD reactor is shown in Figure 5.13 [67], which is significantly different from the one shown in Figure 5.7. Most importantly, the regions of uniaxial

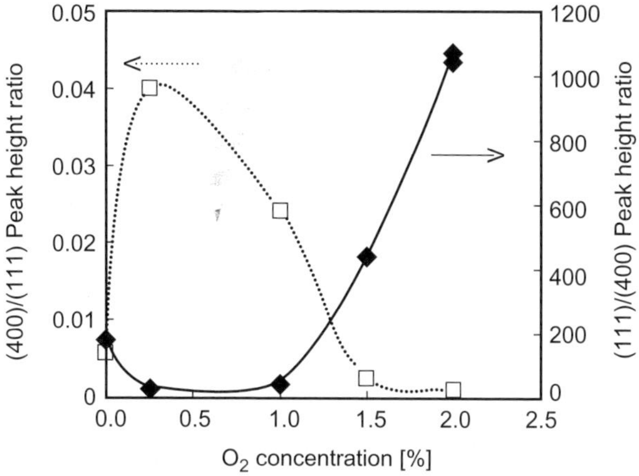

Figure 5.12. Change in the degree of ⟨100⟩-orientation by O_2 addition to the source gas. The CVD conditions are $c = CH_4/H_2$, $P = 100$ Torr, and $P_m = 4$ kW [67].

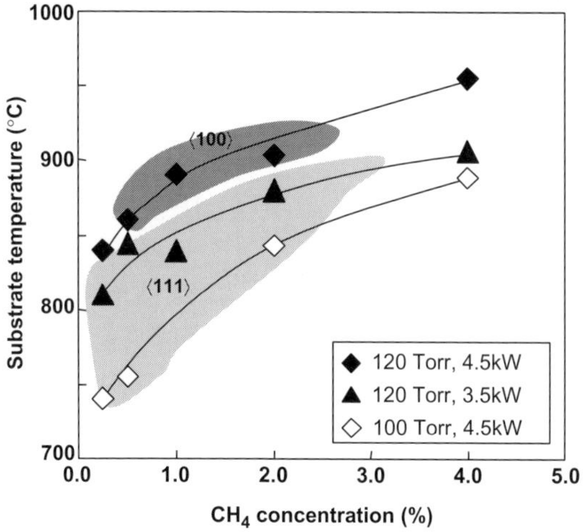

Figure 5.13. The approximate regions of uniaxial ⟨100⟩ and ⟨111⟩ growth obtained using an ASTeX 5-kW MPCVD reactor. The symbols express the data points of T_s and c for given P and P_m [67].

⟨100⟩ and ⟨111⟩ growth are located in totally different regions as compared with the results of Figure 5.7. Figure 5.14 shows the α-parameter curves and the uniaxial growth regions in the c–T_s plane using CH_4 and H_2 as a source gas. The reactor used was the one shown in Figure 3.4, and $P = 90$–140 Torr. The SEM images of Figures 5.14 (a)–(c) show the film morphologies that correspond to the CVD conditions indicated by a–c in Figure 5.14, respectively. Figure 5.15 shows a similar diagram, but a mixture of CH_4, H_2, and CO_2 were used as the source gas. In this diagram, the regions for various α-parameter values are depicted in the C^*–T_s plane, where C^* is an "effective carbon concentration" defined in the figure. The SEM images of (a)–(d) correspond to the CVD conditions indicated by a–d in Figure 5.15, respectively. It should be noted that both $\alpha < 1$ and $\alpha > 3$ were attained, hence the possibility of controlling the α-parameter was expanded in a wider range.

There are some methods to evaluate α-values experimentally [22, 86, 87]. The first method is given by Maeda et al. [86], where the α-parameter for specific CVD conditions is derived from the change in the crystal shape between an initial cubo-octahedral diamond particle of $\alpha = 3/2$ and the same particle after the growth.

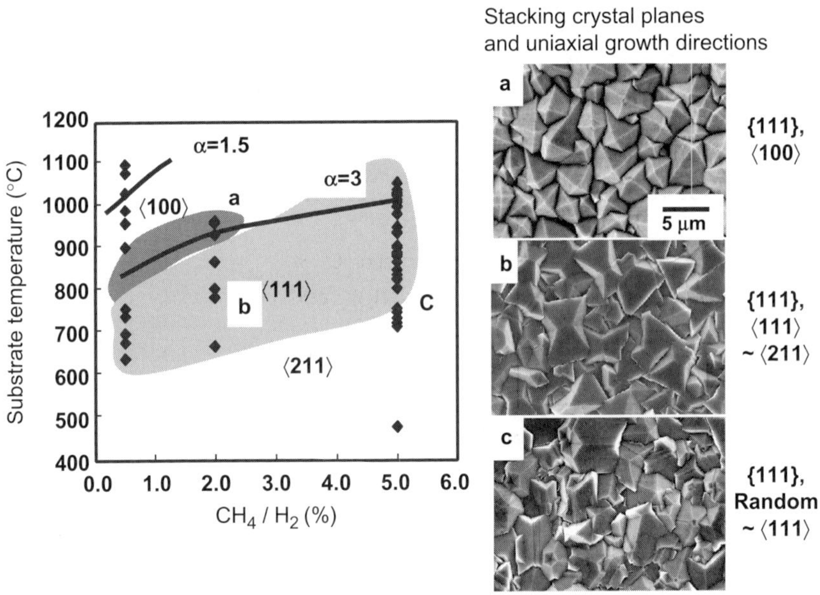

Figure 5.14. Diagram of α-parameter curves, uniaxial film orientation, and morphology. The SEM images (a)–(c) on the right are the film surfaces grown under the CVD conditions indicated by a–c in the c–T_s diagram on the left. The regions ⟨100⟩, ⟨111⟩, and ⟨211⟩ mean the direction of uniaxial growth of the films. For this experiment, the large MPCVD reactor shown in Figure 3.4 was used [68, 69].

Figure 5.16 shows top views of cubo-octahedral diamond particles. The growth rates are calculated separately for $\alpha > 3/2$ and $\alpha < 3/2$. The reason for starting with a diamond particle of $\alpha = 3/2$ is that such particles can be synthesized under the standard conditions of the NIRIM-type MPCVD reactor, i.e. $c = 1\text{–}2\%CH_4/H_2$, $P = 30\text{–}40$ Torr, $T_s \sim 800\,°C$, and $P_m = 300\text{–}400$ W.

Figure 5.17 (a) illustrates a plan view of the change in the crystal shape from $\alpha = 3/2$ (gray lines) to $\alpha > 3/2$ (dark lines). Figure 5.17 (b) is a cross-sectional view

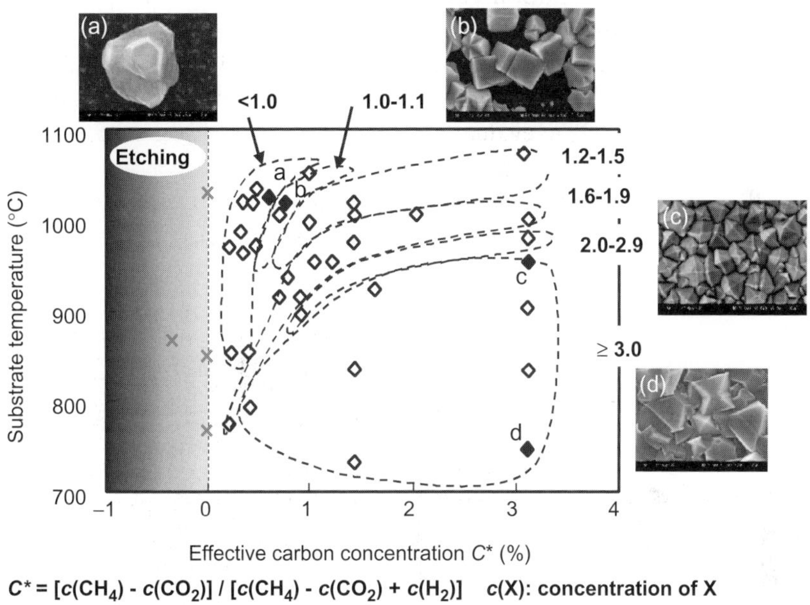

Figure 5.15. Diagram of α-parameter curves and film morphology. The SEM images (a)–(d) are the film surfaces grown under the CVD conditions indicated by a–d in the diagram. The numbers in the diagram are the α-parameter values associated with the regions encircled by the dashed curves. The horizontal axis is an "effective carbon concentration," C^*, defined by the concentrations of CH_4, H_2, and CO_2 in the source gas. For this experiments, the large MPCVD reactor shown in Figure 3.4 was used [68, 70].

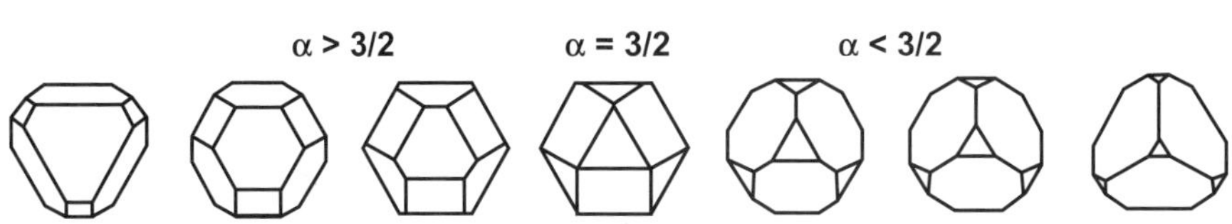

Figure 5.16. Top view of cubo-octahedral crystals [86].

along a vertical plane through A–A* of Figure 5.17 (a). From these figures, the increased thicknesses for the (100) and (111) faces, which are denoted by $T_{\{100\}}$ and $T_{\{111\}}$, respectively, can be expressed using the change in the $\delta_{\{100\}}$ and $\delta_{\{111\}}$, shown in Figure 5.17 (b), as follows:

$$T_{\{111\}} = 1.41\delta_{\{111\}} \tag{5.5}$$

$$T_{\{100\}} = 1.41([\delta_{\{100\}}]^2 + [T_{\{111\}}]^2)^{1/2}\cos\{54.9^\circ - \tan^{-1}[\delta_{\{100\}}/T_{\{111\}}]\} \tag{5.6}$$

Similarly, Figure 5.18 (a) illustrates a plan view of the change in the crystal shape from $\alpha = 3/2$ to $\alpha < 3/2$, and Figure 5.18 (b) is a cross-sectional view along a vertical

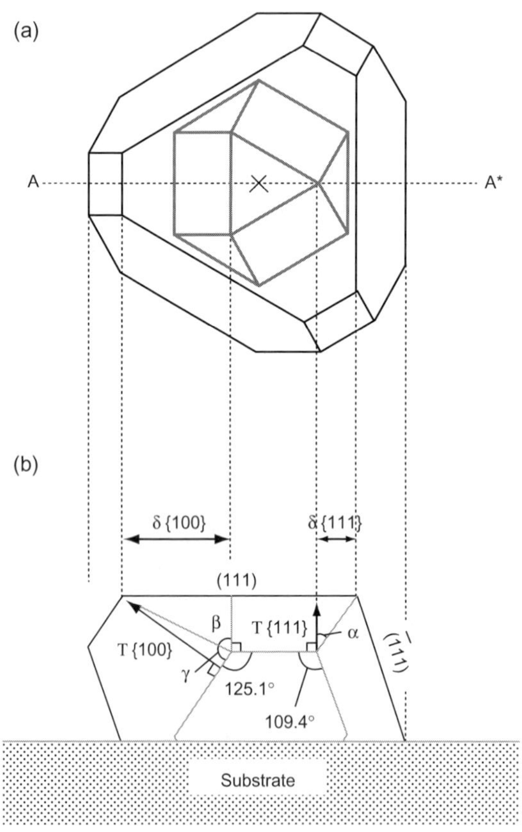

Figure 5.17. Geometrical relationship between an initial crystal ($\alpha = 3/2$) and a grown crystal ($\alpha > 3/2$) [86].

plane through A – A* of Figure 5.18 (a). In this case, $T_{\{111\}}$ and $T_{\{100\}}$ are expressed as follows:

$$T_{\{111\}} = 1.41[0.75D - \delta_{\{111\}}] \quad (5.7)$$

$$T_{\{100\}} = ([\delta_{\{100\}}]^2 + [T_{\{111\}}]^2)^{1/2}\sin\{54.9^\circ - \tan^{-1}[T_{\{111\}}/\delta_{\{100\}}]\} \quad (5.8)$$

Maeda et al. [86] actually measured the homoepitaxial growth rates of (100) and (111) faces at $P = 40$ Torr, and v_{100}/v_{111} $(=\alpha/\sqrt{3})$ is depicted in Figure 5.19. From the figure, it can be seen that α can be controlled approximately between $\sqrt{3}$ and $3\sqrt{3}/2$ by changing the CH_4 concentration between 0.5 and 2% at $T_s = 800\,^\circ$C.

A different method of obtaining the α-parameter is proposed in Ref. [87]. In this case, the α-parameter is evaluated from the shape of a single diamond crystal.

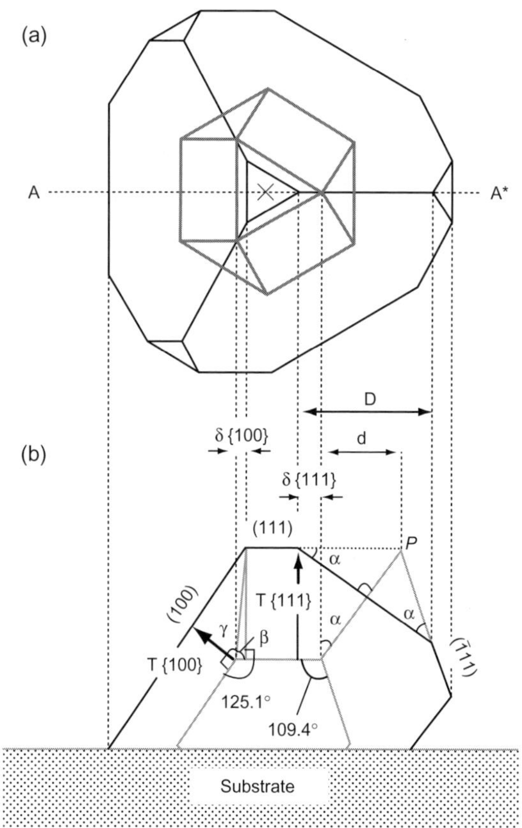

Figure 5.18. Geometrical relationship of an initial crystal ($\alpha = 3/2$) and a grown crystal ($\alpha < 3/2$) [86].

Define notations, *b*, *c*, and *d* for a truncated cube, *g* and *s* for truncated octahedron, as shown in Figure 5.20, and *t* for a growth time, then *v*(111), *v*(100), and the α-parameter are expressed as shown in Table 5.1. This method has a great advantage that no comparison of diamond crystal shapes in an interval of growth time *t* is

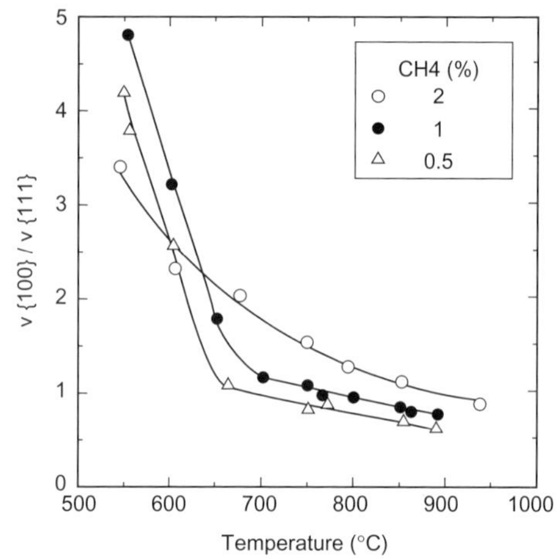

Figure 5.19. Effects of the CH_4 concentration and the substrate temperature on the relative growth rate, v_{100}/v_{111}. Note that $\alpha = \sqrt{3}\,(v_{100}/v_{111})$ [86].

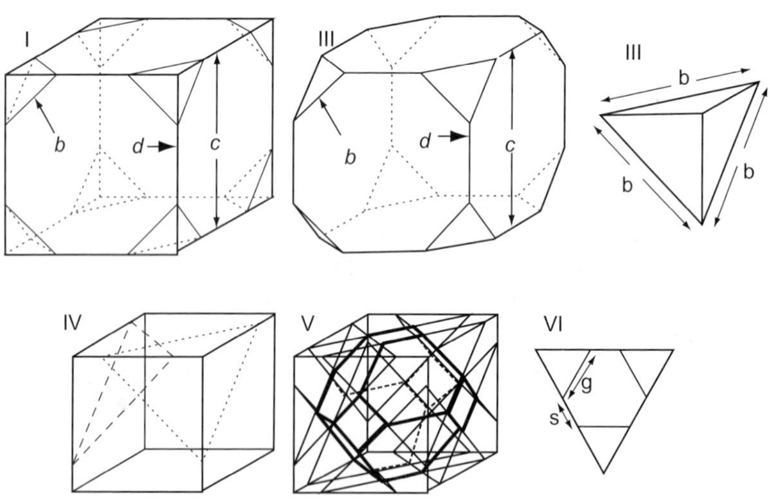

Figure 5.20. A truncated cube before (I) and after (II) a removal of the corners, and one of the prisms removed (III). A cubic crystal (IV) with two triangular prisms removed which impinge on one another. A cubic crystal (V) with all the triangular prisms removed, and one of the hexagonal faces produced (VI) [87].

Table 5.1. Summary of derived equations for v_{111}, v_{100}, and the α-parameter [87].

Variables→ ↓	Truncated cube: Cube dimension, c; Triangular face edge, b; Edge between octagonal faces, d	Truncated octahedron: Cube dimension, c; Square edge length, s; Remaining edge, g
v_{111}	$c\frac{\sqrt{3}}{2t}-\frac{b}{t\sqrt{6}}$	$\frac{s\sqrt{2}}{t\sqrt{3}}+\frac{g}{t\sqrt{6}}$
	or $\frac{2b}{t\sqrt{6}}+\frac{d\sqrt{3}}{2t}$	
	or $\frac{d}{2t\sqrt{3}}+\frac{c}{t\sqrt{3}}$	
v_{100}	$\frac{c}{2t}$	$\frac{s\sqrt{2}}{2t}+\frac{g}{t\sqrt{2}}$
	or $\frac{b}{t\sqrt{2}}+\frac{d}{2t}$	
α	$\frac{c}{c-(b\sqrt{2}/3)}$	$\frac{3(s+g)}{2s+g}$
	or $\frac{3c}{2c+d}$	
	or $\frac{d+b\sqrt{2}}{d+(2b\sqrt{2}/3)}$	

necessary, and the α-value can be evaluated from a single SEM image. However, in the expression of v_{111} and v_{100}, the incubation time for diamond nucleation, which is approximately 20 min and depends on CVD conditions, is included in t so that a certain caution on the definition of t is necessary. To avoid the ambiguity on t, it is perhaps necessary to grow diamond for more than a couple of hours to reduce the effect of the incubation time. Quite often, the crystal shapes on the same substrate are not identical, and there are crystals with different shapes even on the same location of the substrate. These factors might introduce some error bars in the calculated values of v_{100} and v_{111}. On the other hand, t is not included in the expressions of α, and no such problems occur. With these situations being taken into account, this method is very convenient to evaluate α-values.

We shall now go back to Figure 5.7, where curves for different α-values are depicted. This means that the α-parameter depends on the CH_4 concentration and T_s. In fact, it is actually a function of all CVD conditions, and also depends on the details of the reactor structure and setups. Thus, for a more accurate study of the α-parameter, or equivalently the morphology of diamond crystals and films, it is necessary to carefully take into account these factors. The introduction of the concept of the α-parameter made us possible to quantitatively investigate the morphology, and develop the heteroepitaxial growth technology of diamond.

5.6. EFFECTS OF B AND N ADDITION ON SURFACE MORPHOLOGY

It is known that an acceptor level of 0.37 eV above the valence band maximum is formed when diamond is doped with boron (B), and a donor level of 1.7 eV below the conduction band minimum is formed when diamond is doped with nitrogen (N). Apart from the semiconducting characteristics, the crystal habit of diamond grains at the diamond film surface is also influenced by B- and N-dopings. According to Ref. [88] using a NIRIM-type MPCVD reactor, B-doped diamond films preferred (111) facets for low B_2H_6 concentrations. It is also of interest that (100) facets appear by heavy B-doping, when diamond is synthesized in conditions under which otherwise microcrystalline films are formed [89].

On the other hand, the tilt angle τ_{100} (see Section 5.5) changes from $\langle 110 \rangle$ to $\langle 100 \rangle$ as the N_2 concentration was increased, as seen in Figure 5.21 [90]. The growth facet also changed from (111) to (100) as the N_2 concentration was increased from N/C = 0.1 to 10% [91], and yet the film quality, determined by the FWHM of the Raman 1333 cm^{-1} line, was better for N-doped diamond films than undoped films. Also, addition of N_2 markedly increased the growth rate, as shown in Figure 5.22 [92]. It was also confirmed recently that an N_2 addition to the source gas enhanced the $\langle 100 \rangle$-oriented growth and dominantly formed (100) faces at the film surface [93]. For polycrystalline diamond films, the doping efficiency of B is close to unity. It is however less than 10^{-3} for N_2 [94], and in Ref. [91] it was 10^{-4}. Note that the film morphology and the doping efficiency are likely to strongly depend on the CVD reactor design and the growth conditions. Similar effects were confirmed in HFCVD [95, 96]. For N-doping in homoepitaxial growth, see Ref. [97].

5.7. SURFACE ENERGY

The surface energies of (100), (111), and (110) of diamond in a plasma environment (in the presence of H atoms and at high temperature) were evaluated using simple assumptions and an equation [98]. In the surface energy versus T_s diagram for

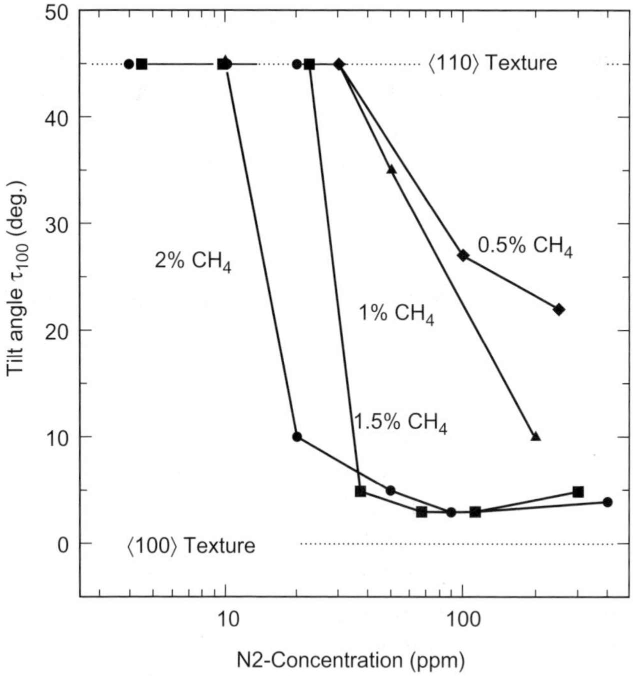

Figure 5.21. Change in the film texture with the N_2 concentration in the source gas [90].

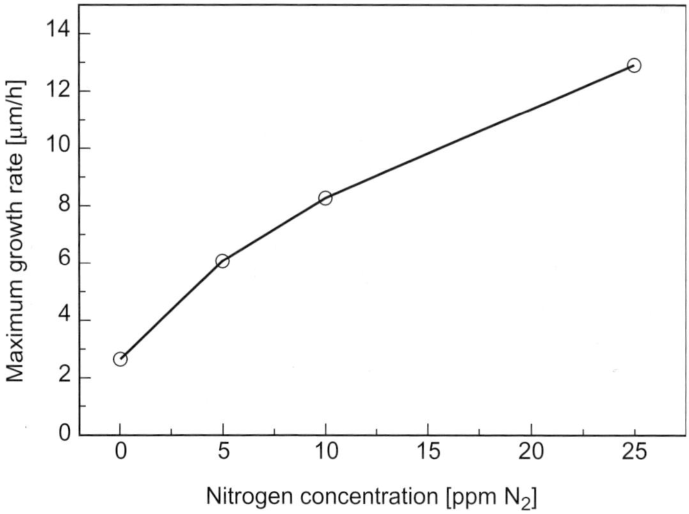

Figure 5.22. Dependence of the maximum growth rate on the N_2 concentration in the source gas [92].

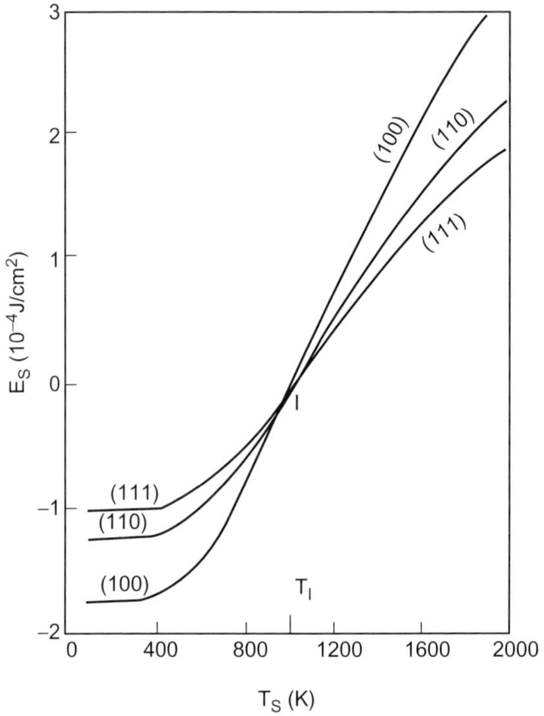

Figure 5.23. Result of theoretical calculation on growth surface energies for (111), (110), and (100). The surface energies intersect at $T = T_I$, when the partial gas pressure of atomic hydrogen $P_H = 100$ Pa (0.75 Torr) [98].

the three surfaces, shown in Figure 5.23, there was a critical temperature T_I, where all surface energies were equal. When the H_2 gas pressure was 2.5×10^4 Pa (188 Torr), the partial pressure of atomic hydrogen was 1×10^2 Pa (0.75 Torr), and T_I was approximately 730 °C. Below T_I, the (100) surface energy was the lowest, while above T_I, the (111) surface energy was the lowest. Below T_I, the diamond surface was almost completely H-terminated. Although the partial pressure of atomic hydrogen and other parameters must be estimated from experiments, this theory is very simple, and will give us an insight into the stability of diamond surfaces, and hence the crystal morphology.

Chapter 6
Formation of Twins

Chapter 6

Formation of Twins

Formation of twin structures on the faces of diamond crystals and films have also been studied by Koidl's group in the early stage of diamond film research [84]. So, we begin with reviewing their works, which is then followed by studies of other groups. To make heteroepitaxial diamond films, it is necessary to avoid the formation of twins, and thus the studies on the formation mechanism and morphology of twins are of great importance.

6.1. TYPES OF TWINS

The temporal evolution of twins has been investigated with an aid of computer simulation by Wild et al. [84]. Figure 6.1 shows a result of a twin evolution on a (111) face of a cubo-octahedral crystal, assuming that $\alpha = 1.75$. Note that time t is in arbitrary units, and the crystal size is normalized. It is seen that a small twin on a (111) face at $t = 0$ laterally increases the area, reaches the adjacent (100) faces, and induces a secondary nucleation on the (100) faces. According to the computer simulation, there are three types of twin evolutions, as shown in Figure 6.2. The types and characteristics of the twins shown in Figure 6.2 are summarized below. Note that the definition $T_{\{ijk\}}$ here is different from that of Section 5.5.

(i) $T_{\{100\}}$: twin on the (100) face
This type of twin evolves and covers part of the (100) face for $\alpha < 2$. It disappears for $\alpha > 2$.

(ii) $T_{\{111\},p}$: twin on the (111) face with the twin plane parallel to the (111) face
For $\alpha < 2$, the twin grows laterally to reach the adjacent (100) faces, and then evolves as $T_{\{100\}}$. For $\alpha > 2$, it grows laterally to reach the adjacent (111) faces, and then evolves as $T_{\{111\},i}$ described below.

(iii) $T_{\{111\},i}$: twin on the (111) face with an inclined twin plane
The twin disappears for $\alpha < 1.5$, while it covers the (111) face for $\alpha > 1.5$.

From the results of Figure 6.2, it is concluded that (111) and (100) faces of diamond are smooth for $\alpha < 1.5$ and $\alpha > 2$, respectively. Qualitatively, when CH_4

and H_2 are used as the source gas, (111) faces are smooth for low CH_4 concentrations and high T_s, while (100) faces are smooth for high CH_4 concentration and low T_s (see Figure 5.7).

It is also of significance that since the shape of twin is sensitively dependent on the α-parameter, as seen in Figures 6.2 and 6.3, the actual α-value under given CVD conditions can also be estimated not only from the crystal morphology,

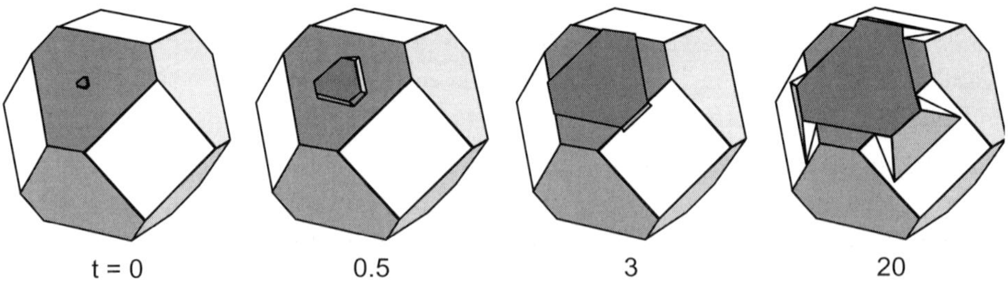

Figure 6.1. Results of computer simulation on the evolution of a twin structure on a (111) face for $\alpha = 1.75$ [84].

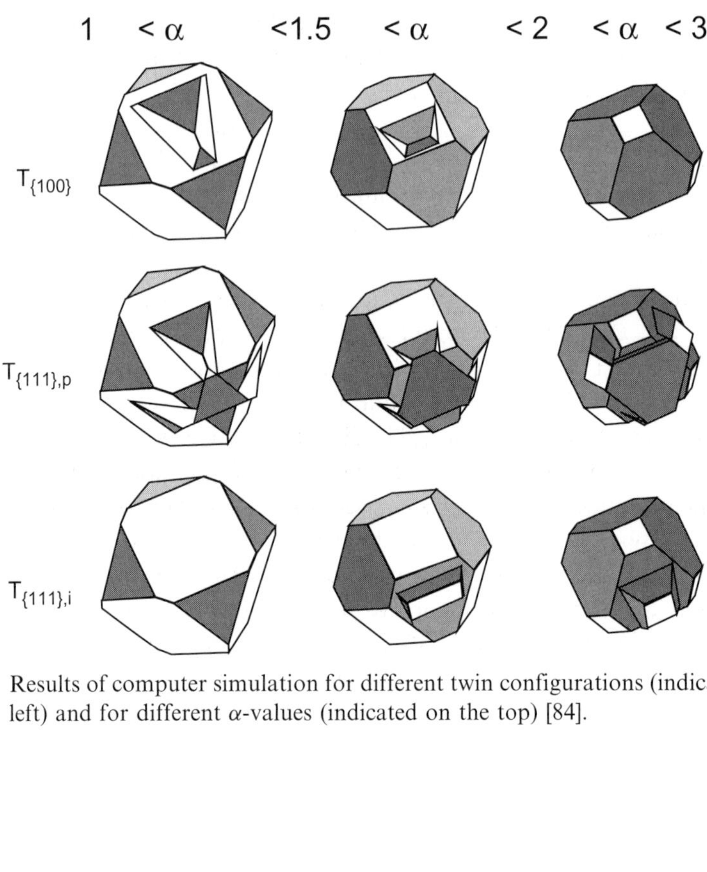

Figure 6.2. Results of computer simulation for different twin configurations (indicated on the left) and for different α-values (indicated on the top) [84].

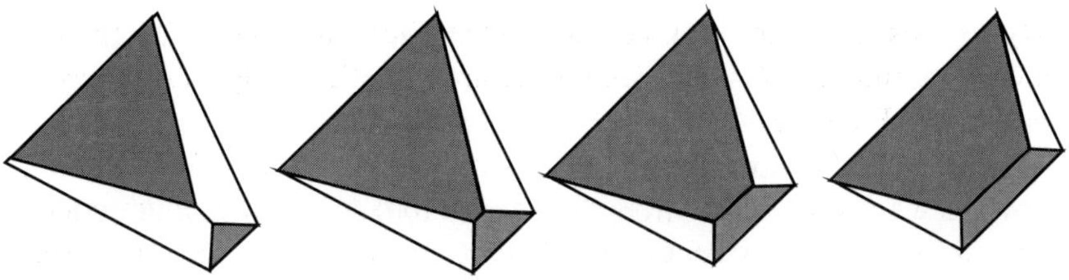

Figure 6.3. Change in the $T_{\{100\}}$ twin shape as a function of α-value [84].

but also from the twin shape at the film surface. Once such data are compiled, it is possible to more precisely control the film morphology, and suppress the secondary nucleation due to twin formation during diamond CVD.

6.2. STRUCTURE OF TWINS

The $T_{\{100\}}$ twin structure was also studied in Ref. [99]. The pyramidal structures, often seen on the (100) faces of HOD films, were identified as penetrating twins, as shown in Figure 6.4 (a). The (111) face of the pyramidal structure tilted by 15.8° with respect to the basal (100) face, and the [221] direction of the twin is parallel to the [100] direction of the basal (100) face. It was inferred that the atomic arrangement of carbon at the beginning of the penetrating twin formation looks like the one shown in Figure 6.4 (b), which is seen from the [110] direction of the basal (100) face.

A morphological study of penetration and contact twins was done by Tamor and Everson in Ref. [100]. Here, the contact twin was defined to always grow on the (111) surface, and shares the [111] direction normal to the twin surface with the parent surface, while the penetration twin was defined to grow on either (111) or (100) surface, and shares the [111] direction that is not normal to the twin's surface with the parent surface. The (100) twin, $T_{\{100\}}$, was inferred to originate from a micro-contact twin on a step, as shown in Figure 6.5, where the dashed lines are grain boundaries. Regarding the penetration twin on the (111) face, $T_{\{111\},i}$, it is like an octahedral pyramid often truncated by a small (100) face, and the [511] direction of the twin is parallel to the [111] direction of the basal (111) face, as shown in Figure 6.6. The (100) face is tilted by 15.8° toward a (110) edge of the basal (111) face. According to their theory [100], for $\alpha < 1.5$, penetration twins persist only on (100) faces, while they are free on (111) faces. For $\alpha > 2$, such twins persist only on (111) faces. In the intermediate range, penetration twins appear on either face.

Formation of hillocks and penetration twins on the (100) surface of single crystal diamond was studied by Tsuno et al. [101]. The misorientation of the (100) surface was less than 3° from the exact (100) lattice plane. A NIRIM-type

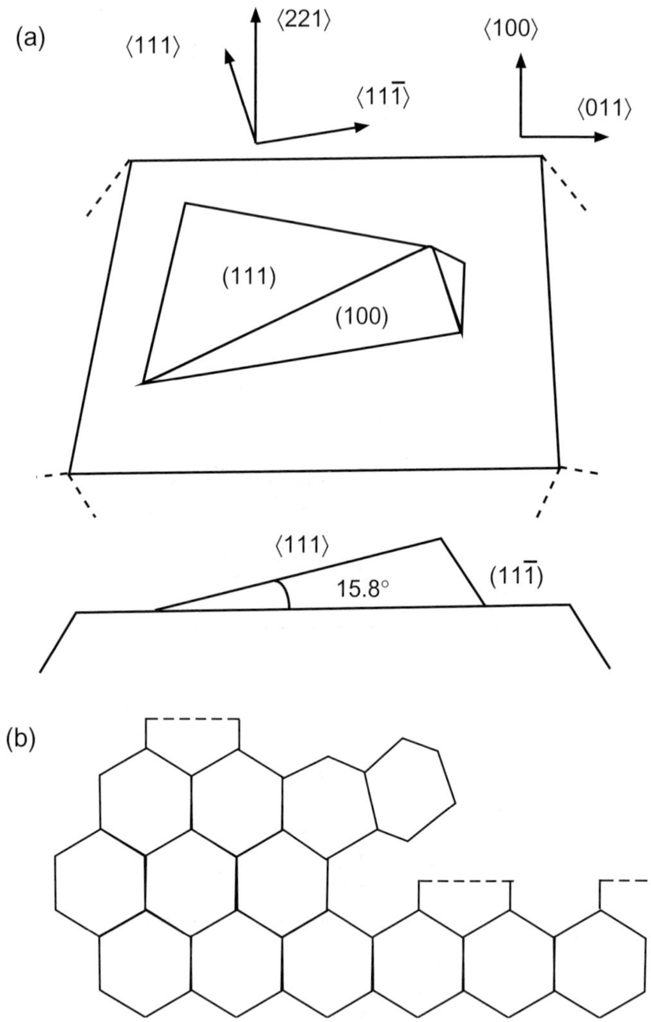

Figure 6.4. (a) Schematic view of a penetrating twin; (b) cross section of the nucleus of a penetrating twin viewed along the [110] direction. The dashed lines are dimer bonds. Surface-bonded hydrogen atoms are not depicted [99].

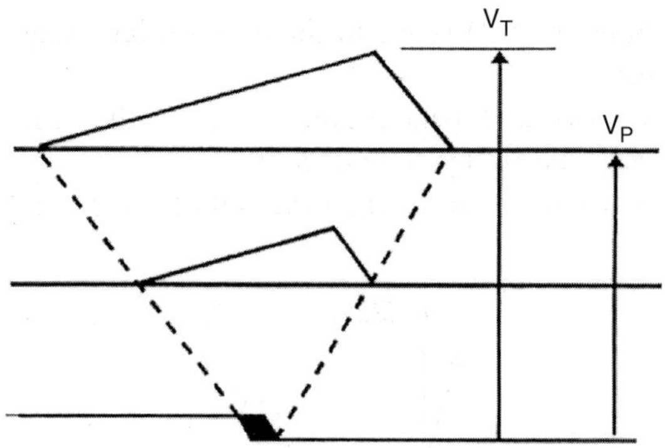

Figure 6.5. Evolution of a micro-contact twin that has been created on a step. The dashed lines are grain boundaries [100].

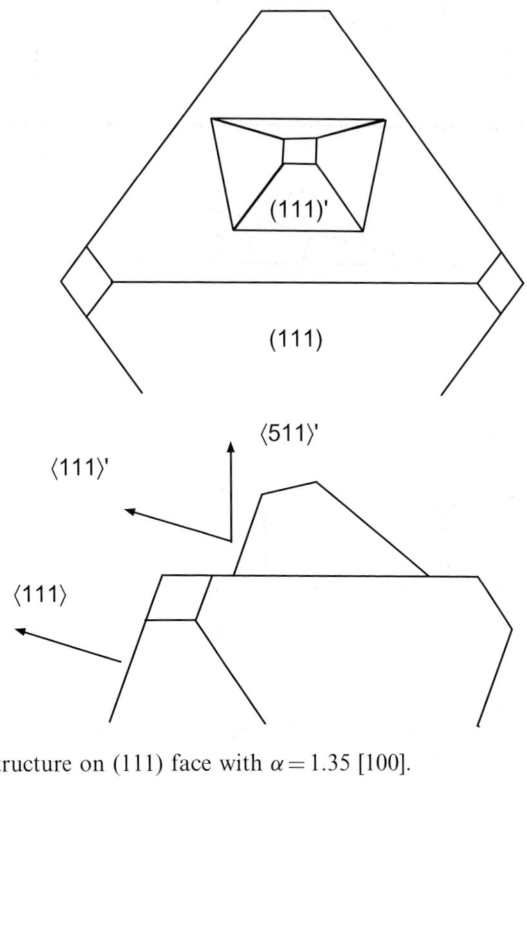

Figure 6.6. Twin structure on (111) face with $\alpha = 1.35$ [100].

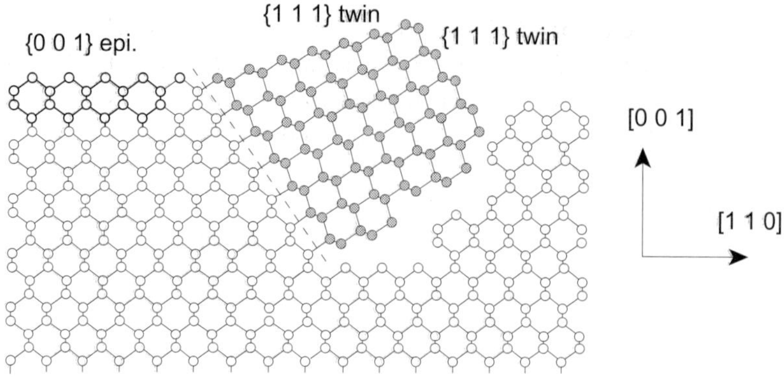

Figure 6.7. Atomic configuration of {111} twin at a single crystal diamond surface [101].

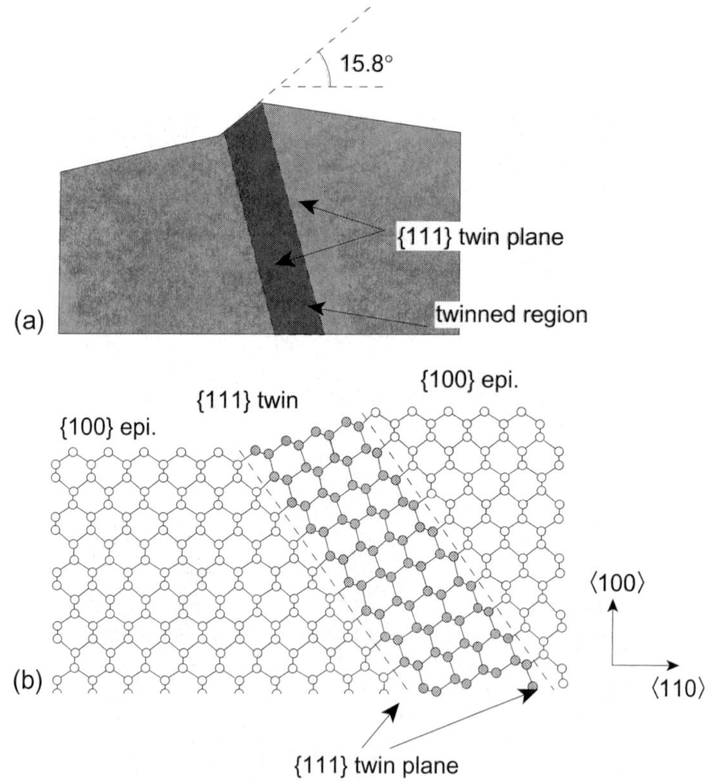

Figure 6.8. (a) Surface structure with a growth hillock and a penetration twin, and (b) its atomic configuration [101].

reactor was used for diamond CVD under conditions of $T_s = 950\,°C$ and $P = 80$ Torr. The source gas was a mixture of CH_4 (3 sccm), H_2O (2 sccm), H_2 (200 sccm), and Ar (50 sccm). As the initial point for penetration twin, it was proposed that a nucleus of the twin was created in an etch pit as depicted in Figure 6.7, similar to Figure 6.5 [100]. In addition, the structure of growth hillocks that were often seen in homoepitaxial growth was understood by the presence of {111} twins, as shown in Figure 6.8. The presence of twin structures was considered to accelerate the diamond growth.

6.3. FIVE-FOLD SYMMETRY

In diamond CVD, diamond crystals with five-fold symmetry are very often seen, as shown in Figure 6.9. The atomic structure in the core region was observed by HRTEM [102], as shown in Figures 6.10 and 6.12. Figure 6.10 shows a core of a diamond crystal with five-fold symmetry. The tetrahedral sectors are bounded by {111} planes, sharing a common [110] axis and twinned relative to

Figure 6.9. Diamond crystals with five-fold symmetry.

each other. The electron beam of TEM is parallel to the [110] axis. Such crystals have a decahedral twin structure with a 7.5° misfit, as schematically shown in Figure 6.11 [102]. Figure 6.12 is a magnified view of the core region of Figure 6.10. It is seen that the core center is not just a merging point

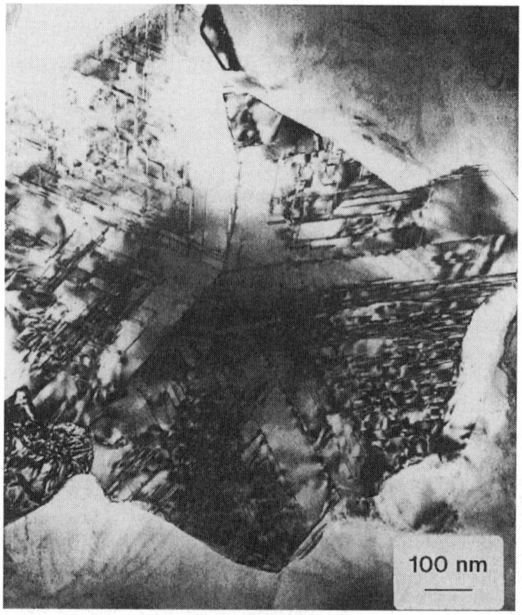

Figure 6.10. TEM micrograph of a multiply twinned particle [102].

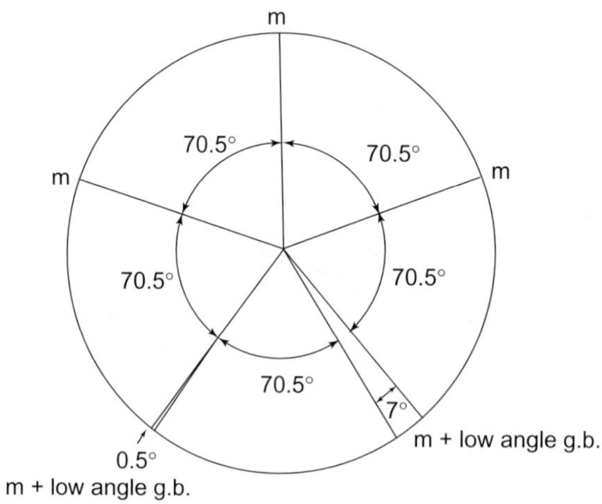

Figure 6.11. Schematic diagram showing defects in a twinned particle. Here, "m" means misfit dislocation, and "g.b." stands for grain boundary [102].

of tetrahedrals, but has a quite complex structure, in spite of the fact that the global view of Figure 6.10 appears to have a rather simple structure. Regarding the five-fold symmetry, see also Ref. [103].

Summarizing the results described so far, whether twins are formed on (100) or (111) faces of diamond crystals, depends on the value of the α-parameter under CVD conditions used. For practical applications, it is often necessary to make diamond films with flat surfaces, and thus one must determine CVD conditions

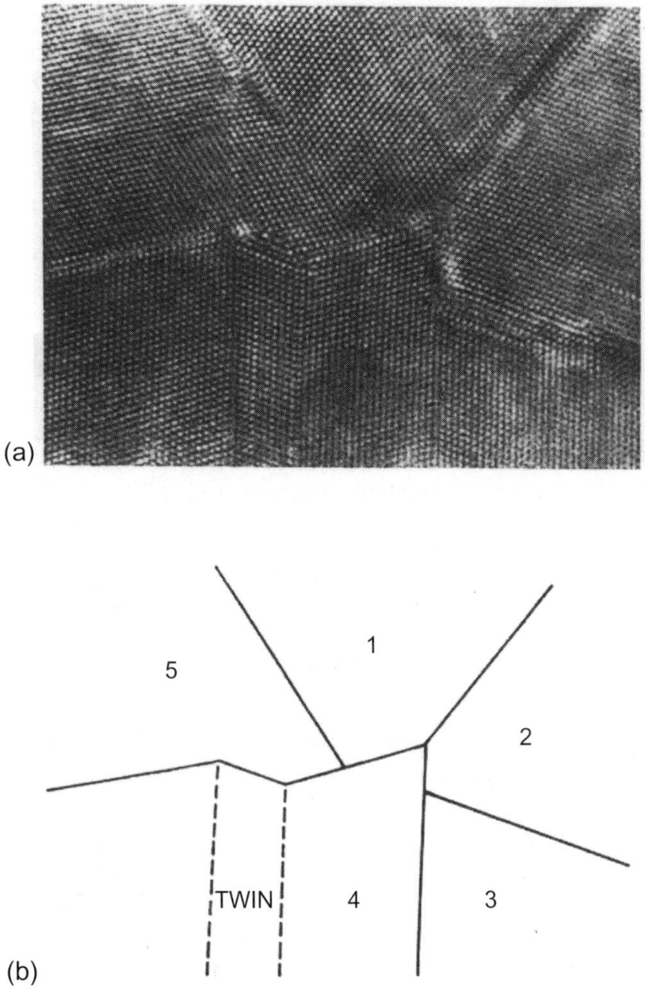

Figure 6.12. (a) HRTEM micrograph of the core of a five-fold multiply twinned crystallite, and (b) an illustration of the growth sectors of (a) [102].

that prevent the twin formation. In Ref. [104], it was demonstrated by HFCVD that two-dimensional secondary nucleation on (111) faces of diamond can be suppressed by using CH_4+O_2/H_2 (O/C $\geq$ 0.8) as the source gas under conditions of $P = 100$ Torr and $T_s = 920\,°C$. To control the film morphology, it is prerequisite to investigate the α-parameter in a wide range of P, T_s, P_m, the source gas composition (e.g. C, H, O, B, N) and other factors such as substrate position. It should be noted that the value of the α-parameter may not be constant even though the CVD conditions are fixed, because α increases as the film is being formed [105].

Chapter 7
Homoepitaxial Growth

Chapter 7

Homoepitaxial Growth

Growth and defect information of homoepitaxial diamond layer is very useful for oriented and heteroepitaxial growth of diamond, because once the non-diamond substrate is covered with a diamond film, it is then nothing but homoepitaxial growth that is occurring. There are a number of works on homoepitaxial growth published so far so that only selected papers that are related with heteroepitaxial growth is reviewed below. For a wider coverage on homoepitaxial diamond growth, the readers can refer to Refs. [19, 24–31]. Most recently, an extremely high growth rate of 50–100 μm/h was achieved [106] under conditions of $c = 12\%CH_4/H_2 + 3\%N_2/CH_4$, $P = 160$ Torr, and $T_s = 1200$–$1220\,°C$. The reactor was Wavemat's 6 kW, and the substrate holder was redesigned for the high growth rate. Such an achievement should be used for oriented and heteroepitaxial growth in the future.

7.1. GROWTH KINETICS

The growth kinetics of (100) and (111) surfaces were studied by Maeda et al. [86] using a NIRIM-type reactor (the source gas was CH_4/H_2 and $P = 40$ Torr), and the results are shown in Figures 7.1 and 7.2, respectively. It is seen from these figures that the growth rate is strongly dependent on the CH_4 concentration for the (100) surface, while it is only weakly dependent for the (111) surface. These results are in strong contrast to those of homoepitaxial diamond growth by MPCVD using CO and H_2 as the source gas [107]. The CVD reactor was a NIRIM-type, and the growth conditions were $c = 5$–$40\%CO/H_2$, $P = 30$ Torr, $T_s = 900\,°C$, and $P_m = 200$ W. Figure 7.3 shows the growth rate of (100) and (111) surfaces. It is seen that unlike the above mentioned results of Figures 7.1 and 7.2 [86], the growth rate of (111) is roughly one order of magnitude higher than that of (100). This is firstly because a high CO concentration was used as oxygen can react and remove non-diamond carbons more effectively than atomic hydrogen. Secondly, oxygen may give a strong influence on surface reactions for diamond growth that depend on the surface atomic structures of diamond.

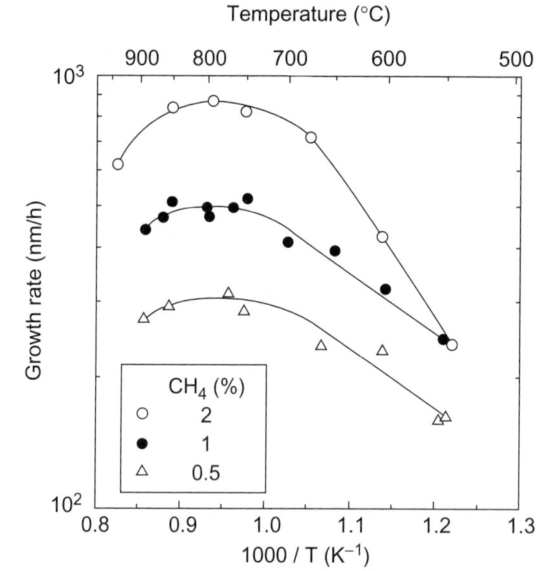

Figure 7.1. Growth rate of (100) surface as a function of CH_4 concentration and T_s. $P = 40$ Torr [86].

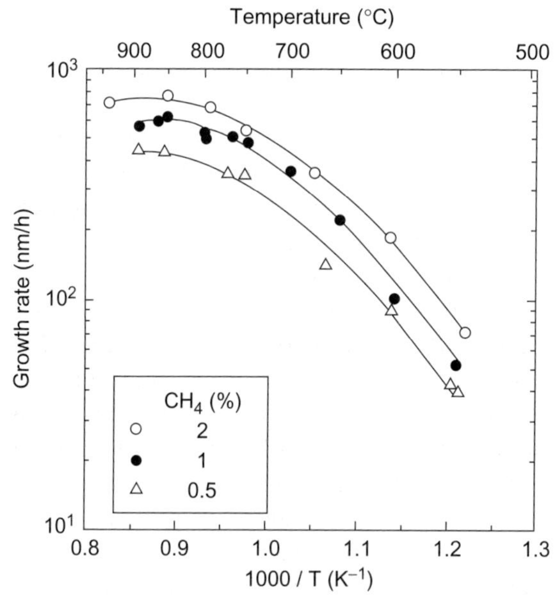

Figure 7.2. Growth rate of (111) surface as a function of CH_4 concentration and T_s. $P = 40$ Torr [86].

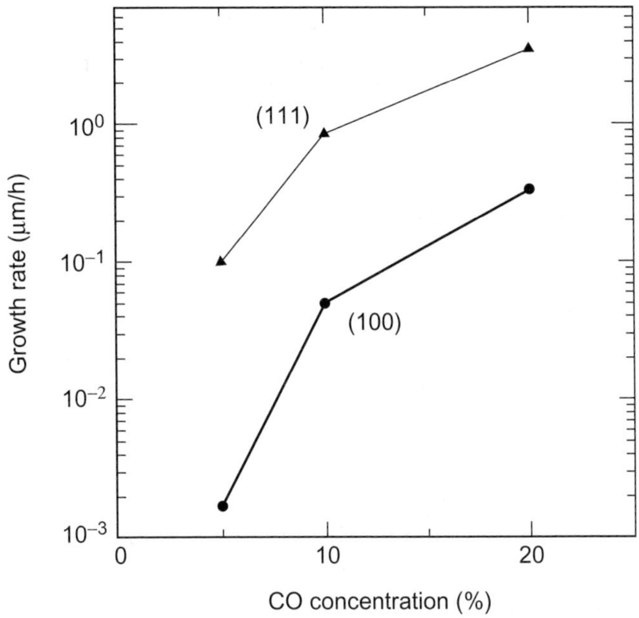

Figure 7.3. Homoepitaxial growth rate using CO/H_2 as the source gas. $P = 30$ Torr [107].

7.2. GROWTH ON OFF-ANGLE DIAMOND SURFACES

Growth of diamond on single crystal diamonds with (100) surfaces with off-angles [108] was undertaken using various CH_4 concentrations and T_s by a NIRIM-type reactor. The results are shown in Figure 7.4. A smooth surface was obtained for $c = 1\%CH_4/H_2$ and $T_s = 1000\,°C$ (see Figure 7.4 (c)).

In Ref. [109], diamond films were deposited using a NIRIM-type reactor on (100) surfaces with off-angles of 0.1°, 3.5°, and 11.0°. The CVD conditions were $c = 1$, 2, and $6\%CH_4/H_2$, $P = 90$ Torr, and $T_s = 875$ and 1200 °C. By increasing the off-angle, the surface morphology changed from hillocks to macro-steps. The hillock growth took place when a two-dimensional nucleation occurred on terraces when the surface step density was low, while the step-flow growth occurred along the [110] direction when the step density was high. The macro-step structures are illustrated in Figure 7.5, and a map of the surface morphology on the plane of the CH_4 concentration and the off-angle is shown in Figure 7.6. It was inferred from the experimental results that the step-flow growth was favored for larger off-angles, lower CH_4 concentrations, and higher substrate temperatures under the CVD conditions used.

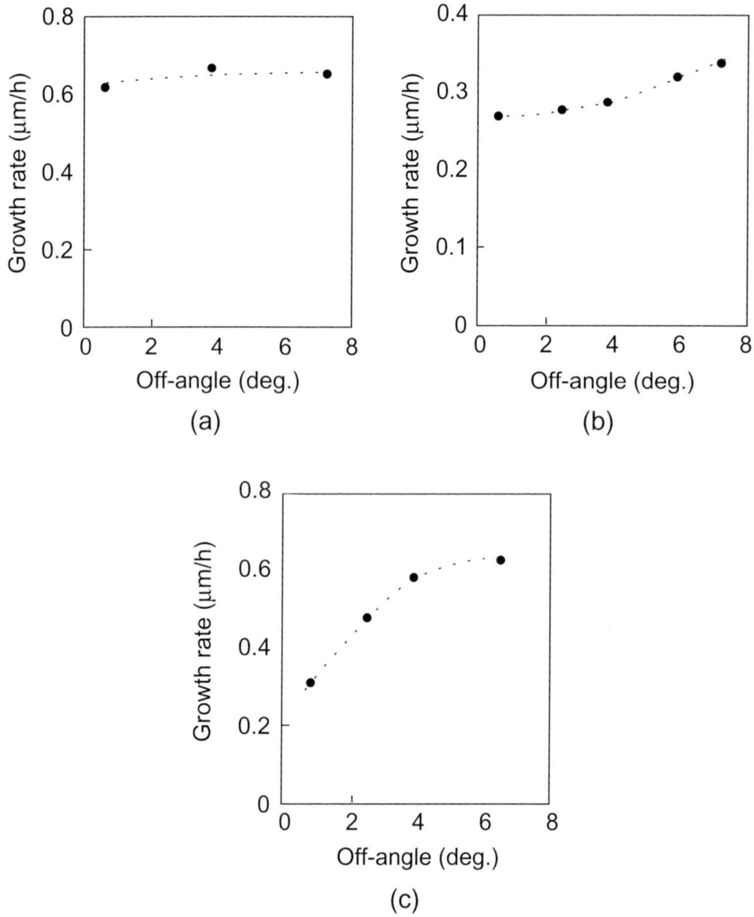

Figure 7.4. Growth rate on off-angle (100) substrates: (a) $c = 6\%CH_4/H_2$ and $T_s = 800\,°C$, (b) $c = 1\%CH_4/H_2$ and $T_s = 800\,°C$, and (c) $c = 1\%CH_4/H_2$ and $T_s = 1000\,°C$. The tilt of the off-angles was toward [110] $\pm 2°$. The gas pressure was $P = 60$ Torr for all cases [108].

An occurrence of a single step-flow growth on a single crystal diamond (100) with a low off-angle was demonstrated using an ASTeX reactor for the first time in Ref. [110]. Successively, Okushi's group has done a number of works related to the single step-flow growth and its characterizations [55–58, 75]. Among these works, Ref. [55] showed that a use of $c = 0.025–0.005\%CH_4/H_2$ for diamond CVD resulted in an atomically flat surface over a $4 \times 4\,mm^2$ area. In this case, a step-flow growth with long ledges occurred when the misorientation angle was small. Although the growth rate was considerably small, the specimens

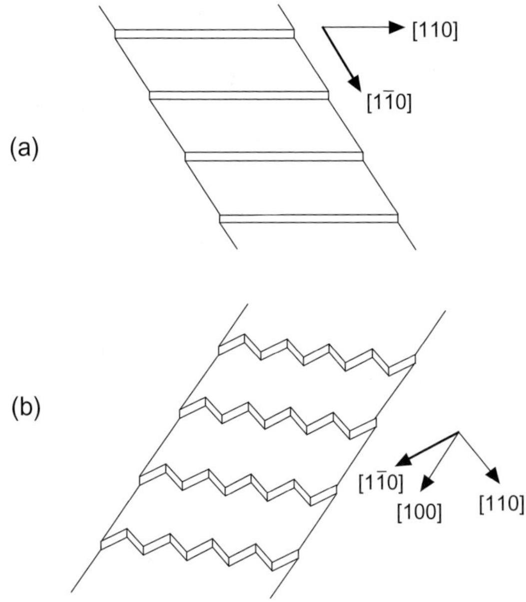

Figure 7.5. Configuration of macro-steps on the off-angle (100) surfaces tilted toward (a) [1$\bar{1}$0] and (b) [100] [109].

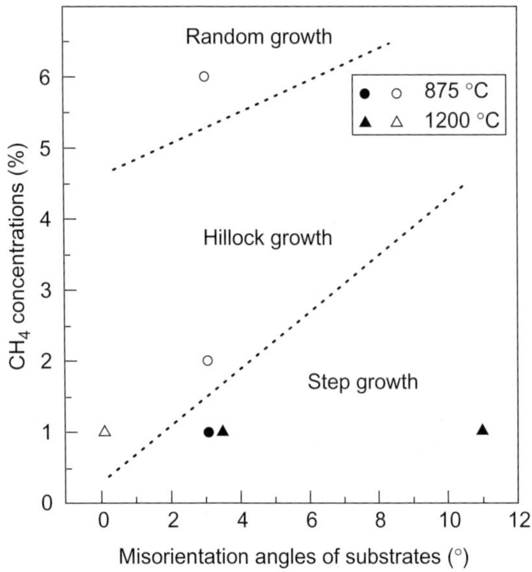

Figure 7.6. Dependence of the surface morphology of diamond films on the off-angle (misorientation angle) of the (100) substrate and the CH_4 concentration. Diamond films were grown at $T_s = 875$ and 1200 °C [109].

thus made had such a low defect density that an intense cathodoluminescence band due to free excitons was observed even at room temperature. Although the CVD conditions are different, these results are totally opposite to the above mentioned inference.

7.3. INTERNAL STRESS

A homoepitaxial growth on (111) surface of natural diamond was undertaken in Ref. [111] by MPCVD for 48 h using CH_4 and H_2 as the source gas at $T_s = 800\,°C$ to deposit an approximately 20-μm thick layer. X-ray double crystal topography indicated that 20-μm deep cracks with triangular symmetry were formed at the layer surface down to the original substrate surface along [110] directions due to the stress between the natural diamond substrate and the deposited layer. The fractional variation of the lattice constant ($\Delta d/d$) due to the strong strain was evaluated to be $(\Delta d/d) = 8.7 \times 10^{-4}$ even after the cracking occurred.

Since diamond CVD is undertaken at $T_s \sim 800\,°C$, and the stress at the interface is generally very high when the specimen is taken out of the reactor to the ambient environment because of the difference in the thermal expansion coefficients between the CVD diamond film and the substrate. In addition, the intrinsic stress within the CVD diamond film is high if it is polycrystalline. The presence of grain boundaries can be the cause of the intrinsic stress. Thus, reducing both interfacial and intrinsic stresses is an important issue for practical use of CVD diamond films. So far, no effective method has been found to solve this problem.

7.4. DEFECT STRUCTURES

Defect structures in a homoepitaxial diamond layer grown on a (100) surface of single crystal diamond were studied in detail by cross-sectional HRTEM. [112,113] The diamond CVD was done by MPCVD under conditions of $c = 1\%CH_4/H_2$, $P = 40$ Torr, and $T_s \sim 850\,°C$. Figure 7.7 (a) shows defects in the homoepitaxial layer on the (100) surface. It is seen that there are needle and oval-shape defects, both of which consist of loops that are stacking parallel to {111} planes and approximately 50-nm wide. The loops are aligned along $[\bar{1}12]$ or $[1\bar{1}2]$ directions, as illustrated in Figure 7.7 (b). Based on the HRTEM images, it was concluded that the defects are interstitial-type Frank dislocation loops with deformed configurations. The horizontal dark line in

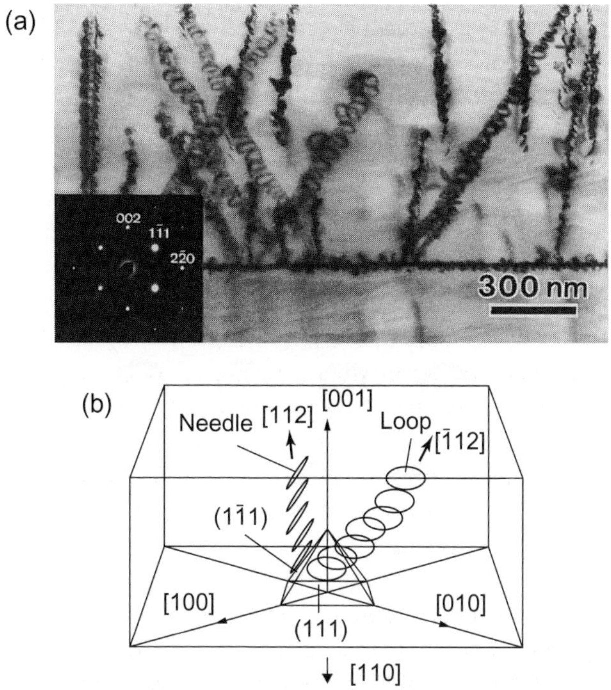

Figure 7.7. Defects in a homoepitaxial layer. (a) HRTEM image and (b) an illustration of the defect structures shown in (a) [112, 113].

Figure 7.7 (a) is the (100) interface between the basal single crystal diamond and the CVD diamond layer. There exist many defects of 5–20 nm in diameter along the interface. Also, there were many stacking faults and microtwins at the $\{1\bar{1}1\}$ corner of the diamond specimen. Similarly, stacking faults and microtwins were present on the $\{1\bar{1}0\}$ corner, and dislocations were dominant in the near surface region.

Defects in CVD diamond were studied in Ref. [102]. Figure 7.8 shows a presence of a mirror plane (indicated by M) in a crystal, and Figure 7.9 includes both mirror planes and {111} stacking faults (indicated by SF). These data suggest that CVD diamond in general contains numerous defects in the atomic scale.

7.5. HIGH QUALITY DIAMOND GROWTH

It may not be an overstatement to say that homoepitaxial growth technology on (100) surface has been revolutionarily changed by Okushi's group by using a very low concentration of CH_4/H_2, such as 0.05%, as the source gas, as described in Section 7.2 [55–58]. First, it was demonstrated that a step-flow growth actually occurs on a less than 3°-off (100) surface [114, 115]. More recently, they were successful in growing a smooth diamond layer without morphological imperfections such as unepitaxial crystallites, pyramidal hillock and macroscopic steps on a 0.4°-off (100) surface using $c = 0.025\%CH_4/H_2$ [116]. In this case, the specimen surface consisted of atomically flat terraces, and single or double-height atomic steps. The band A (400–600 nm) in the CL spectrum is known to arise from defects

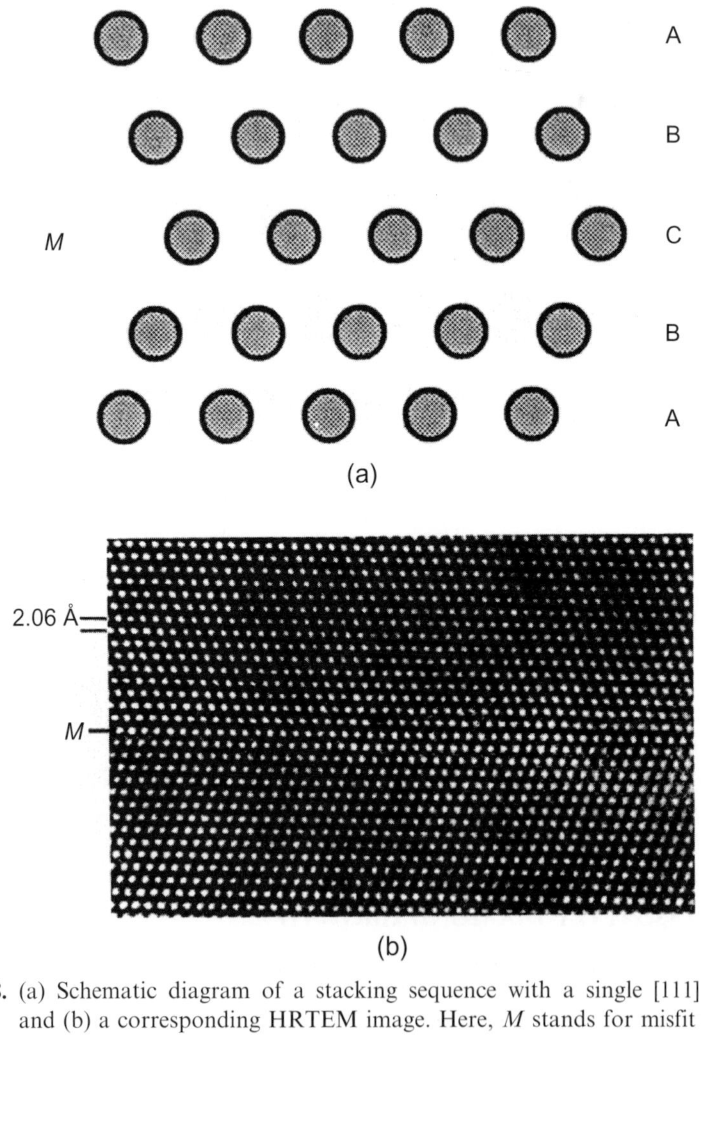

Figure 7.8. (a) Schematic diagram of a stacking sequence with a single [111] twin plane, and (b) a corresponding HRTEM image. Here, *M* stands for misfit [102].

(see Section A.6 in Appendix A on "band A"), but it was absent in their specimens. According to Refs. [117, 118], band A appeared only when unepitaxial crystallites grew on the (100) surface. Thus, a use of (100) single crystal diamond with small off-angles as the substrate and a very low CH_4 concentration as the source gas for diamond CVD lead to a very high quality homoepitaxial CVD diamond layer due to an atomic scale step-flow growth. Electronic properties of such high quality diamond layers were very high and unique [119–128].

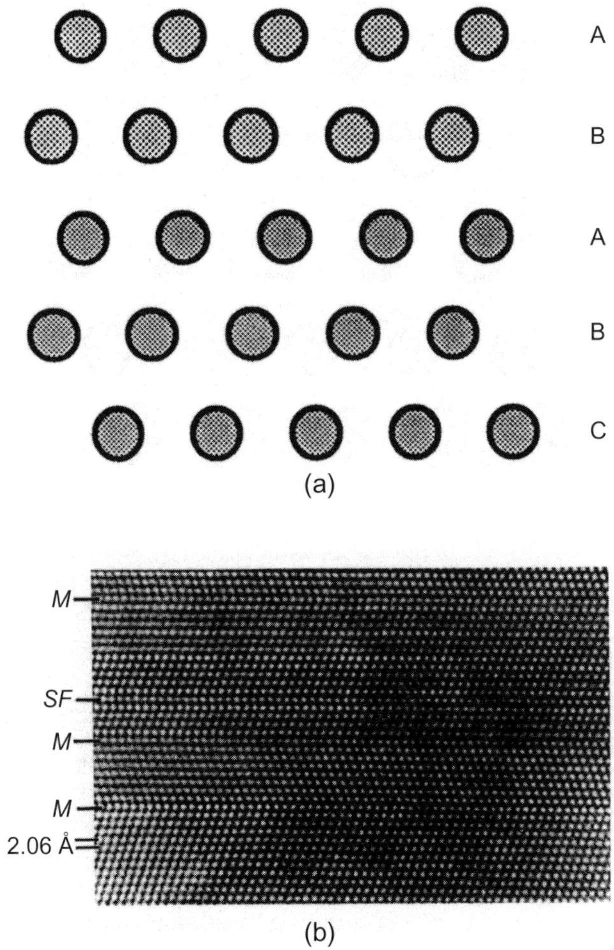

Figure 7.9. (a) Schematic diagram of a {111} stacking fault viewed in a [110] projection and (b) a corresponding HRTEM image. Here, *M* and *SF* stand for misfit and stacking fault, respectively [102].

7.6. ENLARGEMENT OF SINGLE CRYSTAL DIAMOND SURFACE AREA

One of the problems of homoepitaxial technology is that the surface area of diamond is very small, usually less than 4 mm × 4 mm, and hence practical applications such as electronic devices using homoepitaxial diamond layers are very difficult in spite of the fact that the film quality is very high. To increase the surface area of single crystal diamonds, a formation of so-called "mosaic" structures have been attempted in Refs. [129, 130]. For the substrate, two diamond plates with (100) surface were placed side by side so that they formed a flat surface. A diamond deposition was then undertaken by HFCVD for about 16 h and 34 h, and the deposited diamond layers were 14- and 40-μm thick, respectively. After the growth, there was a band of enhanced growth along the contact boundaries between the two diamond plates. This band contained screw dislocations and stacking faults.

A similar work was done in Refs. [131, 132] using less than 7 diamond plates with (100) surfaces as the basal substrate. For the determination of crystallographic orientations of the diamond plates, electron back scatter pattern (EBSP) was used and found to be very useful (the depth resolution was 40 nm). The diamond growth was done by MPCVD using a SAIREM bell-jar type reactor (see Figure 3.7) under conditions of $P = 45$–56 Torr, $P_m = 1100$–1300 W, and $T_s = 790$–$950\,°C$ [71]. An optimized step-flow growth occurred when the off-angle of the diamond (100) surface was 2.5°, $c = 1\%CH_4/H_2$, and $T_s = 800\,°C$. Such conditions for step-flow growth are consistent with those of Badzian et al. [109] described in Section 7.2. To suppress the formation of twins and hillocks, N_2 of 10 ppm was added to the source gas. The diamond growth at the contact boundaries was strongly influenced by (i) the crystallographic orientations of the basal diamond plates, (ii) the difference in heights and orientations between adjacent diamond plates, (iii) the difference in off-angles between adjacent basal diamond plates, and finally (iv) the CVD conditions. Note that the off-angle determines the direction of the step flow (see Figure 7.5). For diamond plates with (100) surfaces, the growth structure at the contact boundaries was categorized into three types, A–C. For type A, diamond plates were completely merged, but a band of enhanced growth was present along the contact boundaries. The band position shifted by 40 μm for a 20-μm thick layer from the original position of the contact boundaries, but the diamond layer was epitaxial. For type B, the shift was small but the diamond layer was not necessarily epitaxial. For type C, one of the layers grew over the other, and macrosteps were present on the whole layer. As a conclusion of their works, to realize a perfect single crystal diamond layer across the different diamond plates, the difference in crystallographic orientations must be less than 2°, and the plates must have the same height.

An interesting attempt was done by Geis et al. [133, 134], to make a single crystal diamond, starting from small diamond crystals. In Ref. [133], diamond crystals of about 100 μm in size were seeded in an array of reversed pyramid-shape pits made on a Si(100) wafer by anisotropic etching using KOH so that the (100) faces of the diamond crystals were aligned parallel to the Si wafer surface. Successively, diamond CVD was undertaken to grow the diamond crystals and make a continuous film. SEM images of the specimen before CVD and after a 240-μm deposition are shown in Figure 7.10.

In Ref. [134], cubic diamond seeds (250 μm on a side) were used. By wetting them, they spontaneously stuck together to form a plate structure of approximately 1 cm^2. Using those seeds as a mosaic substrate, a coalesced single diamond plate was made by depositing a diamond layer of 20-μm thickness, as seen in Figure 7.11. Using the current technology of diamond growth, it is fairly easy to make even a thicker diamond layer. It must be noted, however, that not all diamond crystals were aligned exactly in the same direction, but there existed an angular distribution on their orientations within ±1°. It has not yet been investigated whether this angular distribution is removed by depositing a thicker diamond layer.

Recently, a mosaic technology has been further developed in Refs. [135, 136]. In this case, 16 pieces of diamond Ib (100) plates of 4 mm × 4 mm in size were used, and a diamond layer of 1-mm thickness was deposited on them. Since a custom-made MPCVD reactor was used to achieve a fairly uniform

(a) (b)

Figure 7.10. (a) Diamond crystals seeded on Si and (b) after a growth for 80 h with the overlayer thickness of ~240 μm [133].

plasma distribution and a uniform substrate temperature, there was no abnormal growth in the deposited diamond, and the surfaces at the contact boundaries were smooth. As a result, a diamond plate of 16 mm × 16 mm in size was made. Successively, the basal diamond plates were removed to make a free-standing CVD diamond plate, as seen in Figure 7.12. It seems that the plate is not precisely a

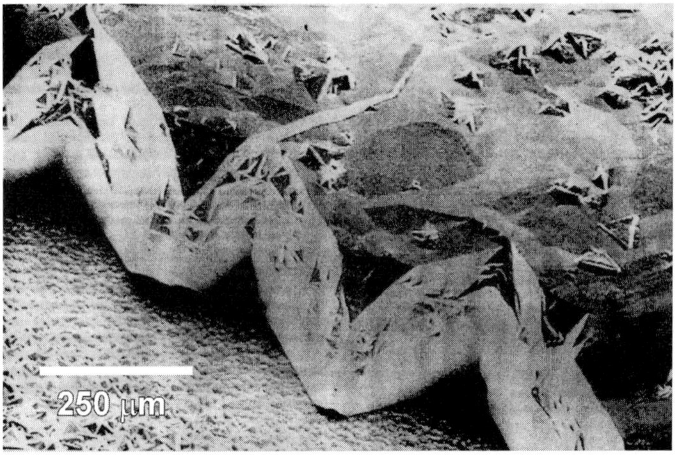

Figure 7.11. Cubic diamond mosaic substrate with an approximately 20-μm thick diamond layer [134].

Figure 7.12. Optical microscopic image of a free-standing, 1-mm thick CVD diamond plate made of 16 pieces of 4 × 4 mm diamond mosaic plates. The basal diamonds were removed [135, 136].

single crystal because of small misorientations between the basal diamond plates, but the quality (nitrogen impurity content, for instance) of the free-standing CVD diamond plate was found to be better than natural IIa diamond. More importantly, it is considerably easier to fabricate diamond devices such as field effect transistors on the large surface of the quasi-single crystal plate than on 3 to 4-mm square single crystal diamonds that are commercially available at present.

Chapter 8
Surface Reconstruction

Chapter 8

Surface Reconstruction

Although this monograph is mainly concerned with the oriented growth of diamond films, it would be worthwhile to briefly review the atomic structures of diamond surfaces studied by STM and AFM. In this regard, Ref. [137] is comprehensive and will be very useful. Additional descriptions on surface reconstruction are given in Appendix D.

8.1. (100) SURFACE

RHEED and STM studies of a homoepitaxial diamond layer grown on a type Ib diamond (100) surface was first carried out by Tsuno et al. [138] by MPCVD using $c = 6\%CH_4/H_2$ at $T_s = 830\,°C$. The RHEED pattern exhibited 2×1 or 1×2 dimer structures along [110] and $[\bar{1}10]$ directions, respectively. The STM images of these structures are shown in Figures 8.1 (a)–(c), where the observed area is 75 Å × 75 Å. In Figure 8.1 (a), the direction of the dimer rows in area A is perpendicular to those in area B, which is lower than area A by 0.9 Å. Similar features are also seen in Figure 8.1 (b). This corresponds to a monatomic step, 0.89 Å. In Figure 8.1 (c), a double height step of 1.8 Å is seen between areas C and D. It was inferred that the atomic arrangement in the dimer rows was such that depicted in Figures 8.2 and 8.3. Indeed, two dimer rows were separated by 5.04 Å in the STM observation, consistent with the theoretical value. The dimer structure, shown in Figure 8.1 (b), was inferred to be C(100) 2×1 : H.

For possible structures of diamond (100) surface, see Appendix D [139]. Also, a study of reconstructed (100) and (111) surfaces was presented in Ref. [140].

A more detailed study of (100) surface was done in Ref. [141]. The diamond CVD was undertaken using $c = 5\%CO/H_2$ on the (100) surfaces of single crystal diamonds. Figures 8.4 (a)–(c) show typical STM images of (100) 2×1 dimer rows. Also, STM images of B-doped diamond layer on (100) single crystal diamond were observed. Figure 8.5 shows a local structure of the reconstructed surface with a S_A step and the unit cell. For the S_A step as well as more information of (100) surface reconstruction, see Appendix D.

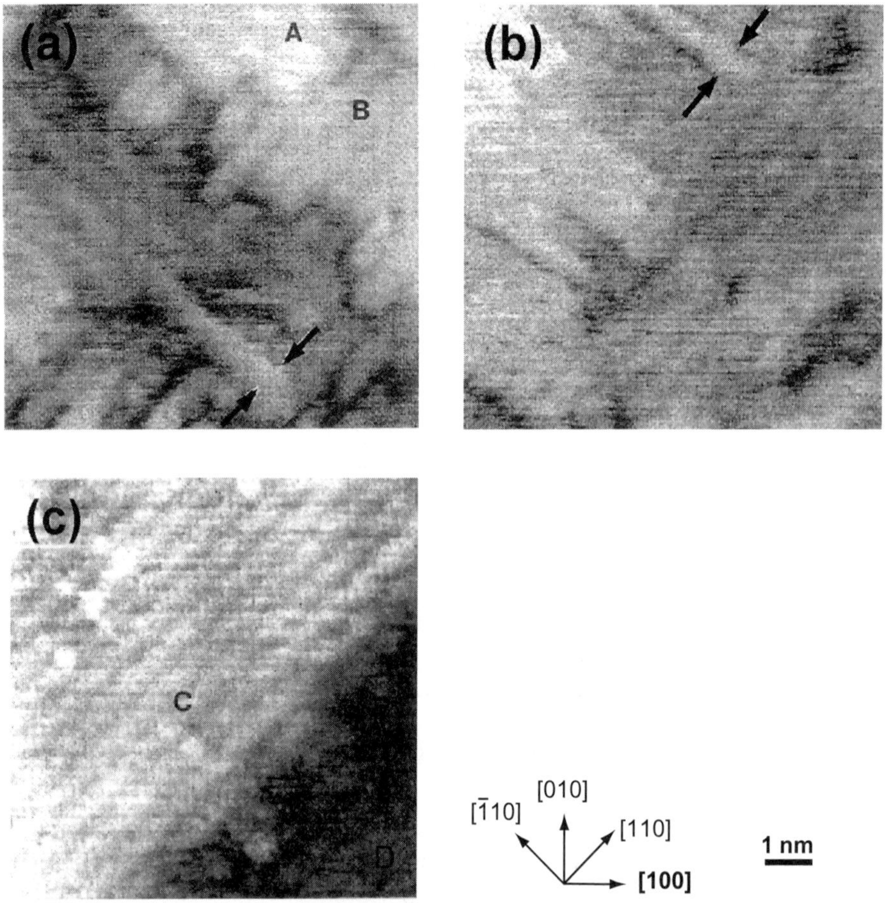

Figure 8.1. STM images of a homoepitaxial diamond layer surface deposited on single crystal diamond (100) surface [138].

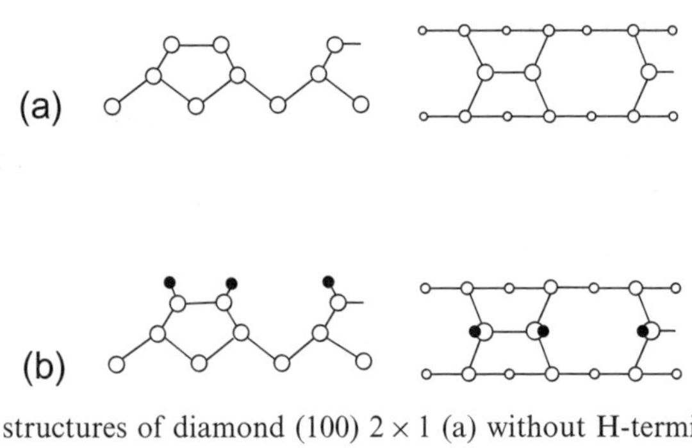

Figure 8.2. Dimer structures of diamond (100) 2×1 (a) without H-termination and (b) with H-termination [138].

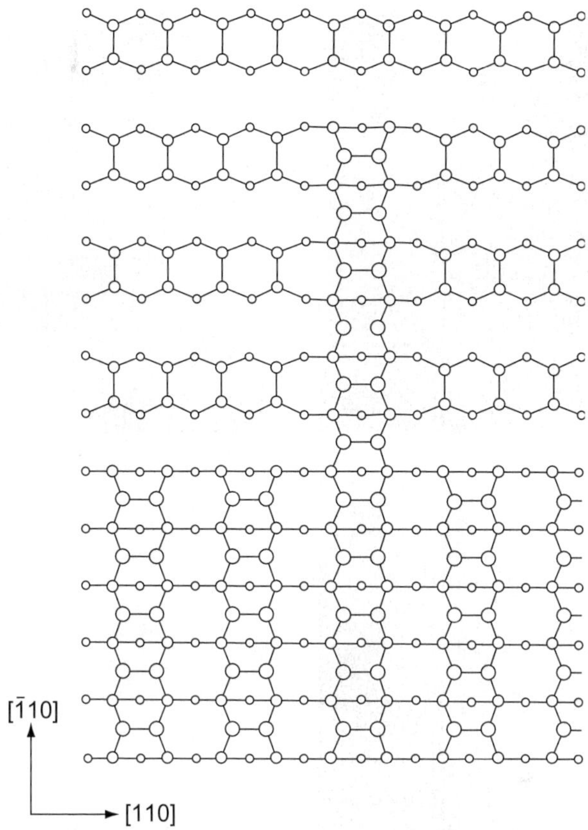

Figure 8.3. Extension of a dimer row. The larger circles represent upper carbon atoms [138].

8.2. (111) AND (110) SURFACES

Reconstruction of (111) surfaces has been investigated by STM [142, 143]. A diamond layer was deposited on bulk diamond Ib (111) under conditions of $c = 1\text{–}2\% CH_4/H_2$, $P = 60$ Torr, and $T_s = 850\,°C$. Consequently, a 1×1 atomic structure was resolved. The observed steps were extended along the [112] direction, and had a height of 0.2 nm, in coincidence with the theoretical value of a single bilayer step, 0.206 nm. The surface should be either H-terminated of CH_3-terminated, but it has not been uniquely determined from STM images. AFM studies of undoped and B-doped polycrystalline diamond films were done in Ref. [144].

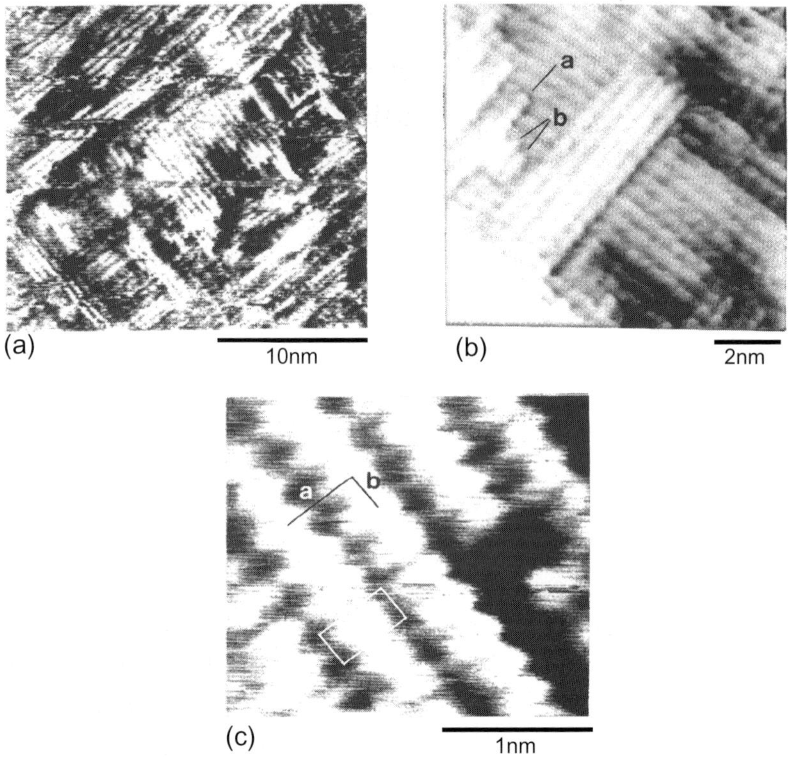

Figure 8.4. STM images of undoped, *as grown* CVD diamond surfaces: (a) current image of 25 nm × 25 nm area, (b) topographic image of 10 nm × 10 nm area, and (c) current image of 2.5 nm × 2.5 nm area [141].

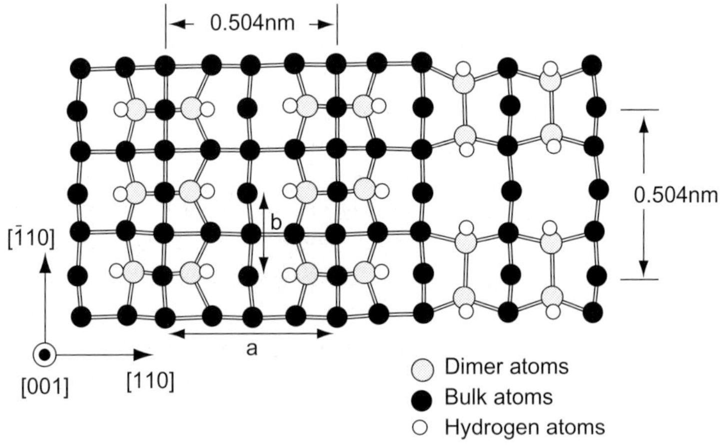

Figure 8.5. Reconstructed (100) 2 × 1 : H surface with a S_A step. The unit cell is a × b [141].

The (100), (110), and (111) surfaces of homoepitaxial diamond layers, grown on type IIa diamonds by HFCVD, were observed by an atmospheric AFM in Ref. [145]. On the (100) surface, there was indication of the presence of 2×1 reconstructed dimers, though no atomic image was observed. Other surfaces were found to be quite rough. A STM study of (111) faces on polycrystalline

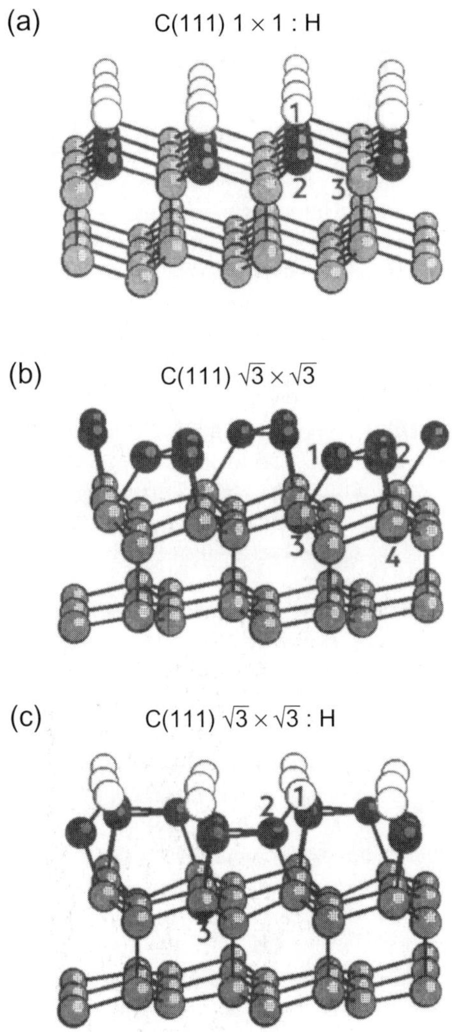

Figure 8.6. Reconstructed (111) surface structure proposed in Ref. [140]. The open circles are H atoms.

diamond films made by HFCVD was presented in Refs. [139, 140, 146]. On the (111) faces of an *as grown* diamond grain, dimer rows were observed to make an angle of about 120° to each other. The possible structures were investigated with an assistance of electronic state calculations. Figures 8.6 (a)–(c) show stable surface structures of (a) (111) 1×1 : H, (b) (111) ($\sqrt{3} \times \sqrt{3}$) R30°, and (c) (111) ($\sqrt{3} \times \sqrt{3}$) R30° : H, respectively. Figures 8.7 (a) and (b) are the top views of Figures 8.6 (b) and (c), respectively.

Other stable reconstructed structures of (111) surface are shown in Figures 8.8 (a) and (b). Figure 8.8 (a) is a 2×1 single chain structure, while Figure 8.8 (b) is Pandy's chain with H termination [140].

A homoepitaxial diamond layer deposited by MPCVD on a diamond Ib (111) surface was studied by STM [147]. See also Ref. [142]. The H-terminated (111) surface had a 1×1 structure with C–H bonds vertically sticking out of the surface.

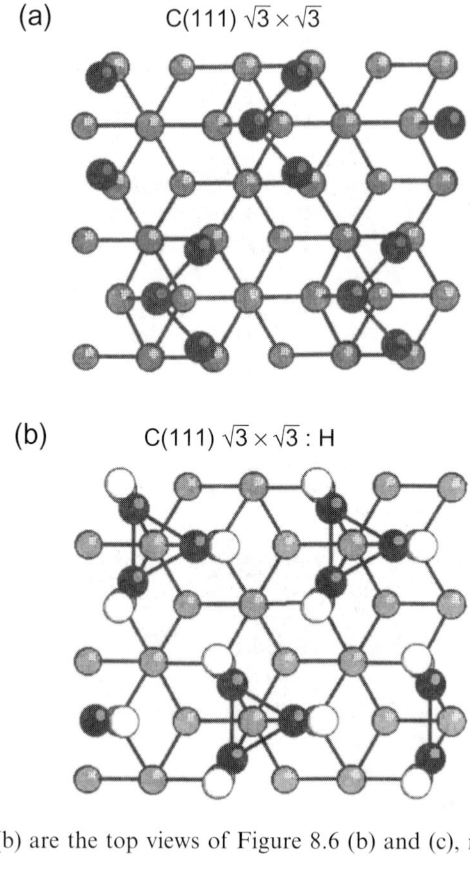

Figure 8.7. (a) and (b) are the top views of Figure 8.6 (b) and (c), respectively [140].

In addition, protrusions were observed, as shown in Figure 8.9, which were considered to have a local structure of (111)-($\sqrt{3} \times \sqrt{3}$) R30°. Unlike the inference given in Ref. [146], the $-CH_3$ units are chemisorbed on every other carbon atoms at the top surface, and there is no possibility of H···H interference in this model. The presence of both C–H and $-CH_3$ terminations at homoepitaxial CVD diamond (111) surface was separately confirmed by the vibrational spectra in high-resolution EELS [148, 149].

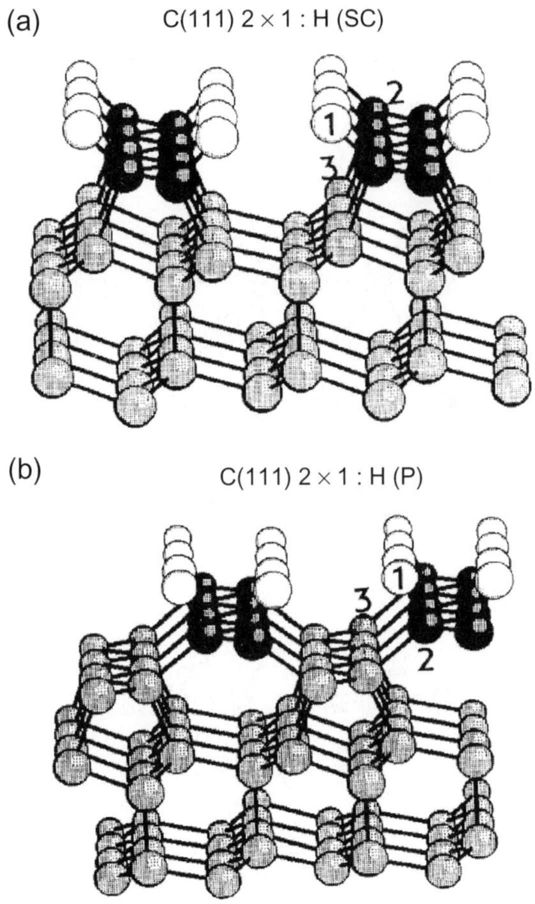

Figure 8.8. Other stable reconstructed structures. (a) 2 × 1 single chain structure, and (b) Pandy's π-chain structure with H termination [140].

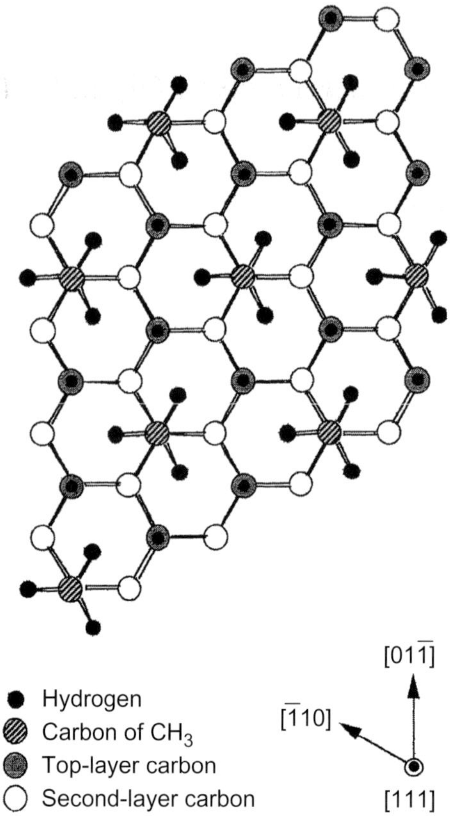

Figure 8.9. A model of local structure at the (111) surface. Other areas are H-terminated [147].

Chapter 9

Heteroepitaxial Growth on cBN, Ni, and Other Substrates

Chapter 9

Heteroepitaxial Growth on cBN, Ni, and Other Substrates

In this chapter, heteroepitaxial growth of diamond particles and films on cubic boron nitride (cBN), Ni, Co, Cu, TiC, BeO, Ni_3Si, graphite, sapphire, and Si will be described. The crystal parameters of these and other materials are listed in Appendix E.

9.1. HETEROEPITAXIAL GROWTH OF DIAMOND ON cBN

Cubic boron nitride (cBN) has a zinc blende-type crystal structure with a lattice constant of 3.615 Å, which is very close to that of diamond (3.567 Å). The difference is only about 1.3%. According to RHEED measurements with the electron beam parallel to the {111} layer of cBN, a growth of diamond by DC plasma CVD on cBN(111) [150] using $c = 0.5\%CH_4/H_2$, $T_s = 900\,°C$, and $P = 180$ Torr led to a result that a smooth {111} layer of diamond was epitaxially deposited in such a way that the [110] direction of diamond was parallel to that of cBN. Namely, D{111}//cBN{111} and D[110]//cBN[110]. In the RHEED pattern, however, extra spots were observed, which were presumably due to the twinnings of {111} diamond layers. In the Raman spectra, there were two lines due to cBN at 1054.5 and 1306.5 cm^{-1}, while there was only a single line due to diamond at 1325 cm^{-1}. This latter value is significantly smaller than that of natural diamond, 1332 cm^{-1}, implying that the deposited diamond layer is subject to a tensile (expansive) stress. This result was confirmed later [151] and the tensile stress was determined to be 22 GPa, which was attributed to the difference in the lattice constants. The polarized Raman spectroscopy of cBN and the diamond layer deposited on cBN(111) indicated that the polarization of cBN was coincident with that of diamond, in consistent with the fact that diamond was heteroepitaxially deposited on cBN.

Diamond growth on cBN(100) was also studied using $c = 2\%CH_4/H_2$, $T_s = 950\,°C$, and $P = 180$ Torr by DC plasma CVD [152]. Judging from the polarized Raman spectroscopy and the SEM image of Figure 9.1, these growth conditions seemed to be appropriate for (100) growth of diamond. The polarized Raman spectroscopy was consistent with the fact that a (100)-oriented diamond layer was heteroepitaxially grown on cBN(100). In this case, the orientational relationship was D{100}//cBN{100} and D[100]//cBN[100].

A more thorough study of diamond growth on cubo-octahedral cBN crystals with diameters of about 500 μm was undertaken by DC plasma CVD in Ref. [153], where B- and N-terminated faces of (111) were distinguished by SEM based on the results of Ref. [154]. See also Ref. [155]. The growth conditions were $c \sim 2\%CH_4/H_2$, $T_s = 950$–$970\,°C$, and $P = 200$ Torr. On the B-terminated cBN(111) face, diamond nuclei were created at the terrace edges, and then laterally grew on the same terrace. At this stage, diamond crystals were already oriented in registry with the basal cBN(111) structure. The diamond crystals grew laterally until the diameter exceeded about 100 nm, and then grew further to be three-dimensional islands with three-fold symmetry. Finally, they started to coalesce to form larger islands with smooth surfaces, as shown in Figure 9.2. The RHEED pattern, however, showed the presence of high density of twins in the diamond crystals.

In contrast to the B-terminated (111) face, only randomly oriented diamond particles were grown on the N-terminated cBN(111) face with a number density of only $\sim 10^7/cm^2$. Thus, diamond nucleation was difficult on the N-terminated cBN(111) face. The difference in nucleation density between the B- and N-terminated cBN(111) faces was attributed to the fact that the formation of energy of the B–C bond, $E_{B\text{–}C}$ (= 348 kJ/mol), is greater than that of the B–H bond, $E_{B\text{–}H}$ (= 320 kJ/mol), while the formation of energy of the N–C bond, $E_{N\text{–}C}$

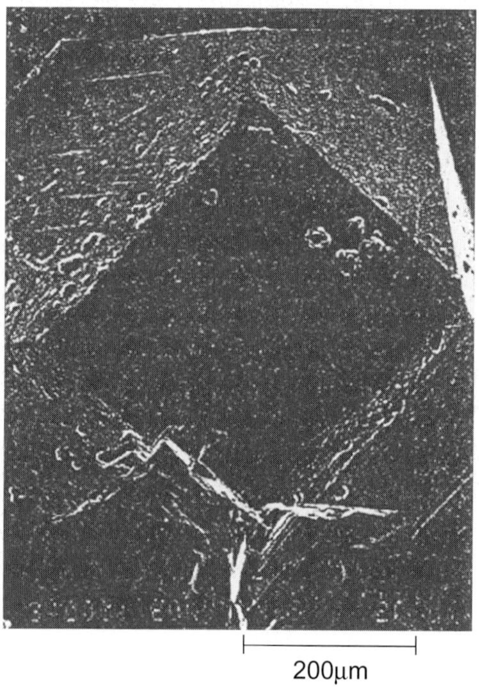

Figure 9.1. SEM image of diamond film on (100) face of cBN [152].

(=291 kJ/mol), is smaller than that of the N–H bond, $E_{N\text{–}H}$ (=391 kJ/mol). For this reason, diamond nucleation on the N-terminated cBN(111) face was difficult because of the stable N–H bonds. For the case of cBN(100) face, it appeared by SEM that the diamond nucleation occurred along the striations, which run along [110] on the cBN(100) face, with a nucleation density of $\sim 10^{10}/cm^2$, and ultimately, a uniform diamond layer with (100) orientation was heteroepitaxially formed with the progress of diamond coalescence. The diamond growth process on cBN(100) is shown in Figure 9.3.

A diamond deposition on cBN powder by HFCVD under conditions of $c \sim 0.4\%CH_4/H_2$, $T_s = 1075\,°C$, and $P = 10$ Torr showed that the (111) face of cBN was covered with tilted (100) faces of diamond, although the {111} lattice layers of diamond were parallel to the {111} lattice layers of cBN [156]. The (111) face of cBN was terminated either by B or N so that B–C and N–C bonds were formed

Figure 9.2. Coalescence stage of epitaxial diamond islands on B-terminated cBN(111) face [153].

at the interface between cBN and diamond. This result is in contradiction with that of Ref. [153] described above, in which DC plasma CVD and cBN crystals were used. It is argued that both cases are possible [156], because surface reactions can take place via -B-H and -N-H surface structures. It seems that the cBN powder surfaces were not so flat as cBN crystal surfaces so that diamond nucleation started from areas that were not N-terminated.

Similar to Ref. [153], a diamond film was deposited on (111) faces of a large cubo-octahedral cBN crystal with a diameter of about 500 μm by HFCVD under conditions of $c\sim 0.5\%CH_4/H_2$, $T_s = 850\,°C$, and $P = 20$ Torr [157, 158]. On B-terminated (111) faces of cBN, a three-dimensional heterogeneous nucleation of diamond particles occurred, which was followed by a lateral growth of the particles, and finally a smooth epitaxial diamond layer was formed as a result of coalescence. On other (111) faces of cBN, however, the deposited diamond surfaces were rough, concave, and highly stepped after a 5-h deposition. Figure 9.4 shows a SEM image of the diamond surface after 40-h deposition. It is seen that the facets

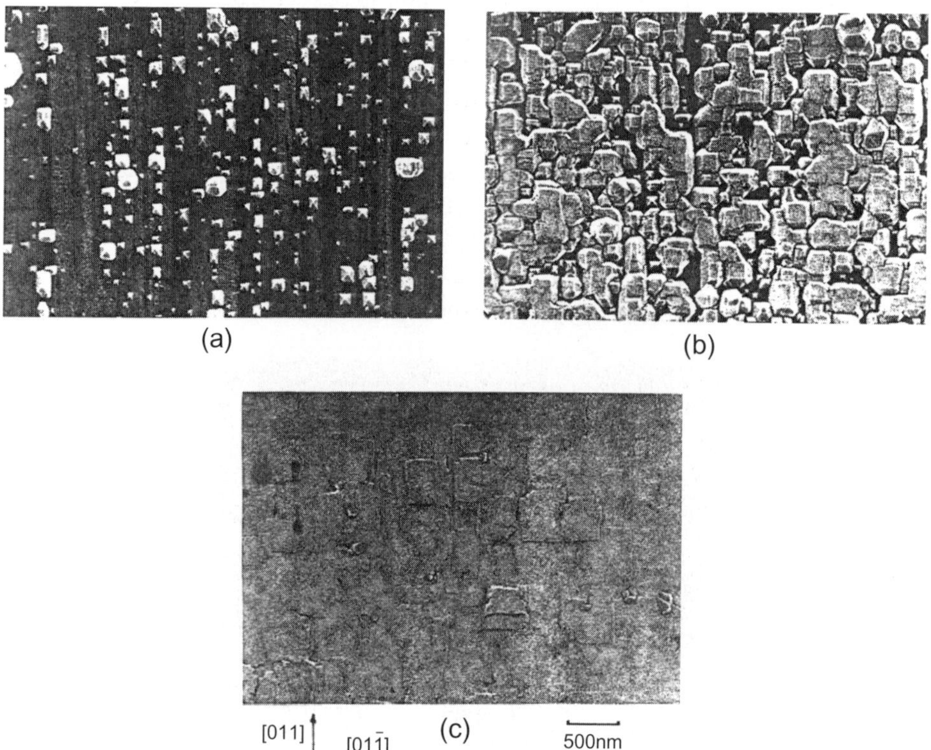

Figure 9.3. Growth process of epitaxial diamond thin film on B-terminated cBN(100) face: (a) 2 min, (b) 4 min, and (c) 5 min [153].

exhibit equilateral triangular and V-shaped features. On N-terminated faces, only randomly oriented diamond crystals were deposited, unlike B-terminated cBN(111) faces, except for the regions where microcracks were present.

In Ref. [158], it was inferred that (i) cBN(111) face is reconstructed so as to achieve charge neutrality due to B- and N-termination, and (ii) if B_2 and N_2 dimers exist on the B- and N-terminated faces, the difficulty of diamond nucleation on N-terminated face is explained that, since the N_2 bonding energy (9.91 eV/molecule) is much greater than both the B_2 bonding energy (3.09 eV/molecule) and the C_2 bonding energy (6.32 eV/molecule), more energy is needed to break the N_2 dimer bonds to form C–N bonds for diamond nucleation. The difficulty of diamond nucleation on N-terminated faces was also attributed to the fact that the formation of C–N bonds may produce volatile reaction products unlike C–B bonds. The Raman lines of diamond layers were located between 1332 and 1333 cm^{-1}, and hence the residual stress in the diamond layer was virtually absent. The FWHM of the Raman line was 8–9 cm^{-1}. In the spectrum of diamond-deposited, N-terminated cBN(111) face, the TO phonon band of cBN was observed at 1054.9 cm^{-1}, indicating that the diamond layer was thinner on this surface than on B-terminated cBN(111) and cBN(100). The fact that the Raman line was observed at 1332 to 1333 cm^{-1} is in contradiction to the previously stated results of Refs. [150, 151], where the Raman line was positioned at 1325 cm^{-1}. The cause for this difference is not clear, but seems to have originated from the different diamond CVD methods used, i.e. DC plasma CVD versus HFCVD.

A cross-sectional HRTEM [157, 158] of the interface between B-terminated cBN(111) and diamond indicated that the interface was clean and free of any other components, in agreement with the results of Ref. [150]. Only a misfit

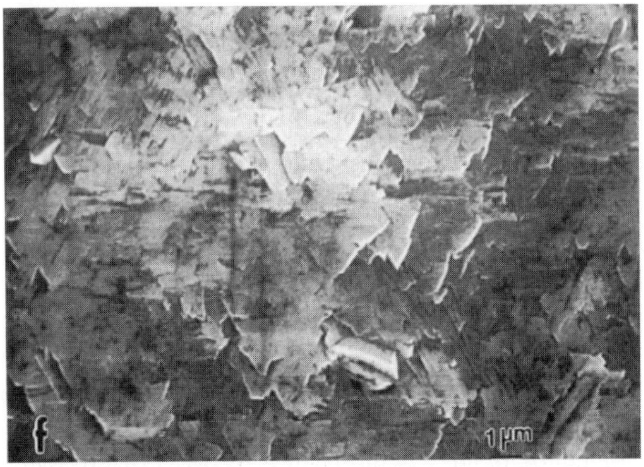

Figure 9.4. Diamond film morphology on B-terminated cBN(111) face [158].

dislocation was observed due the 1.3% lattice mismatch between cBN and diamond. The diamond layer (10-μm thick) contained numerous stacking faults and microtwins, which also were confirmed by electron diffraction (ED). On the other hand, cBN(100) faces contained a number of striations, and initial diamond crystals had a pyramidal shape. Despite heavy striations, a smooth and continuous diamond film was formed after 5 h.

Diamond growth on cBN by MPCVD was done in Ref. [159], where cubo-octahedral cBN crystals of 20–50 μm in size were used as the substrates. For B-terminated cBN(111) faces, epitaxially oriented, partially coalesced diamond particles were grown under conditions of $c = 1\text{–}3\% CH_4/H_2$ (1% was optimum), $P = 38\text{–}113$ Torr, and $T_s = 920\text{–}1050\,°C$. The diamonds had triangular (111) faces. By contrast, the diamond nucleation density on N-terminated cBN(111) faces was much lower, and diamond orientations were random. For cBN(100) faces, the (100) faces of deposited diamond crystals were clearly observed. The growth conditions were $c = 2\text{–}3\% CH_4/H_2$, $P \geq 80$ Torr, and $T_s \geq 950\,°C$, which were significantly narrower than those for B-terminated cBN(111) faces. In the Raman spectrum of a diamond layer deposited on the cBN(111) face mentioned above, the line due to diamond was located at 1335 cm^{-1}, 2 cm^{-1} higher than that of single crystal diamond (1333 cm^{-1}), meaning that a compressive stress remains in diamond. Moreover, there was a broad band around 1550 cm^{-1}, indicating an existence of non-diamond components. The fact that the position of the Raman line for the diamond layer was higher than that for single crystal diamond is in contradiction to the previous two results. It is again speculated that the internal stress/strain of diamond layer deposited on cBN(111) faces depends on the CVD methods and the film quality.

All of the above studies concluded that diamond grows heteroepitaxially on cBN with the orientational relationships that:

$$\mathrm{D}\{111\}//\mathrm{cBN}\{111\} \quad \text{and} \quad \mathrm{D}[\bar{1}10]//\mathrm{cBN}[\bar{1}10] \quad \text{for B-terminated cBN(111)}, \tag{9.1}$$

and

$$\mathrm{D}\{100\}//\mathrm{cBN}\{100\} \quad \text{and} \quad \mathrm{D}[100]//\mathrm{cBN}[100] \quad \text{for cBN(100)}. \tag{9.2}$$

As a standard procedure of substrate pretreatment for diamond deposition, (i) scratching the substrate surface with diamond powder or paste, or (ii) ultrasonic treatment of the substrate in diamond powder suspended in alcohol are widely used. By contrast, for diamond growth on cBN(111), no such pretreatment is necessary, partly because of the close match of lattice constants.

The (111) growth of diamond is achieved at least up to 10 μm [158], demonstrating that a stable growth of (111)-oriented diamond film is possible. Note also that the feature of coalescence for (111) faces of diamond layer observed on cBN(111) is very similar to that for Pt(111) described in Section 12.1.

In Ref. [160], it was predicted that in view of strain, a combination of D(100) and cBN(221) is most favorable for heteroepitaxy. Indeed, heteroepitaxial growth of diamond was later carried out on (221) and (100) faces of cBN by MPCVD under conditions of $c = 1\%CH_4/H_2$, $P = 30$ Torr, and $T_s = 850\,°C$ [161]. In both cases, the specimen surfaces consisted of (100) faces of diamond, but they appeared as if thin rectangular blocks were linked rather than stacked. Regarding the orientational relationship between cBN and diamond, it was D[110]//cBN[110] and D{100}//cBN{221} for (211), while it was D[100]//cBN[100] and D{100}// cBN{100} for (100).

It seems that the industrial use of heteroepitaxial growth of diamond on cBN would be unlikely because cBN substrates with large surface areas are not available, and even small cBN crystals are as costly as diamond. On the other hand, an advantage of using cBN as the substrate is that an n-type doping of cBN is easily done by doping Si in the HP-HT synthesis [162]. Since p-type doping is easy for diamond by B incorporation, p-n junction devices such as ultraviolet (UV) light emitting diode (LED) can be fabricated in the future for non-cost-sensitive applications. To this end, the issue of high density of stacking faults and microtwins in the deposited diamond layer must be solved to deposit a high quality diamond layer.

Finally, it should be mentioned that electronic state calculations of H-terminated cBN surfaces and diamond growth are studied in Refs. [163, 164]. Also, in a recent paper [165], diamond was deposited on large cBN crystals of 200–350 μm in size that were embedded in a Cu plate. It appeared that (i) diamond nuclei were cubo-octahedral crystallites with approximately 100 nm in diameter on the (111) faces of cBN, (ii) in some cases, dense carbon tubes with a diameter of 100 nm and a few micrometer in length were grown, and (iii) diamond crystals grown on Cu had deep holes in the center of the (111) faces. This article also compiled past articles on diamond growth of cBN.

9.2. ORIENTED GROWTH OF DIAMOND ON METALS AND COMPOUNDS

Diamond CVD on group VIII transition metals (Cr, Mn, Fe, Co, and Ni) has been investigated in search of relevant substrates for diamond heteroepitaxy. The material properties are listed in Table 9.1 [166]. See Appendix E for more information. These transition metals can dissolve carbon and hydrogen to form a surface layer of metal (M)–C–H complex. This seems to be common to many metal substrates, including Pt and Ir, as will be seen later.

Table 9.1. Properties of group VIII transition metals and diamond [166].

Material	Crystal structure	Lattice constant (at 20 °C, Å)	Melting point (°C)	Thermal expansion coefficient at 20 °C ($\times 10^{-6}$/°C)
Diamond	fcc	3.56	3830	0.8
Ni	fcc	3.52	1453	13.3
Co	fcc (β-Co, >417 °C)	3.55	1495	12.5
Fe	bcc (α-Fe, < 912 °C)	2.87	1536	12.6
	fcc (γ-Fe, >912 °C)	3.65		
Mn	cubic (β-Mn, >727 °C)	6.29	1245	22.8
Cr	bcc	2.88	1875	6.2

fcc: face centered cubic
bcc: body centered cubic

9.2.1. Ni

The lattice constant of Ni (face centered cubic) is 3.517 Å, which differs only 1.2% from that of diamond (3.567 Å), and Ni is used as a catalytic metal solvent for HP-HT diamond synthesis. Hence, Ni is considered to be an appropriate material for heteroepitaxial diamond growth. Belton and Schmieg [167] used Ni(100) as a substrate for diamond growth by HFCVD. The growth parameters were as follows: $c = 0.2\%CH_4/H_2$, $P = 50$ Torr, the distance between the filament and the Ni substrate was 2 mm, and the filament temperature was 2600 K (2327 °C). After a 2-min growth, microcrystalline graphite islands were formed, which was followed by a growth of glassy carbon. After 109 h, diamond crystals of about 20 μm in diameter were grown, and finally a continuous diamond film was formed after 200 h. The diamond particles were randomly oriented, presumably because the basal graphite and glassy carbon were orientationally random. It seems that a considerably slow growth rate and a random orientation of diamond are partly because the growth parameters were not optimized: the filament/substrate distance was too short, the gas pressure was too high, and the CH_4 concentration was too low.

Sato et al. [168] found that (111)-oriented diamond crystals were deposited on *polycrystalline* Ni surface by MPCVD under conditions of $c = 0.5\%CH_4/H_2$, $T_s = 880$ °C, and $P = 100$ Torr. The number of epitaxially aligned diamond crystals was higher for $T_s > 880$ °C, while the nucleation density was the highest in the temperature range of 800 °C $< T_s <$ 900 °C. A problem was that graphite was deposited when $c \geq 1\%CH_4/H_2$. This is because the catalytic action of Ni is so strong that graphite is formed instead of diamond due to decomposition of hydrocarbon gases. In a successive paper by Sato et al. [169], diamond was deposited by MPCVD on *single crystal* surfaces of Ni(111) and (100). Cubo-octahedral diamond crystals were grown for 700 °C $< T_s <$ 1000 °C, but it was only when CH_4/H_2 was less than 0.9%. For $CH_4/H_2 > 0.9\%$, disordered graphite was deposited. According to the

Laue analysis of X-ray diffraction, the orientations of diamond crystals were in registry with Ni(111) or (100) within an experimental error of 2°. Typical growth conditions of diamond on Ni(111) were $c = 0.5\% CH_4/H_2$, $P = 100$ Torr, and $T_s = 880$ °C. On the other hand, those on Ni(100) were $c = 0.5\% CH_4/H_2$, $P = 60$ Torr, and $T_s = 910$ °C. In both cases, (111) and (100) faces of diamond crystals were parallel to the Ni surfaces, as seen in Figures 9.5 and 9.6, respectively.

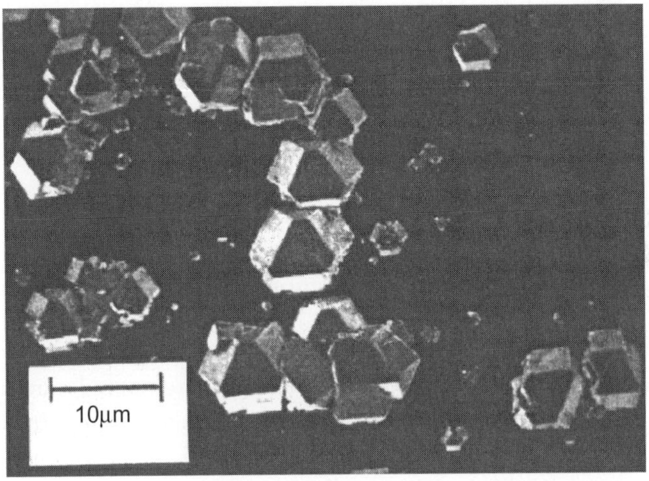

Figure 9.5. SEM image of epitaxial diamond crystals grown on Ni(111) surface: $c = 0.5\% CH_4/H_2$, $P = 100$ Torr, and $T_s = 880$ °C [169].

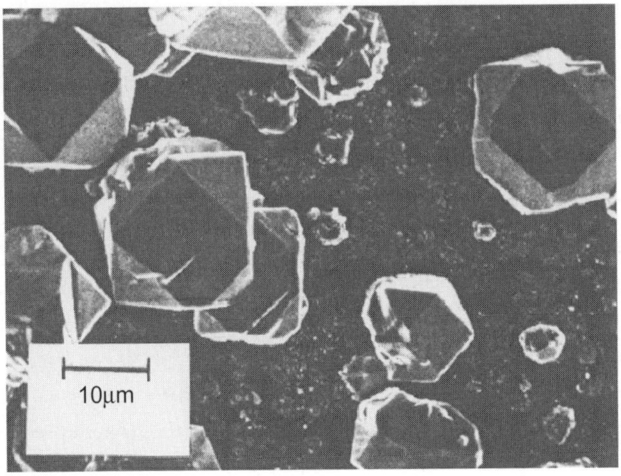

Figure 9.6. SEM image of epitaxial diamond crystals grown on Ni(100) surface: $c = 0.5\% CH_4/H_2$, $P = 60$ Torr, and $T_s = 910$ °C [169].

For the case of Ni(111), it is seen that adjacent diamond crystals have almost the same orientation, and hence partially coalesce to form a continuous surface as they grow. This indicates the possibility that a continuous epitaxial diamond film can be deposited on a Ni single crystal by MPCVD. It was argued [167] that the heteroepitaxial growth of diamond on Ni resulted from the following properties of Ni in the temperature range of $700\,°C \leq T_s \leq 1000\,°C$: (i) catalytic activity to hydrocarbon gases, (ii) high solid solubility of C, (iii) high diffusion rates of C and H, and (iv) little tendency to form stable carbides. An interesting observation was that well-crystallized graphite was deposited on the back side of the Ni substrate when diamond CVD was done under conditions of $c = 0.5\%CH_4/H_2$, $P = 100$ Torr, and $T_s = 980\,°C$, as shown in Figure 9.7. Indeed, there was only a single sharp line at $1580\,cm^{-1}$ in the Raman spectrum that is known to be due to graphite.

After the work of Zhu et al. [166], oriented growth of diamond on Ni(100) was studied extensively by Glass' and Sitar's groups at North Carolina State University (NCSU), and a three-step process was established to suppress graphite formation by HFCVD [170–172]:

Step 1 (pretreatment): The method of seeding for diamond nucleation centers on Ni surface has been optimized as the research was developed. In Ref. [170], the Ni surface was scratched with diamond powder of 0.25 μm in size. In their successive experiments, instead of scratching, diamond powder was sprinkled on the

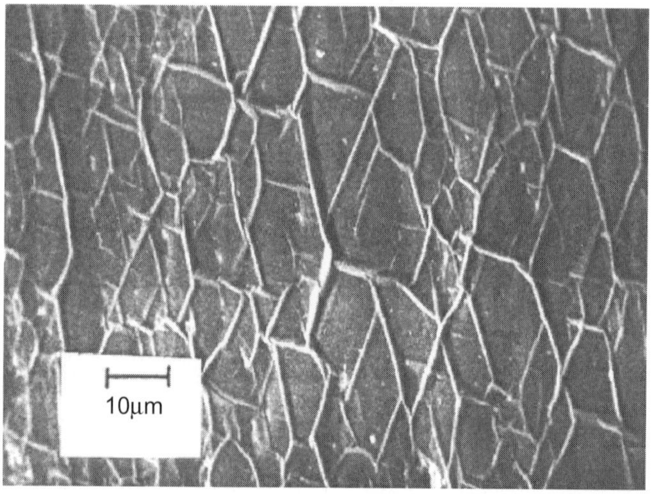

Figure 9.7. A SEM image of well-crystallized graphite layer formed on the back side of Ni substrate: $c = 0.5\%CH_4/H_2$, $P = 100$ Torr, and $T_s = 980\,°C$ [169].

Ni surface [171]. Fullerene (C_{60}) powder of 0.25 μm and graphite powder of 10–15 μm were also used [172]. In Refs. [173, 174], diamond powder of 0.5 μm in size, which was suspended in acetone, was applied on the Ni surface. For heteroepitaxial growth of diamond on Ni, the seeding step is very important to make oriented diamond nuclei and suppress the graphitization of diamond simultaneously.

Step 2 (annealing): The seeded substrate was then annealed in pure hydrogen (i) at 950 °C for 30 min and then the temperature was increased to 1200 °C for 1 min [170], or (ii) at 1100 °C until the color of the specimen surface changed from dark-gray to reflective or shiny [171, 173], or alternatively (iii) at 1050 °C for a short time [174]. This step allows deoxidation of the substrate surface, recrystallization of Ni, and formation of Ni–C–H molten states. It was believed that the seeded diamond particles reorient to be in registry with the Ni surface structure due to interactions between diamond and Ni together with the Ni–C–H complex formation [170, 175]. By monitoring the intensity of the *reflected* and *scattered* lights from the specimen surface [176], one can identify the time of termination of this step. Namely, the high temperature annealing is switched to the diamond growth when the intensity of the *scattered* light drops. This is because toward the end of this step, the seeded diamond powder disappears from the Ni surface as they completely react with Ni, and the specimen surface becomes flat.

Step 3 (growth): The substrate temperature T_s is lowered to 900–950 °C, and diamond CVD is undertaken at $P = 20$ Torr. To make azimuthally (100)- and (111)-oriented diamond films on Ni(100) and Ni(111), $c = 0.5\%$ and $0.3\% CH_4/H_2$ were used, respectively [171].

As a result of the three-step process, heteroepitaxially oriented diamond crystals were formed on Ni substrates, and the coalescences were developed between some adjacent crystals, as seen in Figures 9.8 and 9.9 [171].

Later, Sitar's group [173] was successful in synthesizing both (i) diamond particles in which approximately 90% were well-oriented, and (ii) fully coalesced, continuous diamond films by annealing Ni in $0.5\% CH_4/H_2$ at 1100 °C. Figures 9.10 (a) and (b) show the results of (100)- and (111)-oriented diamond films of 30-μm thickness, respectively, after a 49-h growth. It is seen that the crystals are entirely coalesced at the film surface where marked step bunchings occur. Raman spectra indicated that both films virtually contained no sp^2 carbons. Unfortunately, the film was delaminated and broken into pieces upon cooling. The fracture occurred mostly along the grain boundaries. This seems to be common to all diamond films grown on metal substrates because of the difference in the thermal expansion coefficients and the internal stress within diamond films.

Thus, special care is needed in cooling the specimen at the end of the CVD process. It is of interest that a flow pattern was observed on the Ni surface [170, 177], indicating that a molten phase was formed at the Ni surface during the diamond growth. A rapid dissolution of seeded diamond powder during Step 1 indicates that a molten layer is indeed present at the Ni surface. Since the melting points of Ni–C eutectic and nickel hydride are 1318 and 1150 °C at atmospheric pressure, respectively, which are significantly lower than the melting point of Ni, 1450 °C, it is likely that the molten phase actually exists in Step 2, as the specimen was heated to 1200 °C. It was considered from a depth profile by micro-Auger analysis [170] that diamond crystal surfaces were covered by an Ni–C–H layer, and the diamond growth proceeded by a mechanism similar to the Vapor–Liquid–Solid (VLS) mechanism. An Auger depth analysis of quenched specimen showed that the carbon contents in the top 1-μm thick layer was approximately 6%, which was 3% higher than the carbon solubility in Ni. This also supports the presence of Ni–C–H

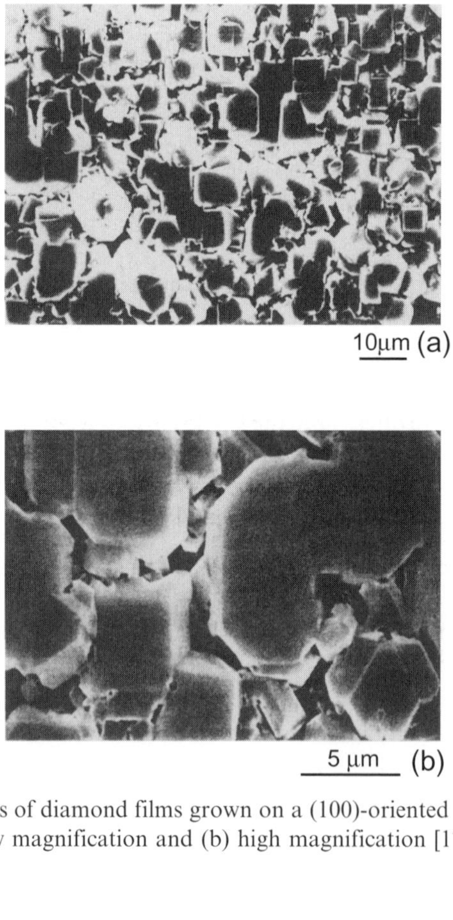

Figure 9.8. SEM images of diamond films grown on a (100)-oriented single crystal Ni surface with (a) low magnification and (b) high magnification [171].

molten layer described above. An *in situ* differential thermal analysis (DTA) indicated that the surface melting takes place at 945 and 1085 °C for Ni–H and Ni–C–H compounds, respectively [177]. These temperatures are significantly lower than the melting point of Ni–C eutectic, 1326 °C.

In Ref. [174], a depth analysis of Auger spectroscopy after Step 1 showed that the top surface of the specimen contained 6% of carbon, which decreased to 3% at a depth greater than about 1 μm. The value, 6%, of carbon at the Ni surface is close to the eutectic composition of 8%, and the value 3% at a depth of about 1 μm is the upper limit of the carbon solubility in Ni according to the Ni–C equilibrium phase diagram (see Appendices F and G). Electron diffraction measurements at the surface layer indicated that the top layer contains Ni_4C ($a = 3.517$ Å), and the TEM image showed that Ni_4C inclusions of about 50 nm in diameter and a planar density of about $10^8/cm^2$ were present. An XTEM [178] showed that a Ni_4C phase was present between diamond crystals and the Ni substrate, as seen in Figure 9.11,

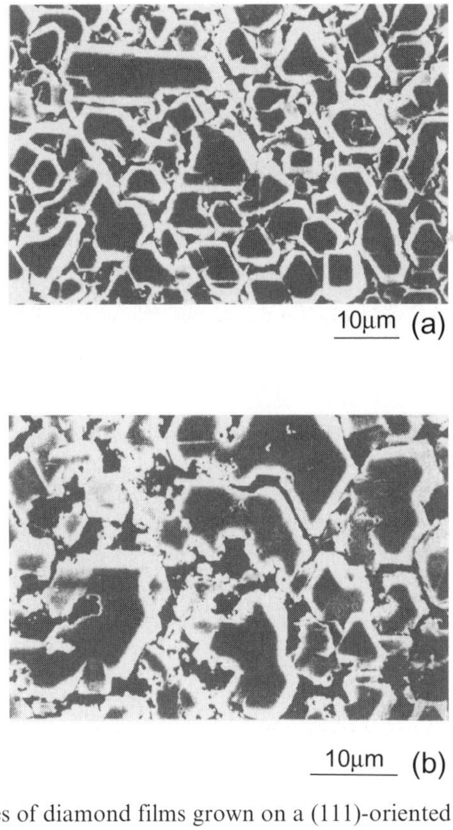

Figure 9.9. SEM images of diamond films grown on a (111)-oriented single crystal Ni surface with (a) low magnification and (b) high magnification [171].

and the thickness was about 500 nm after 7 h of CVD, indicating that the Ni_4C phase continued to grow throughout the diamond CVD in Step 3.

It is also seen in Figure 9.11 that the Ni_4C/diamond interface is faceted. From these results, it was concluded that these Ni_4C inclusions were actually the nucleation sites for diamond growth in Step 2, and the Ni_4C front moves into diamond to react

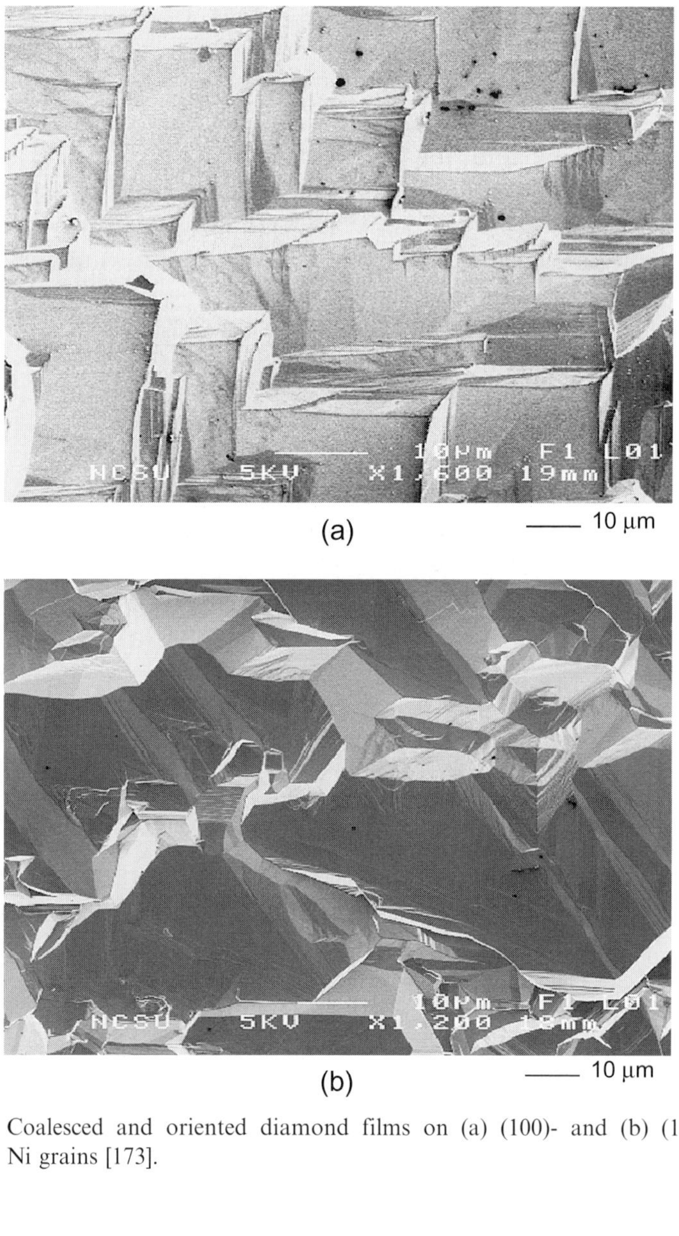

(b) 10 μm

Figure 9.10. Coalesced and oriented diamond films on (a) (100)- and (b) (111)-oriented Ni grains [173].

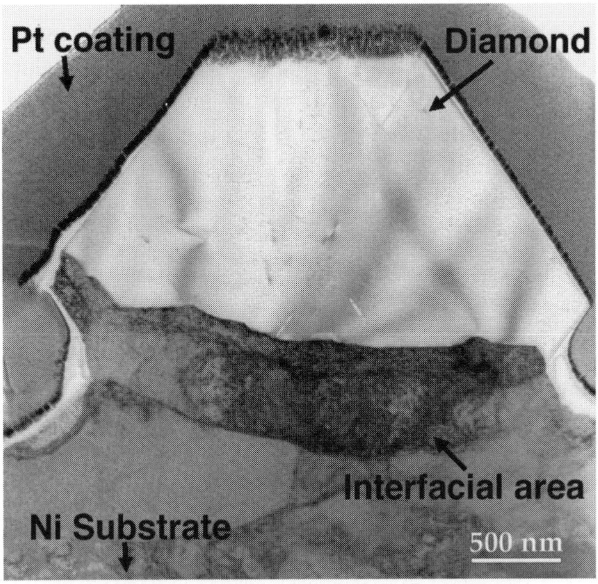

Figure 9.11. Cross-sectional TEM image of azimuthally oriented diamond crystal on Ni. The presence of the interfacial layer, Ni_4C, is apparent [178].

with defective layers of diamond. This is consistent with the observed fact that the defect density in the diamond crystal region in the vicinity of the Ni_4C/diamond interface is very low. The TEM observation also revealed that the difference in the orientational angle between adjacent diamond crystals was only 1.8° on average.

9.2.2. Co

As seen in Table 9.1, cobalt has an fcc structure at temperatures above 417 °C (it is hexagonal below 417 °C [179]) with a lattice constant of 3.554 Å [180], only 0.6% smaller than that of diamond. The melting point of Co–C eutectic is 1318 °C [181], see Appendix F. Thus, the properties of Co are very similar to those of Ni, and hence similar results are anticipated when it is used as the substrate for diamond CVD. Liu et al. [182] used a multi-step process similar to the one described in Section 9.2.1 to grow oriented diamond crystals on Co(0001) by HFCVD. For the seeding, diamond powder of 1–2 μm was used. The seeded substrates were annealed at 900 °C in H_2 for 10–30 min in the CVD chamber using a W filament as a heater. Then, T_s was increased to 1100 °C for 10–60 min, at which point, the seeded material was dissolved into Co to form a Co–C–H complex. Finally, T_s was lowered to 900 °C for diamond growth under conditions of $c = 0.3\%CH_4/H_2$ and $P = 30$ Torr. Consequently, (100)-oriented diamond crystals were grown on Co(0001), as shown in Figure 9.12 (a) and (b) that were obtained by (a) diamond powder seeding, and (b) a high CH_4 pretreatment, respectively. In both cases, (111)-oriented diamond crystals were grown on Co substrates. Similarly, oriented diamond crystals were grown when the Co substrate was seeded with graphite powder of 10–15 μm in size. Interestingly, unusually faceted structures, which had the same morphology as diamond, were observed on the Co surface. It was speculated that they were not diamond but a Co–C phase, and could be the nuclei for diamond growth.

It was inferred from the observed results that carbon seeds and hydrogen formed a ternary, molten, eutectic complex of Co–C–H on the Co surface in the annealing step at 1100 °C. The subsequent lowering of the substrate temperature left supersaturated carbon to be the nuclei for diamond growth. Although the melting point of the Co–C eutectic is 1318 °C, the authors of Ref. [182] inferred that the melting point decreased below 1100 °C because of the presence of hydrogen. Thus, for the diamond growth on Co substrates, it seems that a formation of a

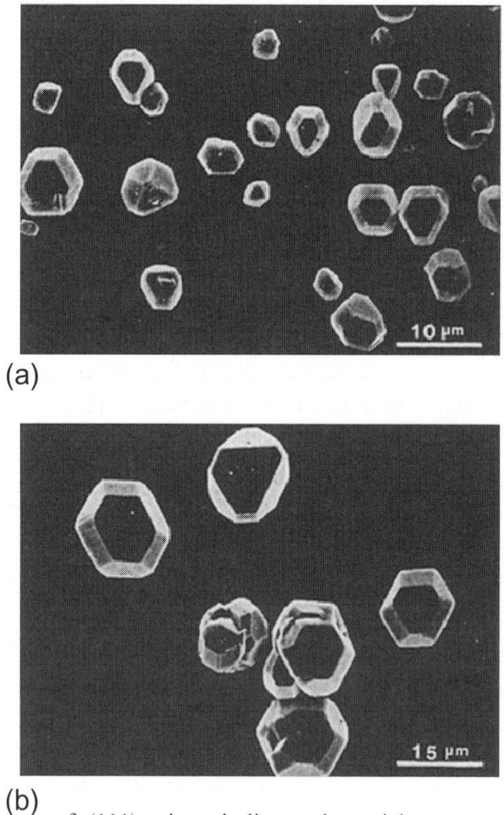

Figure 9.12. SEM images of (111)-oriented diamond particles grown on Co(0001) using (a) diamond powder seeding and (b) a high CH_4 treatment [182].

molten Co–C–H surface layer suppresses the co-deposition of graphite, and plays an important role in diamond nucleation. Such a process seems to be common to diamond growth on other metals.

9.2.3. Cu

Copper (Cu) is also known to be fcc with a lattice constant of 3.608 Å [183] that is only 1.1% larger than diamond (3.567 Å). Hence, Cu is also considered to be a good material for diamond epitaxy. In Ref. [184], a single crystal Cu was implanted by C atoms with a dose of 1×10^{18} ions/cm^2, and annealed at about 800 °C. Diamond CVD was then undertaken for both pure and C-implanted Cu using $c = 3\%CF_4/H_2$ or $0.5\%CH_4 + 0.7\%O_2/H_2$ under conditions of $P = 33$ Torr and $T_s = 800$ °C. Unfortunately, there was no indication of epitaxial growth or preferred orientation of diamond. However, a high quality graphite layer of ~900 Å thickness was formed on the Cu surface, which seemed to enhance diamond nucleation. This is schematically shown in Figure 9.13. The orientational relationship between diamond and graphite was D[111]//G[0001] and D[110]//G[$11\bar{2}0$], as also shown in Figure 9.13.

In Ref. [185], a Cu surface was abraded with diamond powder, and the diamond deposition was done by HFCVD under conditions of $c \sim 1\%CH_4/H_2$, $P = 20$ Torr, and $T_s = 800$ °C for 2 h. As a result, a diamond film, consisting of diamond grains of 0.5 μm in diameter, was deposited. A patterning by selective deposition was found to be possible by irradiating KrF or XeCl excimer laser pulses on unnecessary areas.

Diamond growth on polycrystalline Cu by electron cyclotron resonance (ECR) plasma CVD was investigated in Ref. [186]. For diamond nucleation, the BEN technique was used under conditions of $c = 60\%CH_4/H_2$, $P = 0.1$ Torr, $T_s = 500$ °C, while $V_b = -50$ V was applied to the Cu substrate for 30 min. This was followed by a diamond growth under conditions of $c = 1\%CH_4$, $1\%O_2$, and 49%Ar diluted with H_2, $P = 0.1$ Torr, $T_s = 700$ °C, and $V_b = +30$ V for 15 to 20 h. Consequently, diamond crystals were grown on the Cu substrate, where they were oriented in such a way that the (111) faces of diamonds were nearly parallel to the substrate surface. A more detailed examination of the diamond growth on poly- and (111) single-crystal Cu substrates were carried out in Ref. [187]. In this case, the BEN treatment was undertaken under conditions of $c = 30 \sim 67\%CH_4/H_2$, $P = 0.1$ Torr, $T_s = 750$ °C, and $V_b = -50$ V for 10–30 min. The diamond growth was done under conditions of 6.5–7%CH_4, 6.5%O_2, and 61%Ar diluted with H_2, $P = 0.25$ Torr, $T_s = 460$ °C, and $V_b = +30$ V. On the Cu substrate, there were diamond crystals whose (111) faces were perfectly coalesced. A SEM observation of diamond growth indicated that (i) diamond particles of roughly 100 nm in diameter can migrate on the Cu surface to coalesce, and (ii) the clusters of diamond crystals undergo rotations and changes in relative positions during CVD. Such a curious behavior seems to be attributed to the fact that Cu does not form carbide, and thus the interaction between diamond and Cu is weak.

In Refs. [188, 189], diamond growth was done by MPCVD on polycrystalline Cu using the BEN technique, but the nucleation density was only $\sim 10^6/\text{cm}^2$, significantly smaller than that on Si, $\sim 10^{11}/\text{cm}^2$. The nucleation density was $\sim 10^7/\text{cm}^2$ on Cu that had been scratched with a 0.25-μm diamond paste. A surface analysis indicated that there existed a thin graphitic layer of 5–10 Å thickness on Cu. The work of Ref. [183] presents presumably the best-optimized diamond

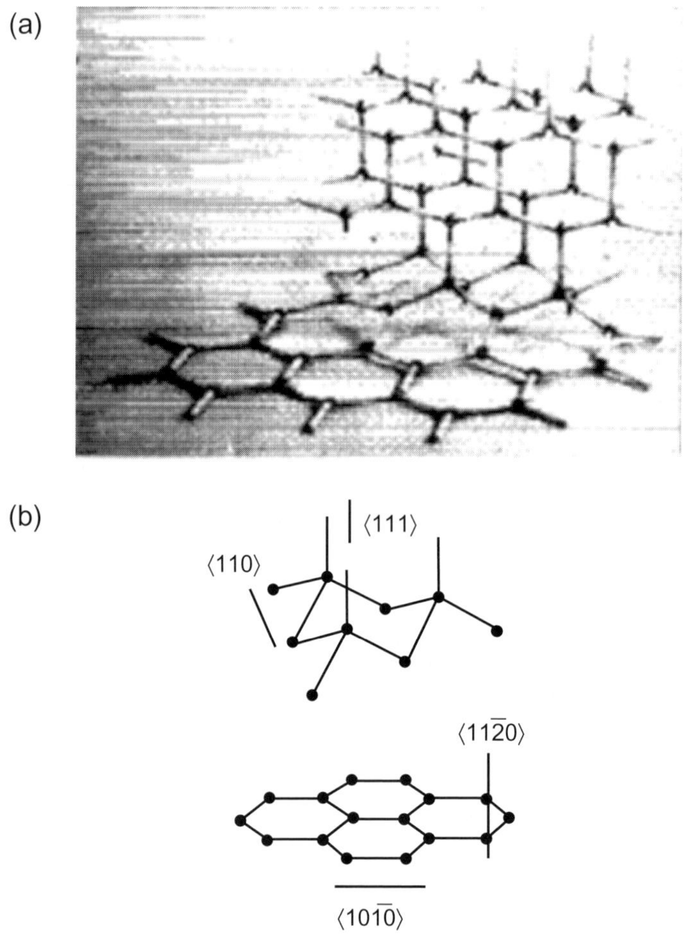

Figure 9.13. A simple model of diamond nucleus on the edge of graphite basal plane on Cu: (a) atomic arrangement and (b) diamond and graphite structures [184].

growth method on Cu. The substrates were polycrystalline Cu foils that had been abraded by a SiC paper and a 0.25-μm diamond paste. To relieve the stress in the diamond film due to the mismatch of thermal expansion coefficients between diamond ($1.5–4.8 \times 10^{-6}$/K) and Cu (17.7×10^{-6}/K), the Cu foils were scribed with a grid pattern. The diamond deposition was done by a NIRIM-type MPCVD reactor using a mixture of CH_4, O_2, and H_2 as the source gas under conditions of $P = 60$ Torr, $T_s = 650–825$ °C. Consequently, a continuous film was grown over 1 cm^2-area of Cu. The adhesion of the films to the Cu substrates was so weak that it was possible to obtain freestanding diamond films. The growth experiments were repeated by changing the CH_4 concentration and O_2/C ratio, and the results are shown in Figure 9.14.

A SEM image of diamond particles is shown in Figure 9.15. Unlike past works, diamond film surfaces were well facetted with (111) and (100) faces, or consisted of cubo-octahedrons. Under certain conditions, either (111) or (100) faces of diamond particles were nearly parallel to the substrate surface. It is of intrigue that the (111)-oriented diamond grains have hexagonal faces, as seen in Figure 9.15, rather than triangles that were seen in Refs. [186, 187]. Thus, both ⟨111⟩- and ⟨100⟩-textured diamond films were demonstrated to be synthesized on polycrystalline Cu foils.

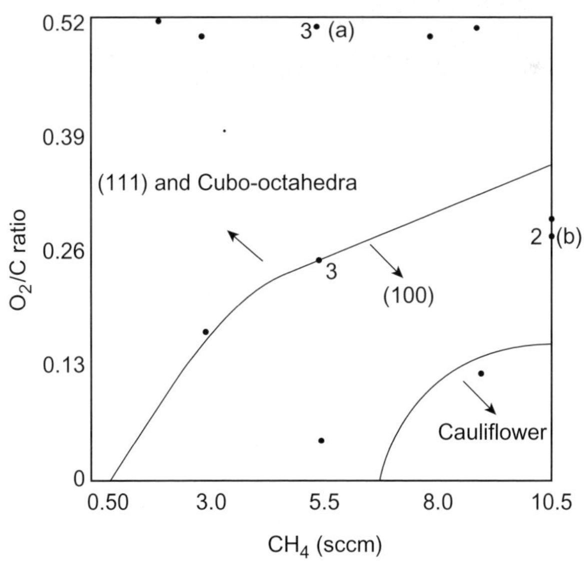

Figure 9.14. Morphology of diamond crystals grown on Cu as a function of the O_2/C ratio and the CH_4 flow rate (in units of sccm). The numbers next to closed circles indicate how many times the run was repeated [183].

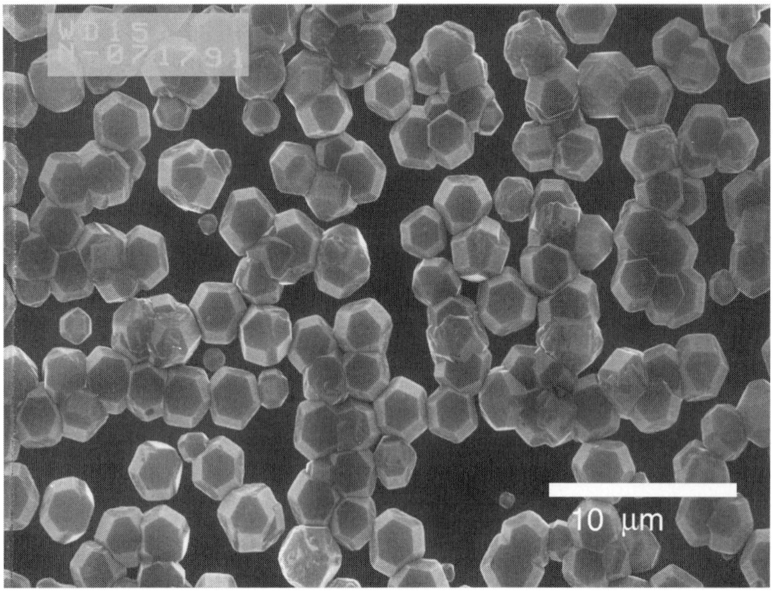

Figure 9.15. ⟨111⟩-oriented diamond crystals grown on Cu foil [191].

The interface between diamond and Cu was studied in detail by HRTEM [190]. The substrate was pretreated by the ultrasonic treatment with diamond powder suspended in alcohol. For diamond deposition, 10%CO/H_2 was used at $P = 35$ Torr. The substrate was a TEM grid of Cu, and the cross-sectional TEM was observed without special pretreatment, if diamond was small enough. There existed (0002) planes of the intermediate graphite layer approximately parallel to the Cu surface. Its thickness was 4 to 5 nm. Since diamonds are ⟨111⟩-textured in this case, it was inferred that the presence of intermediate graphite layers helped grow ⟨111⟩-oriented diamonds.

9.2.4. TiC

An oriented growth of diamond crystals on TiC(111) was done in Ref. [192] using BEN. The substrate used was a single crystal TiC, which had been mechanically polished with diamond powder of 0.1–30 µm in size. This was followed by re-polishing with 0.05-µm alumina powder to remove diamonds embedded in TiC in the first polishing. It was then processed with hydrogen plasma, followed by the BEN process for diamond nucleation in a MPCVD reactor. The diamond deposition was done using $c = 0.2\%CH_4/H_2$, $T_s = 900\,°C$, and $P = 40$ Torr for 8 h. The nucleation density was $1.5 \times 10^8\,cm^{-2}$, and 10–15% of the diamond particles were oriented with respect to each other. The (111) faces of diamond crystals were almost parallel to the substrate surface, and the orientational relationship was $D[01\bar{1}]//TiC[01\bar{1}]$ and $D[10\bar{1}]//TiC[10\bar{1}]$. It appeared that there existed small misorientations between the (111) faces of diamond and TiC, presumably because of the 21% mismatch in the lattice constants between diamond and TiC (TiC is fcc with a lattice constant of 4.32 Å, while the lattice constant of diamond is 3.567 Å).

9.2.5. *BeO*

The diamond/beryllium oxide (BeO) interface has been theoretically studied in Ref. [193], which was experimentally examined by Angus' group [194]. BeO is hexagonal (wurzite), which is geometrically similar to lonsdaleite (hexagonal diamond), and the Be–O bond length, 1.65 Å, is close to the C–C bond length, 1.54 Å. The lattice parameters of the wurzite structure are $a = 2.696$ Å, and $c = 4.379$ Å. An estimated BeO/diamond adhesion energy was 4.6 J/m^2, only slightly smaller than that of cBN/diamond, 5.4 J/m^2. The diamond deposition was done by HFCVD for 8 h using $c = 0.5\%CH_4/H_2$ at $P = 20$ Torr, and $T_s = 850$ °C on BeO(0001) surface that had been polished with a SiC paper. As a result, hexagonal-shape diamond crystals were grown on the BeO(0001) surface. The orientation of the diamond crystals was heteroepitaxially aligned on BeO(0001) in such a way that D{111}//BeO{0001}, and the $[1\bar{1}0]$ axis of diamond was rotated within ±6° about the $[11\bar{2}0]$ axis of BeO from the ideal contact, as schematically shown in Figure 9.16. Interestingly, no interface layer was detected by HRTEM. It was inferred that diamond crystals were grown on the Be-terminated surface of BeO(0001) rather than the O-terminated surface. From a structural consideration, it was expected that hexagonal diamond could be grown on $BeO(11\bar{2}0)$ surface, but no epitaxial relationship was confirmed in this case.

9.2.6. *Ni_3Si*

In Ref. [195], an oriented diamond growth was attempted on (100) and (111) domains of polycrystalline Ni_3Si surfaces with an average grain size of approximately 20 μm. It was considered to be a good material for oriented diamond growth as it is cubic with the Cu_3Au-type structure, and its (100) surface has only a 1.8% lattice mismatch (the lattice constant $a = 3.504$ Å) with diamond (100). It was also expected that the presence of Si in Ni_3Si suppressed the formation of graphite during diamond CVD, which had been a major problem in depositing diamond on Ni. Prior to diamond deposition, the Ni_3Si specimen was polished with a 0.01-mm alumina powder, and the diamond CVD was undertaken by HFCVD under

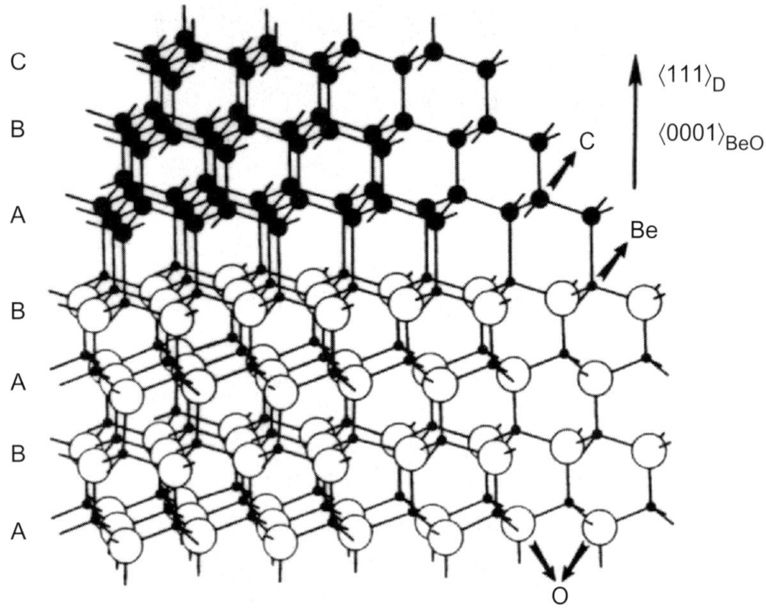

Figure 9.16. Crystallographic model of the ideal superposition of (111) diamond on top of the (0001) basal plane of BeO. Dark, white, and small dark spheres are C, O, and Be atoms, respectively [194].

conditions of $c = 1\%CH_4/H_2$, $P = 20$ Torr, and $T_s = 850\,°C$. After a 16-h growth, oriented diamond crystals were deposited on Ni_3Si with little graphite co-deposition, as shown in Figure 9.17. All crystals seemed to be aligned with the triangular (111) faces almost parallel to the substrate surface, implying that they grew on $Ni_3Si(111)$ domain surfaces. To further develop this technique, it will be necessary to establish a method of depositing a $Ni_3Si(111)$ single crystal film on a large substrate.

9.3. GRAPHITE

Apart from minor differences in position and bond lengths, the (0001) surface of graphite and the {111} layer of diamond have a very similar hexagonal structure, as shown in Figure 9.18 [196]. It may thus be supposed that diamond can easily grow epitaxially on graphite. Unfortunately, it is actually very difficult to deposit diamond on graphite, because graphite can be rapidly etched by hydrogen plasma under conditions of diamond growth. It is only recently that polycrystalline diamond films

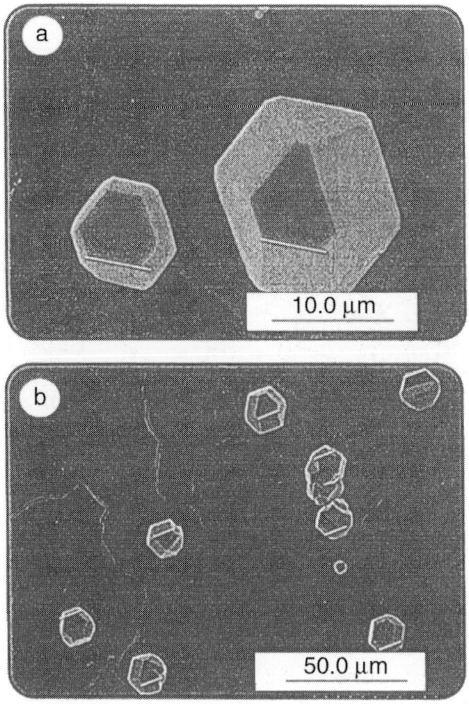

Figure 9.17. SEM images of diamond crystals deposited on Ni_3Si for 16 h under conditions of $c = 1\%CH_4/H_2$ and $T_s = 850\,°C$. The parallel lines on the images mark crystals that seemed to be unidirectionally oriented [195].

can be deposited by CVD on graphite. In 1993, Li et al. [196] used graphite-seeded Si substrates, graphite fibers, and highly oriented pyrolytic graphite (HOPG) to study diamond growth on them by HFCVD. The diamond nuclei preferentially grew epitaxially on the prism plane of graphite in such a way that G(0001) // D(111) and G[11$\bar{2}$0] // D[1$\bar{1}$0]. A theoretical investigation [197] determined the atomic configuration at the graphite/diamond interface, as shown in Figure 9.19. There was a few degree deviation ($\pm 3°$) of diamond orientation from the exact epitaxial arrangement.

Later, diamond growth on graphite flakes was done by MPCVD using $50\%CO/H_2$ [198]. The results indicated the presence of coalesced diamond crystals at the edge of the graphite flakes, consistent with the theory [197]. Also found were some mutually oriented diamond crystals with their (111) faces parallel to the G(0001) surface. So far, however, no epitaxial diamond films have been made on graphite.

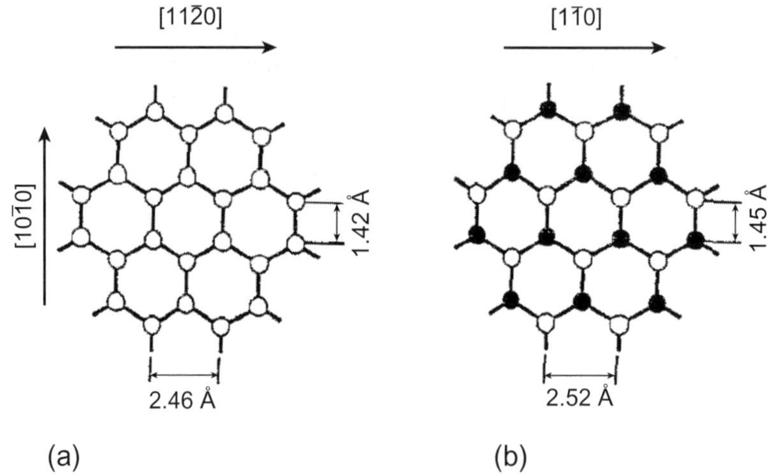

Figure 9.18. Structure of graphite {0001} basal plane and diamond {111} plane: (a) a view of graphite basal plane along the [0001] direction and (b) a projected view of diamond {111} plane along the [111] direction. The filled circles represent raised atoms [196].

9.4. SAPPHIRE

A unique attempt to heteroepitaxially grow diamond directly on sapphire was done in Ref. [199] using a pulsed excimer laser ablation. The laser used was a KrF excimer laser (wavelength: 248 nm, pulse duration: 20 ns, frequency: 5 Hz), and the target was graphite that was about 2 cm away from the sapphire substrate. The substrate was heated to 600 °C, and O_2 of 0.1–0.2 Torr was introduced in the chamber. As a result, very well-oriented diamond crystals with (111) faces were grown, as seen in Figure 9.20. The orientational relationship between diamond and sapphire was D[$\bar{1}$10]//sapphire[1$\bar{1}$00].

9.5. LOCAL EPITAXY ON Si

A local epitaxy of diamond crystallites has been reported in Ref. [200], where Si(100) substrate was pretreated in microwave plasma under conditions of $c = 0.1\%CH_4/H_2$, $P = 100$–120 Torr, and $T_s = 1040$–1050 °C for 10 h, and successively, diamond growth was done under conditions of $c = 1\%CH_4/H_2$, $P = 50$–60 Torr, and $T_s = 850$–880 °C for 40 h. On the Si substrate, there were two mutually oriented large diamond crystals of $120 \times 150\,\mu m$ in size and a number of small oriented crystallites with a pyramid shape of 1–2 μm in size around the large crystals, as seen in Figure 9.21. Both large crystals and small crystallites were oriented 45° off the substrate cleavage planes, i.e. {110}. According to X-ray diffraction measurements, the diamonds were ⟨100⟩-textured. It was not understood at that time what changes occurred by the pretreatment, and why oriented diamond crystals were grown. Now this is understood in such a way that the surface of the Si substrate was locally

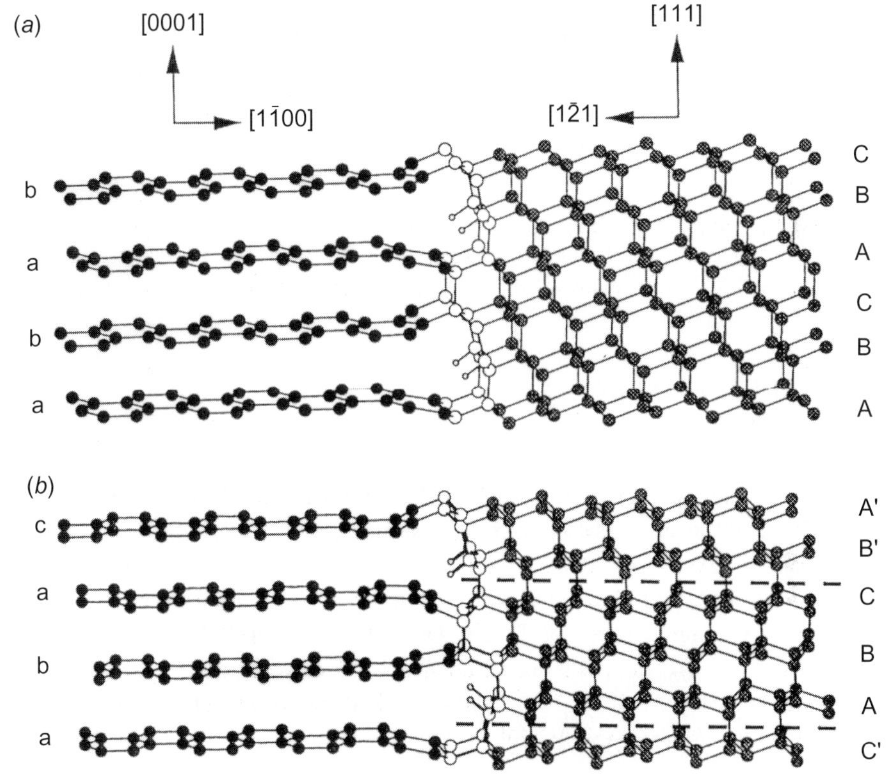

Figure 9.19. Side view (approximately parallel to the graphite [11$\bar{2}$0] and diamond [10$\bar{1}$] directions) of the relaxed structure models of diamond nucleus on the (1$\bar{1}$00) prism plane of graphite: (a) cubic diamond on perfect hexagonal graphite and (b) twinned diamond nucleus adjoining a graphite stacking fault. The three-dimensional simulation cell is periodic in the vertical direction and normal to the page, but finite in the horizontal direction. Twin boundaries in diamond are indicated by the dashed lines, and hydrogen by the small open spheres. The carbon atoms in diamond are indicated in dark gray, the carbon atoms in graphite in black. The atoms shown by the larger open circles indicate the initial nucleus formed at the interface to link the graphite layers with tetrahedrally bonded carbon [197].

carburized by the pretreatment to form a β-SiC(100) layer, and (100)-oriented diamonds were grown on it. This paper concluded from the orientation of the square faces on the diamond crystallites that they were 45° off the Si[110] direction. It is assumed that this conclusion was obtained just from the appearance of the diamond crystals, and on the contrary, an orientational relationship of D[110]// Si[110] may hold, just like the cases of HOD films [1] that will be described in later sections. It seems that the result of Ref. [200] occurred fortuitously, but was a prophecy for the reproducible invention of HOD films that came just two years later.

In Ref. [201], it was claimed that diamond films would be heteroepitaxially grown on mirror-polished (not scratched) Si(100) by thermal CVD at $T_s = 1150$–$1250\,°C$ using 1%alcohol/H_2. In the successive paper [54], it was demonstrated that diamond was deposited heteroepitaxially on Si(100) using $c = 0.5\%$alcohol (water-free)/H_2 at $T_s = 1100$–$1250\,°C$. The growth rate was only 33 nm/h. The monocrystalline nature of the film was confirmed by electron diffraction. There have been only few works on thermal CVD of diamond so that more investigations are needed for establishing this technique.

9.6. INTERFACE LAYERS

At this stage, it may be relevant to briefly describe the interface between diamond and Si substrate. The major issue is whether an interface layer is present, and what are the interface material and its atomic structure.

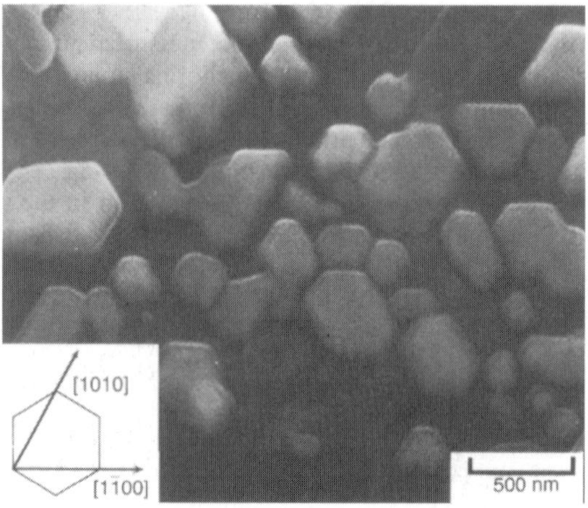

Figure 9.20. SEM image of oriented diamond crystals grown by laser ablation of graphite in an O_2 environment [199].

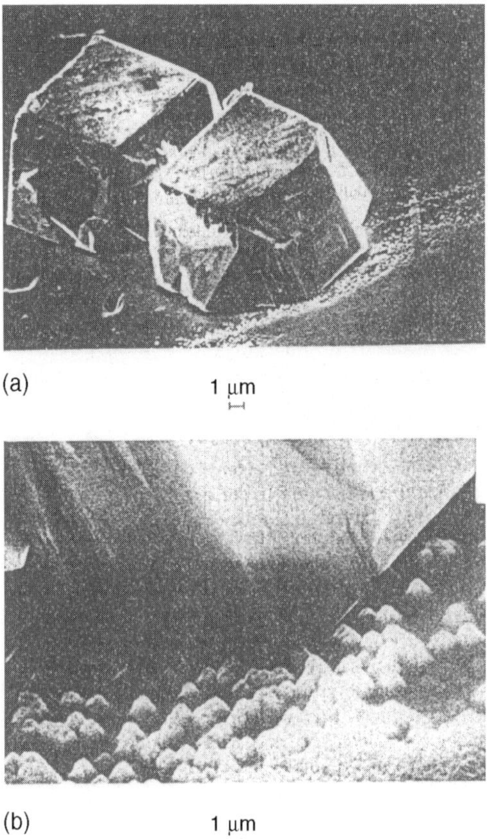

Figure 9.21. (a) Two neighboring locally epitaxial crystals of about $120 \times 150\,\mu m$ in size with each edge 45° off the {110} substrate cleavage plane, and a cluster of oriented cubic nucleations grown on the lower right corner and (b) enlarged view at the lower right corner of the locally epitaxial crystals [200].

In Ref. [202], it was reported by XPS measurements that when Si was used as a substrate for diamond growth by HFCVD, the native oxide, SiO_2, was removed, and a SiC layer was formed before diamond started to grow.

A detailed interface analysis of polycrystalline diamond films grown on Si has been done by Williams and Glass [203] using AES, EELS, TEM, and ED. The Si substrates were scratched prior to diamond CVD, and the diamond films were deposited by a NIRIM-type MPCVD reactor. For a specimen synthesized using $c = 0.3\%CH_4/H_2$, a 50 Å-thick β-SiC layer was present at the Si/diamond interface (see Figure 9.22 [203]), while it was absent for the sample grown using $c = 2.0\%CH_4/H_2$. The β-SiC layer of the former sample was a single

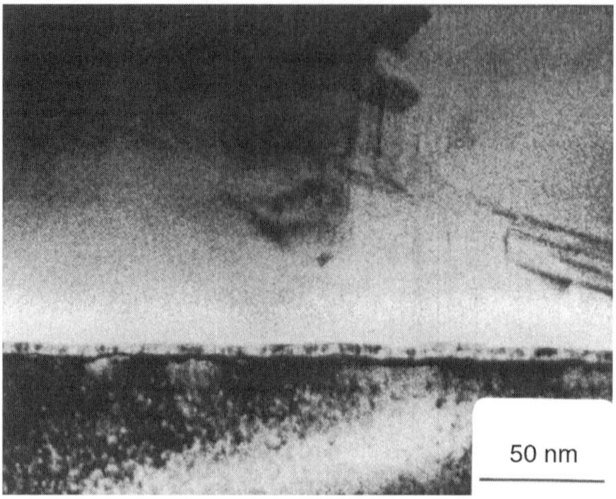

Figure 9.22. XTEM image of diamond film deposited on Si using $0.3\% CH_4/H_2$, showing the presence of a 50 Å-thick β-SiC layer at the interface [203].

crystal with an epitaxial relationship with Si, i.e. β-SiC{111} // Si{111} and β-SiC[110] // Si[110].

The atomic images at diamond/substrate interfaces is shown later. The presence or absence and the structures of the interface layers depend on the method of substrate pretreatment, the diamond CVD method and the growth conditions used. The investigation of the atomic interface structures will give us useful insight into the nucleation and heteroepitaxial growth of diamond.

Chapter 10

Diamond Nucleation

Chapter 10

Diamond Nucleation

10.1. METHODS OF DIAMOND NUCLEATION

The outline of known diamond nucleation methods and process has been described in Section. 2.2, and it is briefly summarized below:

In order to achieve a high nucleation density of diamonds on substrates, higher than $10^8/cm^2$, the following three methods are most frequently used:

(1) Polishing the substrate surface with diamond powder or paste of 0.1–30 μm in size. The nucleation density does not increase so much if SiC or Al_2O_3 powders are used.
(2) Ultrasonically treating the substrate in alcohol in which diamond powder is suspended.
(3) Applying negative direct current (DC) bias voltage on the substrate in the plasma of hydrocarbon/H_2 gas.

The nucleation density of diamond on *as received* Si wafer is only in the order of $10^6\,cm^{-2}$ by a NIRIM-type MPCVD reactor under standard CVD conditions, e.g. $c = 0.5\%CH_4/H_2$, $P = 30$ Torr, and $T_s = 800\,°C$. To make a continuous diamond film within one hour or so under these CVD conditions, a nucleation density of $\geq 10^8\,cm^{-2}$ is necessary. To achieve this, the Si surface is scratched with diamond powder or paste. The powder size is usually 0.1–30 μm. The nucleation density is increased to approximately $10^8\,cm^{-2}$ by this treatment. Alternatively, the Si wafer is ultrasonically treated in alcohol with diamond powder suspension for several minutes. The nucleation density can be increased to 10^9–$10^{11}\,cm^{-2}$ by this treatment. In both cases, the Si surface is roughened.

There are two possibilities on the creation of nucleation sites by scratching. The first is a creation of microedges of Si by scratching or the ultrasonic treatment of the Si surface, as it is known that diamonds tend to nucleate on sharp edges. However, similar treatments using non-diamond powders such as SiC and Al_2O_3 result in a lower nucleation density than $10^8\,cm^{-2}$ that is normally attained by diamond scratching. Therefore, solely the creation of microedges is not able to account for the nucleation density attained by scratching with diamond powder. The second is the embedded diamond nanoparticles or fragments below the Si surface during the polishing or the ultrasonic treatments [62]. It may be a little difficult to imagine that super-hard diamond particles are broken when they crash to a softer Si surface during the ultrasonic treatment, leaving nanosize diamond fragments in Si. Observations of the ultrasonicated Si surface by TEM and ED, however, revealed the existence of nanosize diamond fragments indeed embedded in Si with a density of 10^{10} to $10^{11}/cm^2$.

Since the Si substrate surface is roughened by the scratching, and the embedded diamond fragments are randomly oriented, it seems difficult for diamond to grow in registry with the Si atom arrangement at the substrate surface. Thus, a variety of attempts were made to enhance nucleation: carbon ion implantation, painting with a carbon-containing ink and anneal it [204], depositing a thin layer of metals, etc. For a review of diamond nucleation up to 1995, the reader can refer to Ref. [205].

The most successful method of diamond nucleation was (3), mentioned above, that was invented by Yugo et al. [3, 206–210], using a negative DC bias on the Si substrate in hydrocarbon plasma, which is later called as bias enhanced nucleation (BEN). In the first paper by Yugo et al. [3], a nucleation density of $10^{10}/cm^2$ was achieved. It was successively found by Stoner and Glass [1] that (100)-oriented, azimuthally aligned diamond crystals can be grown on a β-SiC(100) substrate that had been pretreated by BEN. The synthesized film was named as highly oriented diamond (HOD) film, and this method was later developed by a number of researchers is described below. Today, HOD films can be grown on 2–3 inch Si wafers. For a theoretical study of BEN, see Ref. [210].

10.2. BEN METHOD

Since the BEN technique was found by Yugo et al. [3], numerous works have been done on BEN. The establishment of the HOD film growth technique was one of the most important motivations for the BEN studies. Even so, there are still controversy over fundamental issues among researchers: the first is how diamond nuclei are formed by BEN, and the second is whether an interfacial layer exists, or an interfacial layer is necessary to grow HOD films.

10.2.1. Yugo's method

Since the method of BEN was first found by Yugo et al. [3], works of his group till date is first reviewed in the following. The reactor used was a NIRIM-type, as shown in Figure 3.1, where a negative DC bias of $V_b = -100$ V was applied to an *as received* Si substrate with a resistivity of 1–5 Ω · cm and placed on a molybdenum (Mo) holder. The opposing electrode was a tantalum (Ta) wire immersed at the opposite end of the plasma above the substrate. The biasing conditions used are listed in Table H.1. Figure 10.1 is a specimen that has been BEN-treated using 10%CH_4/H_2 for 2 to 5 min, followed by diamond CVD under conditions of $c = 0.5\%CH_4/H_2$ and $P = 30$ Torr for 30 min. As a result, a nucleation density of $10^{10}/cm^2$ was achieved.

To study the nucleation mechanism of BEN, a process gas of 40%CH_4/H_2 and $V_b = -100$ V were employed for 5 min [206]. Note that since this work is prehistoric to the establishment of HOD film growth technology, these conditions are not optimized for HOD film growth, as seen in Table H.1. The specimen was analyzed by RHEED, XPS, and Raman spectroscopy. Particles deposited on the Si substrate, like the one shown in Figure 10.1, had a "star-ball" shape. The RHEED measurements indicated that they consisted of *a*-C, SiC, and diamond. Furthermore, XPS and Raman spectroscopy measurements revealed a presence of glass-like carbon. A hydrogen plasma treatment of those particles resulted in residues with indefinite shapes, consisting of SiC and diamond, according to XPS measurements. Yugo et al. [206] argued that since the mean-free-path of (carbon containing) ions is approximately 5 μm, which is nearly equal to the width of the plasma sheath under the gas pressure of 30 Torr used in their experiments, the kinetic energies of impinging ions to the substrate were nearly equal to the bias voltage, 100 eV. According to the Kinchin-Peace equation [211], the average numbers of atoms displaced by a 100-eV impinging carbon ion are 3.5 for Si, 2–5 for graphite, and 0.5 for diamond. Furthermore, the penetration depth of carbon ions with 100-eV kinetic energy in Si is approximately 0.6 nm with a standard deviation of 0.3 nm. These data suggest that during the BEN treatment under −100 V bias to the substrate, (i) the impinging carbon atoms from the plasma sputter mostly graphitic carbons, or sp^2-bonded carbons, from the deposited carbonaceous particles, and (ii) a carbon ion mixing with Si to form SiC simultaneously takes place to form

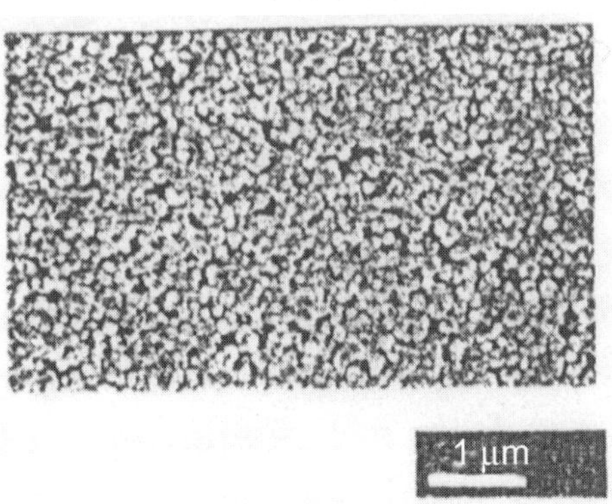

Figure 10.1. SEM image of diamond particles grown on a Si substrate using 10%CH_4/H_2 by applying $V_b = -100$ V to the substrate [3].

a shallow carbon-saturated layer at the Si surface, and (iii) hydrogen atoms are considered to also etch graphitic carbons, and terminate dangling bonds of carbon atoms, converting sp^2 bonds to sp^3. Thus, the BEN treatment has these three effects.

Cross-sectional TEM observations were undertaken for a BEN-treated Si(100) substrate under conditions of $V_b = -100$ V, $c = 40\%CH_4/H_2$, $P = 30$ Torr, $P_m = 400$ W, and $T_s = 850\,^\circ C$ for 5 min (sample A), which was then followed by an etching in pure H_2 plasma for 5 min (sample B), and finally diamond CVD was done for 5 min using $0.5\%CH_4/H_2$ (sample C) [208]. The XTEM micrographs are shown in Figures 10.2, 10.3, and 10.4 for samples A, B, and C, respectively. In Figure 10.2 (a)

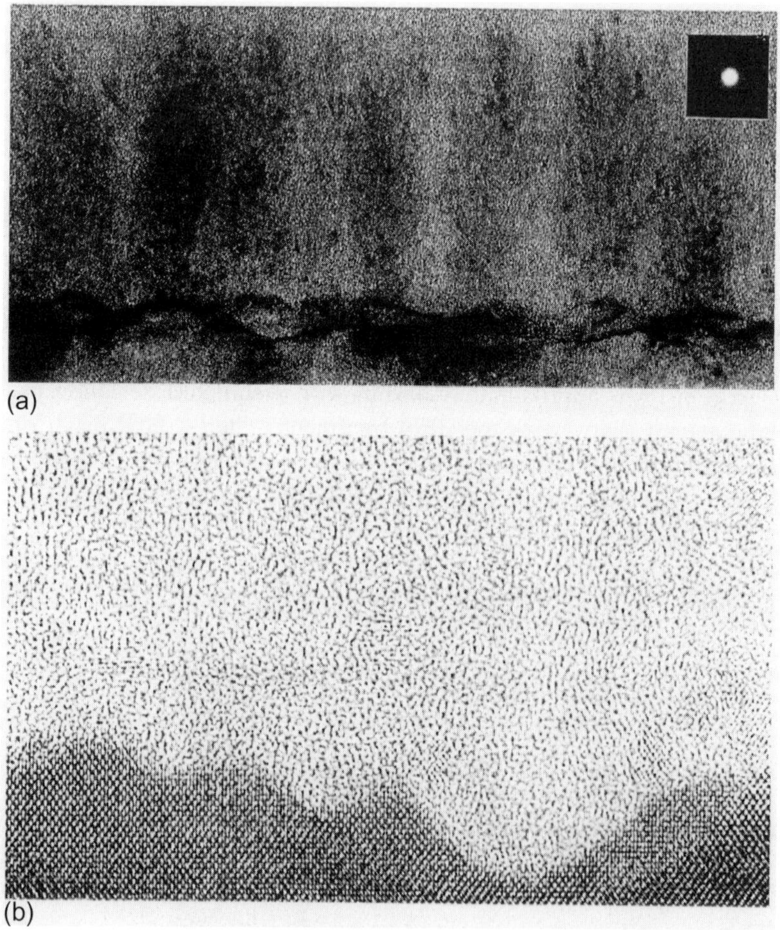

Figure 10.2. XTEM image of the deposited material on Si by BEN ($c = 40\%CH_4/H_2$ and $V_b = -100$ V) for 5 min (sample A). Magnification: (a) × 400 000 and (b) × 3 200 000 [208].

(sample A after BEN), tree-shape structures are standing on the Si surface with a density of $10^{10}/cm^2$. Its magnified view, Figure 10.2 (b), indicates that the tree-shape structures consist of fibrous graphite that contains diamond grains. In Figure 10.3 (a) (sample B after hydrogen plasma etching), the lengths of the tree-shape structures become less than half of those in Figure 10.2 (a), but there are more diamond grains, as seen in Figure 10.3 (b). It is of interest that many of the diamond grains are agglomerations of diamond crystals of about 1 nm in diameter. Finally, in Figure 10.4 (a) (sample B after diamond CVD), the tree-shape

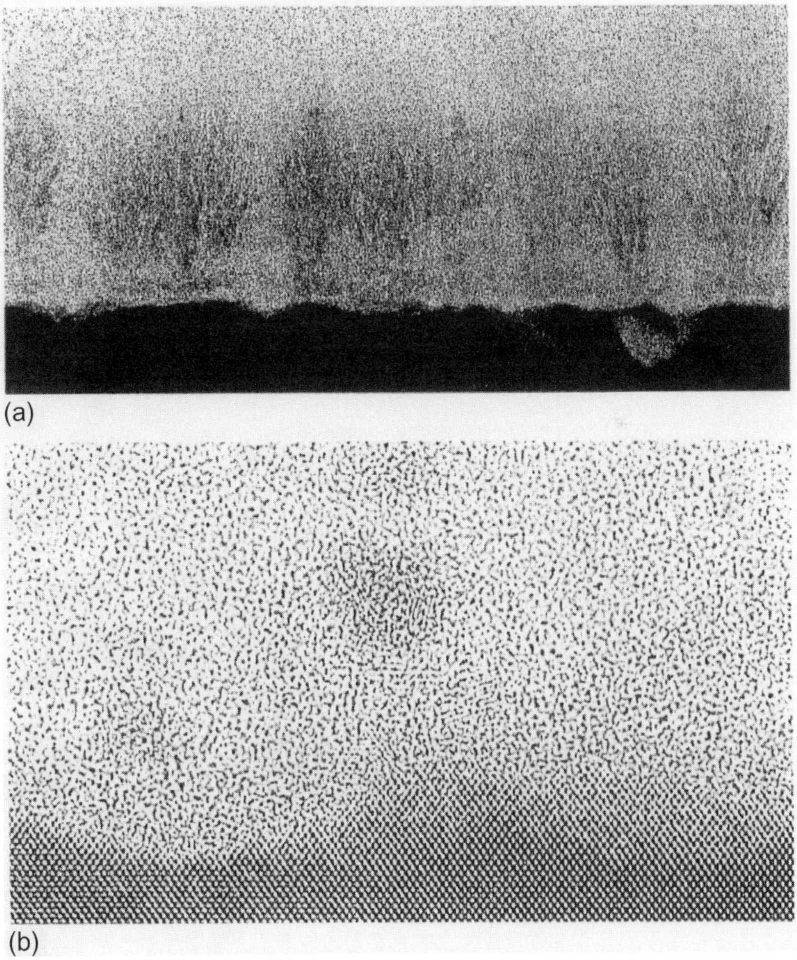

Figure 10.3. XTEM images of sample B prepared by etching sample A using hydrogen plasma for 5 min. Magnification: (a) × 400 000 and (b) × 3 200 000 [208].

structures have disappeared and only their root parts remain. A magnified view of the interface region by TEM, Figure 10.4 (b), indicates a presence of a 0.5-nm thick, random structures, but the materials have not yet been identified.

A cross-sectional TEM observation was also undertaken for a BEN-treated specimen under conditions of $V_b = -60$ V, $c = 2\%CH_4/H_2$, $T_s = 850\,°C$, and $P =$ 15 Torr for 10 min [207]. These are nearly consistent with the HOD film growth conditions, as seen in Table H.3. No carburization process of the Si substrate was done. These BEN conditions are rather milder than other works. As a result,

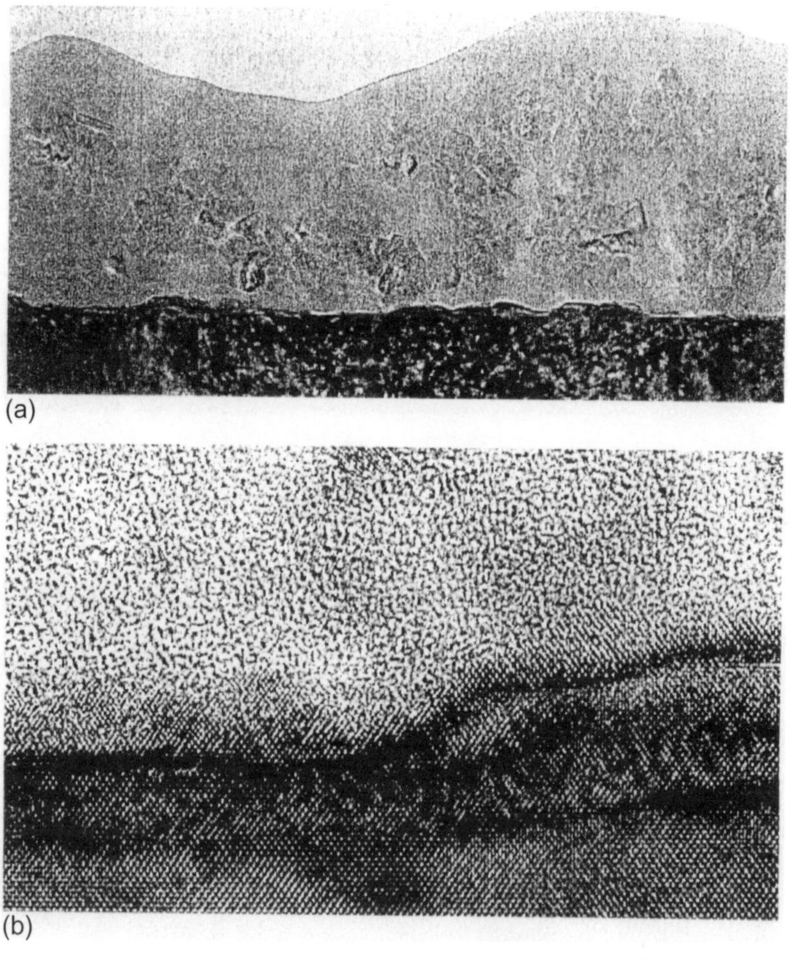

Figure 10.4. XTEM images of sample C prepared by using sample B as a substrate and deposited diamond for 5 min using $c = 0.5\%CH_4/H_2$. (a) × 250 000 and (b) × 4 000 000 [208].

a rectangular mesh structure with 100 Å separation was created at the Si surface, and particles of 10–100 nm in size were deposited with a density of $10^{10}/cm^2$. The height and the diameter of the particles were 100–150 nm and 100 nm, respectively. The HRTEM indicated that the core region of the particle consisted of β-SiC, diamond, and amorphous layers, where in some areas, diamond and β-SiC were epitaxial with each other. It was argued that the presence or absence of the amorphous carbon layer between diamond crystallites and the Si substrate depended on the BEN conditions, i.e. the CH_4 concentration c, the ion energy (or equivalently the bias voltage V_b), and the substrate temperature T_s. It was also speculated that the embryonic carbon clusters underwent a liquid–solid transition to epitaxially align with the substrate structure. However, this model needs experimental evidence. In Figure 10.5, one can see diamond particles with a typical diameter and a height of 100 and 100–150 nm, respectively. A 10-times magnified view of the interface, Figure 10.6, shows that there exists an intermediate layer of roughly 1-nm thickness between Si and diamond particles. This work is direct evidence that diamond particles can be formed on an *as received* Si substrate by BEN.

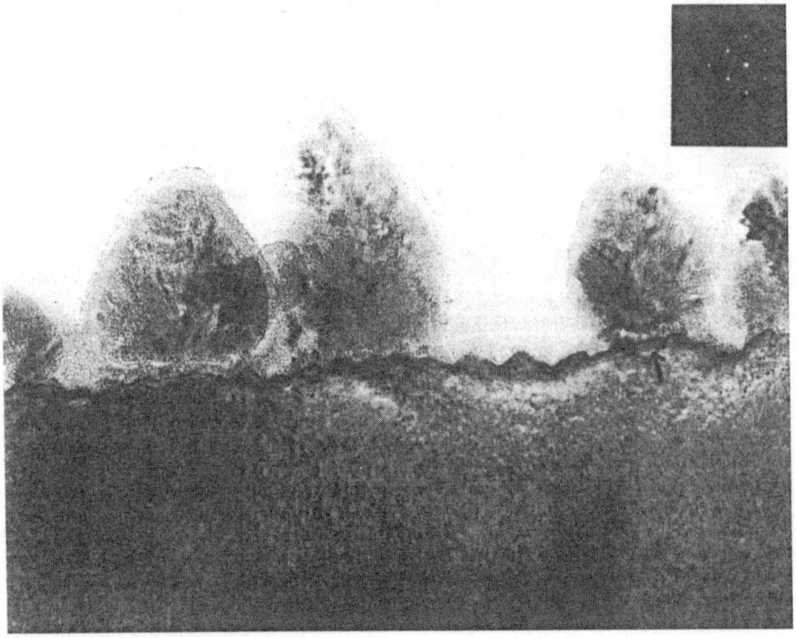

Figure 10.5. XTEM image of diamond crystallites on Si substrate after BEN (× 400 000). The inset is an ED pattern [207].

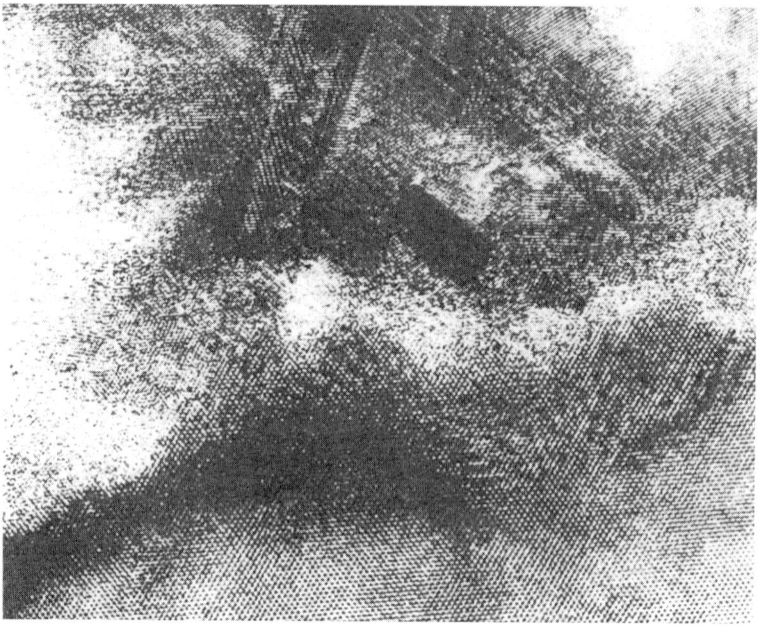

Figure 10.6. XTEM image of diamond crystallites on Si substrate after BEN (×4 000 000) [207].

The distribution of SiC interlayer over a 76-mm Si(111) wafer was examined after a 30-min BEN treatment at $V_b = -125\,V$, followed by diamond growth (MPCVD) using $c = 1.5\%CH_4/H_2$ for 2 h. The thickness of SiC layer was markedly different between z- and x-directions, as observed by FT-IR [212].

10.2.2. Various aspects of BEN

Stoner et al. [2] examined BEN in great detail by XPS, AES, XPS-EELS, Raman spectroscopy, and TEM. The reactor used was an ASTeX-type. The substrates were Si placed on a heated Mo holder. The BEN conditions are listed in Table H.1, and the biasing time was up to 2 h. The substrate temperature was not mentioned in the paper, but assumed to be around 650 °C from other papers. The highest nucleation density obtained by the BEN treatment was $10^{11}/cm^2$, in strong contrast to the nucleation density of $10^8/cm^2$ on scratched Si substrates. A Raman spectrum of the BEN-treated specimen had a higher background and a broad band around $1550\,cm^{-1}$, indicating the presence of high grain boundary density in the specimen. As seen in Figure 10.7 (a), the XPS spectra of C(1s) showed that as the biasing time was increased, the C–Si band intensity, relative to the C–C band intensity, increased

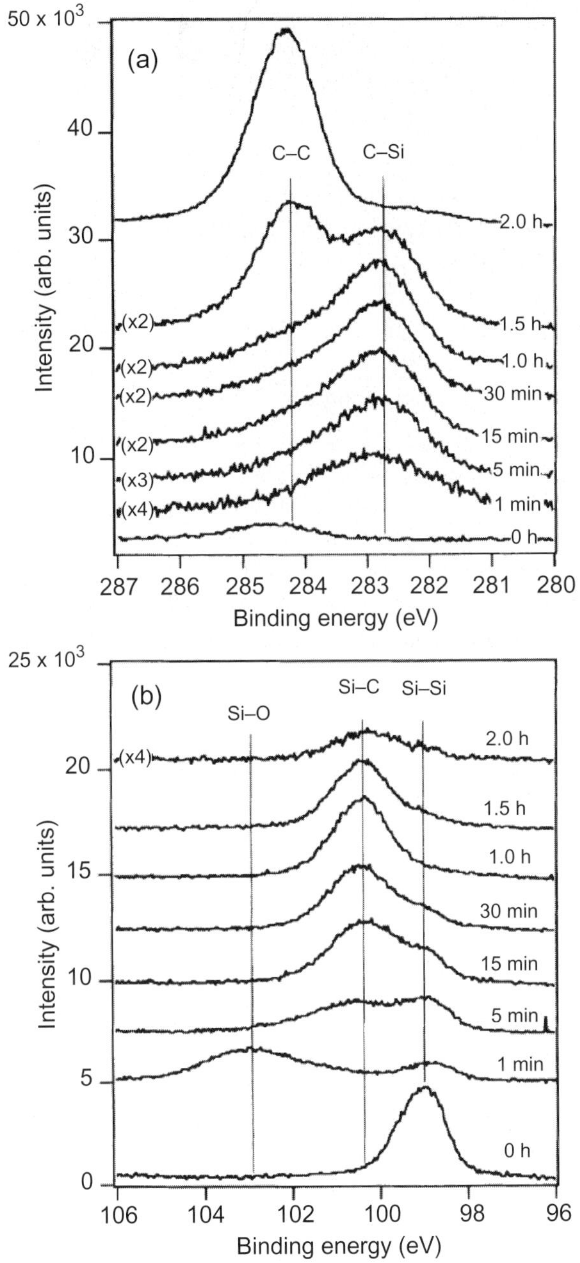

Figure 10.7. XPS analysis as a function of BEN treatment time: (a) C(1s) and (b) Si(2p) progressions with time [2].

about four times up to the biasing time of 1 h, and then rapidly decreased. On the other hand, in the XPS spectra of Si(2p) (Figure 10.7 (b)), the intensity of the Si–C band became more dominant with the biasing time, although the Si–O band was present in the initial stage. Note that the binding energies of C(1s) and Si(2p) in Figure 10.7 are 284.3 and 282.8 eV, respectively. On the other hand, those of Si–Si, Si–C, and Si–O are 99.0, 100.3, and 102.7 eV, respectively. Figure 10.7 thus shows that up to the biasing time of 1 h, the atomic concentration ratio of C/Si was less than 0.37. A deconvolution of the C(1s) spectra showed that up to 1 h, there was a C–C bonding component that was roughly 20% of the total C(1s) intensity. It was concluded by XPS of a BEN-treated and surface-sputtered specimen that the C–C bonds on the specimen surface was created by the biasing process in such a way that either Si atoms were etched or sublimated from Si–C or a flux of hydrocarbon ions to the specimen surface increased. In Figure 10.7 (b), it is seen that the Si–C band increases at 1.5 h, when the C–C band also increases rapidly, as seen in Figure 10.7 (a). This was attributed to the etching of the interfacial SiC layer, which brings Si atoms close to the surface, thus increasing the SiC band intensity. The specimen surface became carbon-rich after a 2-h biasing, and the transition from SiC to diamond was clearly observed by XPS-EELS spectra.

In Figure 10.8, reference spectra of AES and XPS-EELS from various materials observed by Stoner et al. [2] are shown. Based on these data, it is clearly seen that the specimen spectra of Figure 10.7 exhibit a transition process from β-SiC formed by CVD to diamond. This transition process was also confirmed by Raman spectroscopy. According to an XTEM observation for the specimen after a 1-h biasing followed by a 5-h diamond CVD, an *a*-SiC layer of 6-nm (maximum 10-nm) thickness was present between the Si substrate surface and the diamond layer. An HRTEM indicated that diamonds nucleated within the interfacial layer but above the Si substrate. From the observed data, a model of diamond nucleation by BEN was proposed, as shown in Figure 10.9.

Initially, the substrate surface is adsorbed by oxygen and hydrocarbons (Figure 10.9 (a)), but they are soon removed from the surface, or converted to Si–O and Si–C at the surface after the BEN is initiated (Figure 10.9 (b)). As the biasing is continued, a preferential etching of Si and a deposition of carbon takes place, resulting in a formation of *a*-SiC layer and carbon islands (Figure 10.9 (c)). A continued flux of carbon atoms onto the surface, giving energies to the carbon islands, as well as the hydrogen plasma environment, converts the carbon clusters to diamond nuclei (Figure 10.9 (d)). This process is enhanced when the surface layer is saturated with carbon, while the *a*-SiC layer reaches a critical (maximum) thickness of about 90 Å (Figure 10.9 (e)). The surface is roughened by plasma etching as the biasing is continued, and the formation of carbon clusters and their conversion to diamond nuclei proceed (Figure 10.9 (e)). Eventually the surface

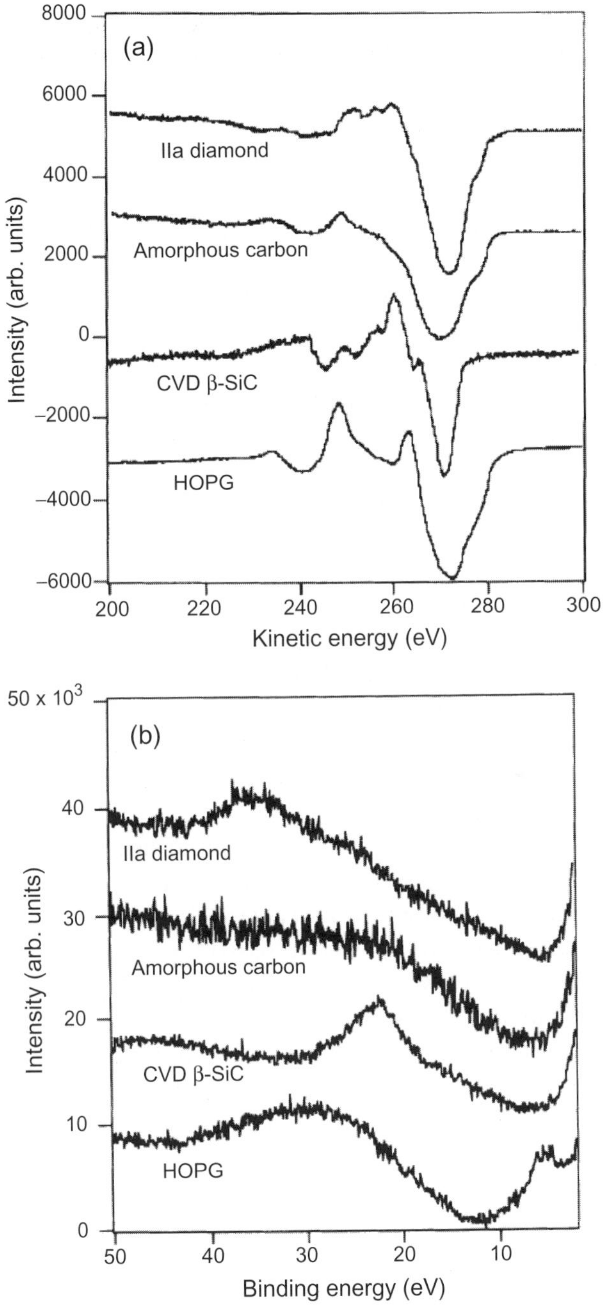

Figure 10.8. Reference (a) AES and (b) XPS-EELS spectra of single-crystal diamond, *a*-C, single-crystal SiC, and highly oriented pyrolytic graphite (HOPG) [2].

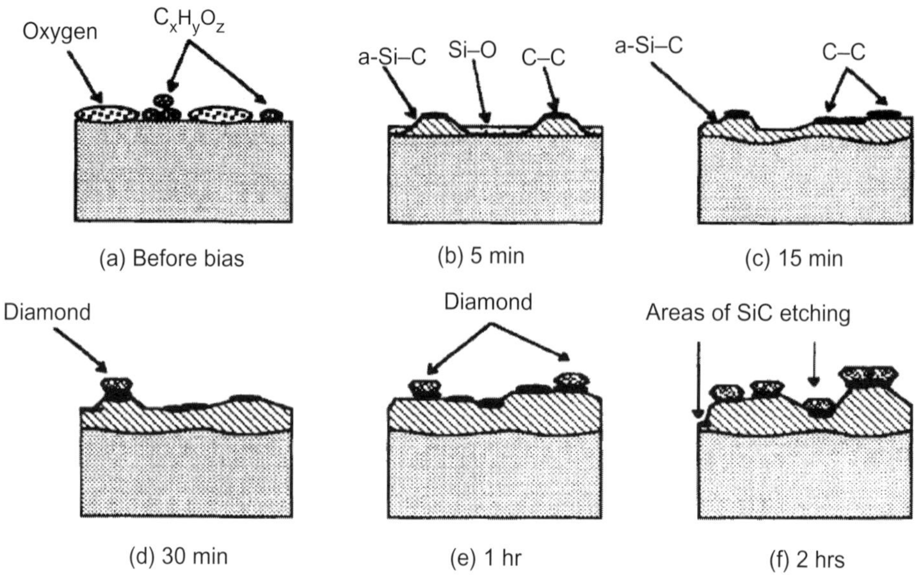

Figure 10.9. Model of diamond nucleation by BEN on Si [2].

is entirely covered continuously with diamond (Figure 10.9 (f)). The diamond nuclei are thus originated within the interfacial layer above the Si substrate surface. There was, however, no indication in the HRTEM that oriented nanocrystalline graphite was generated within the interfacial SiC layer in contact with the Si substrate surface.

The effects of bias voltage V_b on the nucleation density of diamond on Si(111), (110), and (100) substrates have been investigated by Jiang et al. [213] for a wide range of parameter space, as shown in Table H.1 and Figure 10.10 (a)–(c). The reactor used was an 1.5-kW ASTeX reactor for MPCVD. The nucleation density was highest ($\sim 10^9/cm^2$) for $c = 4\%CH_4/H_2$, $P = 14$ Torr, $V_b \leq -120$ V, and $T_s = 850\,°C$, when the biasing time was 20 min.

Bar-Yam and Moustakas [214, 215] (see also Ref. [216]) suggested that because of the vacancies at the deposition surface, diamond was more stable than other forms of carbon at the growth surface during CVD. They argued that the diamond growth surface with a density of less than 1% vacancies was more stable than graphite with the same density of vacancies, because the formation energy of vacancies is 0.4 eV higher for graphite than diamond. Based on this premise, it was predicted that the negative bias would not affect the formation energy of vacancies for graphite, but significantly decrease the formation energy of vacancies for diamond as the vacancies in diamond are negatively charged. If this is the case, the biasing (i) increases the defect density in diamond and (ii) makes diamond more

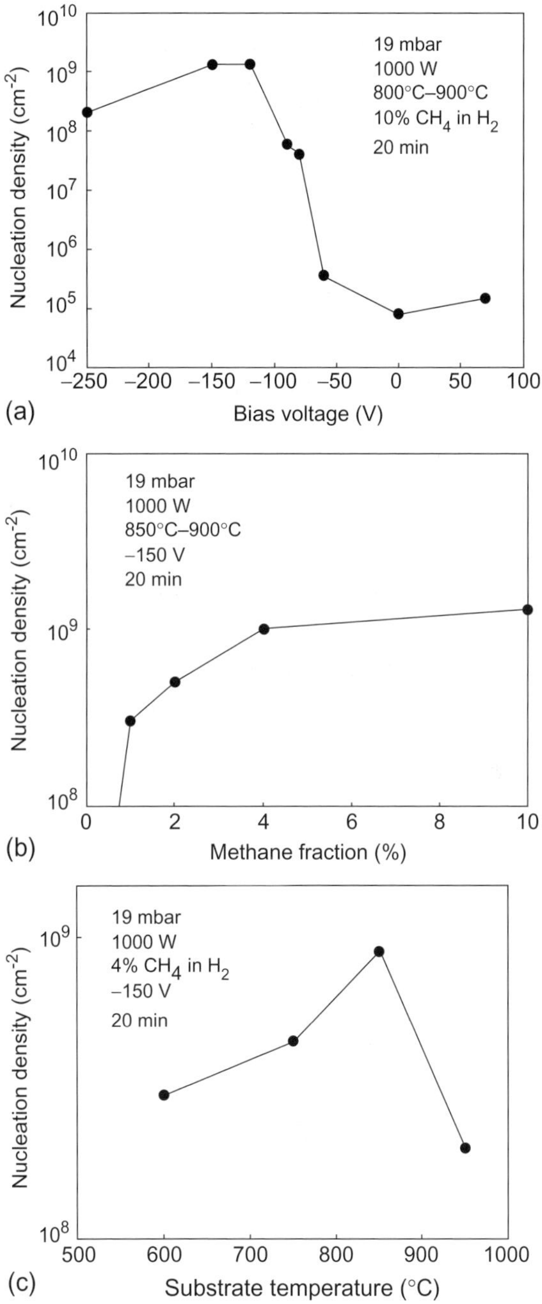

Figure 10.10. Diamond nucleation density as a function of (a) V_b, (b) the CH_4/H_2 concentration, and (c) T_s [213].

stable than graphite as defects increase the free energy of formation of graphite more than diamond. It seems, however, difficult to explain that the diamond growth solely relies upon the defect formation energy, ignoring the traditional understanding that graphite is more easily etched by chemically active atomic hydrogen than diamond. Moreover, the presence of 1% defect density, necessary to stabilize diamond, is extraordinarily high, and such a high density of defects is not present in natural diamond crystals. Indeed, Sheldon et al. [216] pointed out that their Raman results appeared to contradict with this argument, because the broad band at $\sim$1580 cm^{-1} indicated that a significant amount of sp^2 carbon were deposited during the BEN process.

Sheldon et al. [216] studied the BEN mechanism using an ASTeX reactor. Note, however, that the conditions used for BEN do not lead to HOD films. The substrate used was Si(100) that was placed on a heated graphite stage. The Si substrate was treated with hydrogen plasma for 5 min prior to BEN to remove the native silicon oxide layer. The BEN conditions were listed in Table H.1, and T_s, measured by a thermocouple, was usually 865 °C. Under these conditions, the Si substrate surface was entirely covered by a film after biasing for 3 h. The deposited films were examined by Raman spectroscopy, SEM, TEM, and EELS. The results of EELS and XTEM indicated that for a specimen with a 90-min biasing, an *a*-C layer of 4-nm thickness was present at the Si surface, while for a specimen with a 180-min biasing, a coexisting layer of *a*-C, SiC, and diamond was detected, where SiC and diamond were only a few nanometer in diameter. Namely, nanometer-size SiC and diamond crystallites were embedded in the *a*-C matrix. Although there was no indication on the presence of crystalline graphite in the *a*-C matrix, Raman spectra indicated that disordered sp^2 carbons were present. The *a*-C layer also was present when the BEN treatment was done at $T_s = 400$ and 600 °C for 90 min. These results are different from Stoner et al. [2], where a Si–C layer of less than 10-nm thickness was present after the BEN treatment. This difference may be attributed to the difference in the BEN conditions used. It seems that this also arises from the difference in the substrate holder material, Mo in Ref. [2] but graphite in this work [216]. It was noticed that the growth rate of the *a*-C layer was only 0.4–0.6 Å/min. This was either because (i) the concentration of positive carbon or hydrocarbon ions in the plasma was low, or (ii) the ion energies due to biasing were still low.

The average kinetic energy of positive ions was estimated as follows: assuming that the mean-free-paths Λ of the ions are equal for CH_4 and H_2, then Λ is expressed by:

$$\Lambda = \frac{k_B T}{\pi \sigma^2 P} \tag{10.1}$$

where σ is the average collision diameter (~0.3 nm), P the total gas pressure, and k_B the Boltzmann constant. The kinetic energy of a monovalent positive ion $E(+1)$ is then given by:

$$E(+1) = \Lambda \frac{\Delta V}{Z_s} \tag{10.2}$$

where ΔV is the potential gradient across the plasma sheath, and Z_s is the thickness of the plasma sheath, which is in the order of 0.1–1 mm. From Eq. (10.2), $E(+1)$ was estimated to be in the range of 2–20 eV, which was considered to be the lower limit of the kinetic energy.

It is of interest [216] that it was *a*-C but not diamond that was deposited under the bias voltage. This means that the biasing has a detrimental effect on diamond growth. The growth rate of *a*-C film deposited under the bias voltage was more than two orders of magnitude smaller than the growth rate of diamond film under the same conditions but without bias voltage. Also, the nucleation sites of diamond created by the biasing did not depend on the presence of nanocrystalline SiC, because diamonds were directly formed on the *a*-C film that had been bias-deposited for 90 min. In the *a*-C film, there was no sign on the presence of SiC, despite the reasonable inference that Si, which is preferentially etched or sublimated from the deposited film, might be pushed back into the film under the bias voltage.

On the contrary, in Ref. [217], diamond growth was undertaken under a negative DC bias voltage on scratched 3-inch Si wafers. The growth conditions were $V_b \geq -450$ V, $c = 1.3\%CH_4/H_2$, $P = 30$ Torr, $T_s = 300$ °C, and $P_m = 3.25$ kW. For $V_b \leq -150$ V, well-facetted diamond films were grown, and the crystal morphology was dependent upon the applied bias voltage. This means that the α-parameter could be controlled by V_b. So far, it has been understood that a continuous application of bias voltage is detrimental, and generates ball-like deposits rather than well-facetted diamond crystals. Thus, this work is in contradiction with past works, and a further study is necessary to make clear the reason.

In Ref. [218], Kawarada et al. studied the polarity effects in BEN on the nucleation density using a NIRIM-type reactor, a Mo substrate holder, and Si(111) substrates under conditions of $P = 0.2$, 2, and 15 Torr, $c = 2$, 10, and $40\%CH_4/H_2$, and V_b was between −100 and +100 V. After the BEN treatment, diamond CVD was done for 3 h under conditions of $c = 0.5\%CH_4/H_2$, $P = 30$ Torr, and $T_s = 950$ °C to evaluate the nucleation density. Interestingly, it was found that:

(i) The nucleation occurred for both positive and negative bias voltages to the substrate, although the nucleation density was $10^8/cm^2$ for +20 V, while it was $10^9/cm^2$ for −20 V,

(ii) The positively biased specimen surface was smooth, while the negatively biased surface was rough, presumably due to ion damage,
(iii) The RHEED measurements showed no indication on the presence of diamond phase in the BEN-treated specimens,
(iv) The nucleation density was higher when the gas pressure was low (0.2 Torr), the CH_4 concentration was high (40%), and the absolute value of the bias voltage also was high (±100 V), and
(v) The nucleation occurred for $c = 40\%CH_4/H_2$ and $P = 15$ Torr even without bias voltage, presumably because plasma itself is positively charged to create the electric field between the substrate.

The deposition of carbon particles under a positive bias voltage to the substrate was attributed to the electron flux to the substrate, giving the same effects as the electron-assisted HFCVD [219], in which a negative bias was applied to the filament and a high density of diamond nucleation occurred even without any pretreatment of Si substrates. Figure 10.11 shows the *I–V* characteristics of the plasma without substrate, indicating that the electron flux was approximately 10 times higher than the ion flux.

To investigate the role of hydrogen in BEN, Lannon et al. [220] undertook an etching of a SiC film of 100-nm thickness using H^+ and H_2^+ ions (the ratio of $H^+/H_2^+ \sim 2\%$) with energies of 10, 100, 500, and 2000 eV. When the ion energy was 500 eV, Si atoms were most preferentially removed out of the SiC film, leaving 30% of C atoms in the sp^3-bonding state. Based on their experimental results, it was argued about the BEN process proposed by Stoner et al. [2] that both (i) SiC provided by segregation and out-diffusion of Si from the substrate and

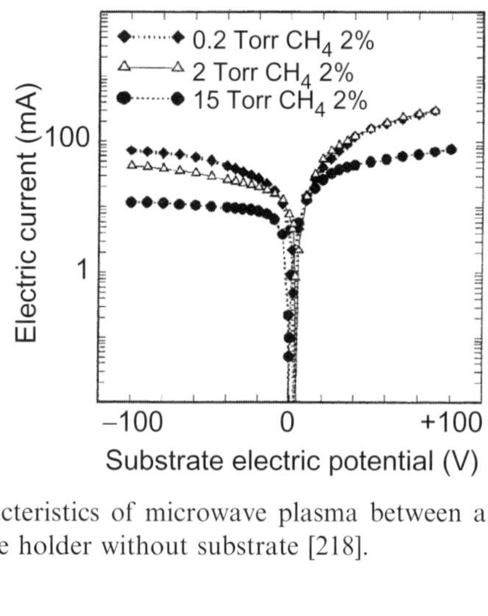

Figure 10.11. *I–V* characteristics of microwave plasma between a cylindrical electrode and a substrate holder without substrate [218].

(ii) hydrocarbon species provided from the plasma are nearly offset by the removal of Si and C by hydrogen ions. Since the removal rate of Si atoms by hydrogen ions is higher than that of C atoms, C–C species appear early in the nucleation stage. It was then inferred that the C–C species are sp^3-bonded carbons: in fact, diamonds. The formation of the 6-nm thick SiC layer in Ref. [2] is a result that the Si supply is transport-limited at that thickness and at the substrate temperature T_s ($\sim 650\,°C$) used. This inference is consistent with the results obtained by Yugo's group [3, 206–210].

Gerber et al. [221] investigated the nucleation density of diamond on Si, which was placed on a heated stage, by changing the BEN parameters: the CH_4 concentration $c \leq 15\%CH_4/H_2$, V_b, the biasing time, and T_s. The nucleation density became maximum of approximately $10^{10}/cm^2$ for $T_s \geq 740\,°C$ at $V_b = -250\,V$, and did not depend on the CH_4 concentration. Indeed, $c = 5$ and $15\%CH_4/H_2$ resulted in the same saturated nucleation density. An estimated ion energy for optimum nucleation was 70–80 eV, and hence it was concluded that the BEN process proceeds by a subplantation of hydrocarbon ions.

In Ref. [222], researches of BEN are reviewed with an emphasis on their works presented in Refs. [223, 224]. Since Ref. [224] will be reviewed later, that part is left out below. In electrical and optical studies of BEN using an ASTeX reactor, where a 3-inch Si wafer was placed on a graphite holder on top of a common 4-inch substrate holder, as shown in Figure 10.12, a secondary plasma was generated between the main plasma and the substrate center when the bias voltage was applied (for example, see Figure 11.51). There was also a plasma at the outer rim of the substrate holder, which had been deposited with diamond [225]. This was considered to be the cause of the poor reproducibility and the limited homogeneous nucleation area by the BEN process. It might well be that this result originated from the specific structures of the reactor chamber and the substrate holder (shown in Figure 10.12), because the presence of the secondary plasma was also seen in Refs. [226, 227] using a bell-jar type ASTeX reactor, but the plasma was disk-like and existed right above the substrate only. The Si substrate was placed also on a graphite block. Thus, the generation of the microwave plasma and the secondary plasma seems to be very sensitive to the reactor design as well as the structure of the substrate holder.

Figure 10.13 shows I_b–V_b characteristics for the plasmas using a bare Si substrate and a Si substrate already covered with a polycrystalline diamond film. The diagram of the I_b–V_b characteristics was divided into three regions, I to III. Note that the current is higher for the diamond-covered substrate, which is attributed to an enhanced secondary electron emission from diamond [224, 225, 228].

Figure 10.14 shows the time dependence of the bias current $I_b(t)$, which is divided into three stages: (i) in the first stage, SiC is formed on the Si surface; (ii) in the

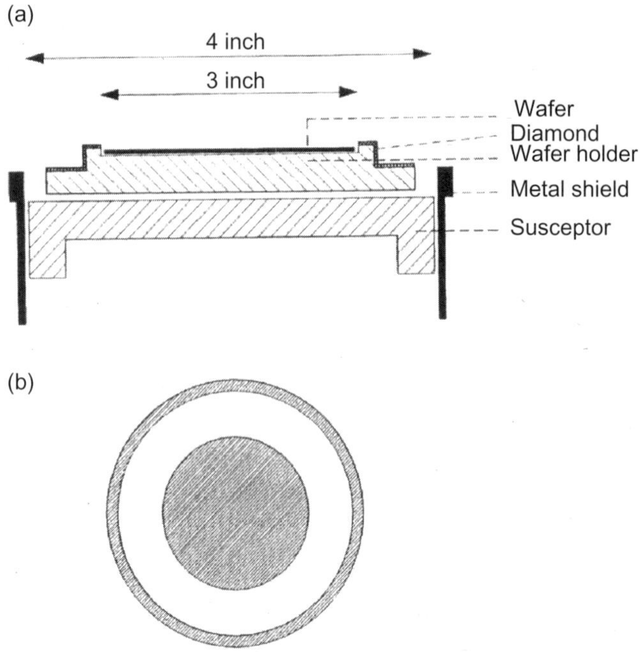

Figure 10.12. (a) Structure of the substrate holder and (b) the areas on the wafer covered by the films [222].

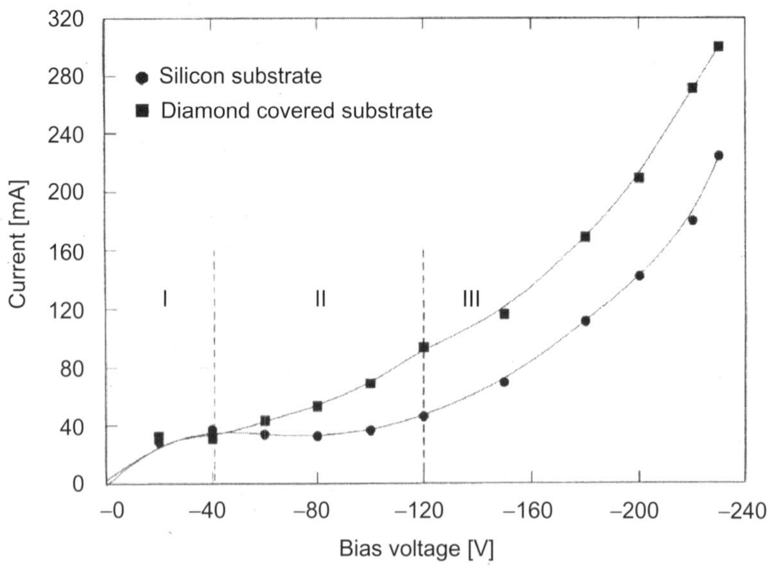

Figure 10.13. I_b–V_b characteristics for a blank and a diamond-coated Si substrate [222].

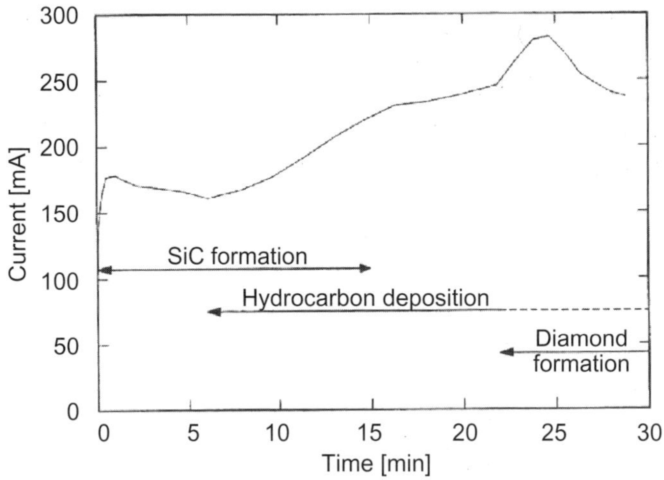

Figure 10.14. Change in the bias current $I_b(t)$ with time. Also indicated is the nucleation to diamond growth sequence [222].

second stage, hydrocarbon deposition occurs; and (iii) in the third stage, the current increases rapidly and diamond formation begins. The fact that the current decrease after $t=25$ min was ascribed to an increase in etching and damages of already formed diamond crystallites by prolonged ion bombardment. Since the final increase in $I_b(t)$ for $t=22$–25 min corresponds to the beginning of diamond formation, this point can be used as the termination point of BEN. The general behavior of I_b–V_b was characteristics for BEN, but the absolute values and the time scale of $I_b(t)$ were poorly reproducible. This is a serious problem of diamond nucleation using BEN. Furthermore, the extension of the secondary plasma changed during the BEN process, and this also caused a problem of reproducibility.

Figure 10.15 shows a typical result of the nucleation density as a function of the bias voltage using $c=2\%CH_4/H_2$. It is of interest to note that the voltage at which the nucleation density saturates, $V_b=-140$ V, roughly corresponds to the voltage of phase III onset in Figure 10.13, $V_b=-120$ V. This figure indicates that there exists a threshold voltage for diamond nucleation by BEN, although the value depends on the experimental setup and the BEN conditions.

Regarding the atomic hydrogen concentration, it is stated [222] that unlike the result of Ref. [229], the atomic hydrogen concentration decreased by about 25%, if the substrate surface was covered with diamond. Such a small change in the atomic hydrogen concentration was very unlikely to change the nucleation density that actually increased by six orders of magnitude by biasing. By contrast, Ref. [216] supported Ref. [229], because the diamond film growth rate increased as much as $\sim$2 orders of magnitude by turning off the bias voltage, and no other factors but the

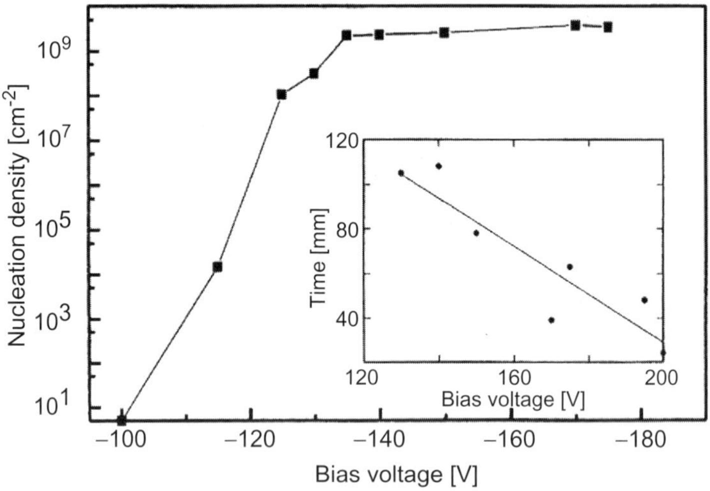

Figure 10.15. Dependence of the nucleation density on bias voltage V_b. The insert shows the pretreatment time as a function of V_b that was determined by the onset of the rapid current increase in Figure 10.14 [222].

atomic hydrogen concentration were likely to cause such a large change. Thus, Ref. [216] concluded that the bias voltage affects chemical reactions as well as physical processes such as ion deposition, and the increase in the atomic hydrogen concentration affects the initial stage of diamond formation. It must, however, be noted that the source gas of Ref. [222] contained as much as 10% Ar for actinometric measurements but that of Ref. [229] used $CH_4/H_2/Ar = 4/496/30$ sccm (Ar is 5.6%). Thus, it should be understood to what extent Ar influences the spectra before a unanimous conclusion is obtained.

Based on a Monte Carlo calculation, Kulisch et al. [222] stated that the electron temperature was hardly changed by increasing the bias voltage. The calculation showed a small average and a maximum carbon ion energies of 15–20 eV and 40–50 eV, respectively, so that the subplantation model of Ref. [221] was not accepted. The marked increase in the nucleation density at about $V_b = -120$ V in Figure 10.15 was attributed to the presence of either a threshold energy of carbon ion formation or a threshold ion dose necessary to create diamond nuclei, and the real cause is still inconclusive [222]. It was stated that the carburization of Si to form SiC plays a secondary role, and is not necessary for enhanced nucleation. This is firstly because according to Milne et al. [228], the carburization only accelerated BEN, and neither increased the nucleation density nor unidirectionally aligned the orientations of diamond nuclei. Secondly, in the TEM study by Jiang and Lia [230] there was no SiC interface layer between Si and diamond. Since SiC particles observed by TEM were strongly misoriented or amorphous, they appeared to be rather detrimental for oriented growth of diamond. Indeed, the (100)-oriented growth of diamond takes place on Si or β-SiC, but the lattice mismatches are so large: D:Si = 2:3 and D:SiC = 4:5.

FT-IR spectra [222] indicated that in the CH_x stretching region, there is a known intense band at 2927 cm^{-1}, and a new band at 2827 cm^{-1} started to grow with the biasing time, as seen in Figure 10.16. This latter band was assigned to D(111)-H vibrational modes. From the data obtained so far [222], it was inferred that the BEN process proceeds in such a way as (i) SiC layer formation, (ii) hydrocarbon layer formation, (iii) development of internal stress, and (iv) diamond nucleation. FT-IR spectra were also observed in Ref. [231] which is shown in Figure 10.17 (a)–(c), where Figure 10.17 (a) is just a spectrum from the Si substrate, Figure 10.17 (b) is the spectrum after a BEN treatment for 15 min under conditions of $c = 5.5\%CH_4/H_2$, $P = 20$ Torr, $P_m = 650$ W, $V_b = -80$ V, and Figure 10.17 (c) is the spectrum of the diamond film deposited for 8 h under conditions given in Table H.1. In Figure 10.17 (b), there exist three sharp absorption bands at 2850, 2920, and 2960 cm^{-1} due to symmetric stretching of CH_2, asymmetric stretching of CH_2, and asymmetric stretching CH_3, respectively. All bands are due to sp^3-bonded groups. By contrast, only a broad band was observed around 2920 cm^{-1} for a CVD-grown diamond film, as seen in Figure 10.17 (c).

In Figure 10.18, the nucleation density due to BEN is shown as a function of the substrate temperature T_s [232] under conditions given in Table H.1. Note, however, that since T_s was dependent on the input microwave power in the reactor used, the plasma condition was different for different T_s. It should also be noted

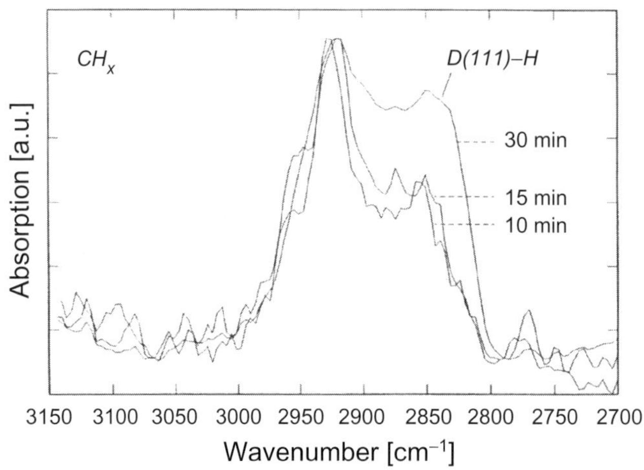

Figure 10.16. Normalized FT-IR spectra in the range of CH_x stretch vibrations after BEN treatments for 10, 15, and 30 min [222].

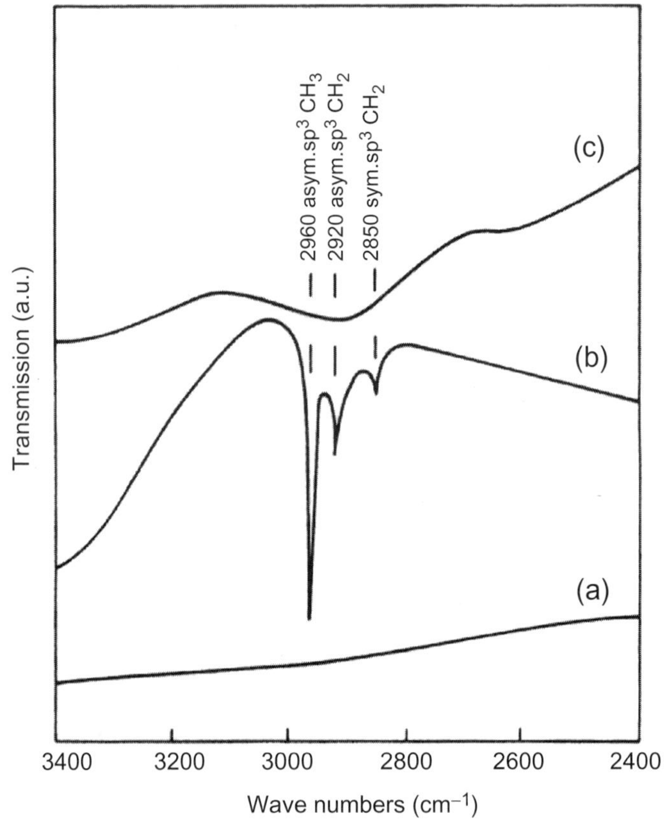

Figure 10.17. FT-IR spectra of (a) Si substrate, (b) *as grown* film after 15 min of BEN ($V_b = -80$ V, $c = 5.8\%CH_4/H_2$, $P = 20$ Torr), and (c) after 480 min of growth ($V_b = 0$ V, $c = 0.96\%CH_4/H_2$, $P = 40$ Torr) [231].

that the nucleation density on Si reached a maximum at $T_s = 780\,°C$. In the EELS spectra (see Figure 11.55) below 46 eV, there was a band at 34 eV that was attributed to the plasmon of diamond, and a band at 24 eV due to other forms of carbon (e.g. *a*-C and SiC). The 34 eV band became more prominent for $T_s \geq 770\,°C$. From these results, the authors of Ref. [232] assumed that the diamond nuclei were formed in the amorphous carbon layer due to the subplantation of carbon ions into the layer, similar to a solid-state recrystallization. Thus, this work supports the results of Ref. [221], but not those of Ref. [222].

Under the BEN conditions of Ref. [233], shown in Table H.1, the nucleation density increased rapidly from less than $10^6/cm^2$ to $10^{10}/cm^2$, when the substrate temperature was increased from 670 to 700 °C. Although the temperature range is different, this result is consistent with Figure 10.18. In the EELS spectra,

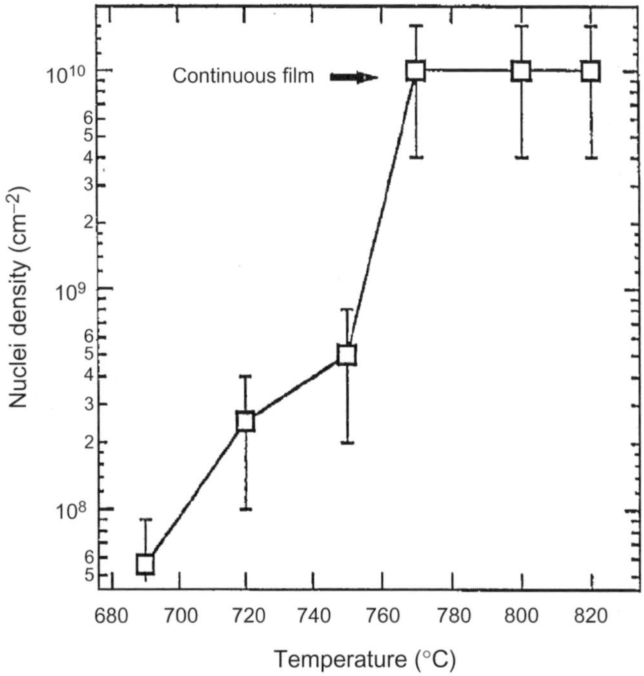

Figure 10.18. Nucleation density on Si by BEN as a function of T_s. Nucleation densities for continuous films were evaluated from the average crystal diameters [232].

$(\sigma + \pi)$ plasmon band exists at 24 eV (which is attributed to *a*-C, SiC, and diamond surface plasmon), diamond band at 34 eV, and second harmonic of $(\sigma + \pi)$ plasmon of *a*-C/SiC at 45 eV. No $(\pi$–$\pi^*)$ band was observed at 5.5 eV. These results indicated that diamond crystallites were embedded in an amorphous carbonaceous matrix. Thus, the formation of diamond crystallites was also explained by the subplantation model.

10.2.3. Optical emission from plasma

In optical emission spectroscopy (OES) from plasma, the emission spectrum near the substrate (1 mm above an Si(100) substrate surface) changed significantly by a negative bias voltage applied to the substrate, while that of the plasma center, about 20 mm above the substrate surface, was virtually unaffected by biasing [229]. The intensities of hydrogen Balmer α and β lines, H_α (656.3 nm) and H_β (486.1 nm), respectively, and the Ar emission line (750.4 nm) were observed near the substrate under the bias voltage to quantify the electron temperature and the atomic hydrogen concentration in the plasma. The CVD reactor used was an ASTeX reactor, which was run under conditions of $T_s = 865\,°C$ and $P_m = 1063\,W$. For pure hydrogen plasma without bias voltage, emission intensities of both H_α and H_β increased with the gas pressure P from 25 to 65 Torr. The intensity ratio of H_β/H_α, which corresponds to the electron temperature, decreased by about 10% from $P = 45$–65 Torr. This is primarily due to the decrease of the mean-free-path for electrons at higher pressure. When a source gas of $CH_4/H_2/Ar$ (=4/496/30 sccm) was used under the bias voltage at $P = 26$, 38, and 52 Torr, the emission intensities of both H_α and H_β lines increased for all pressures with V_b from 0 to −150 V. This indicates that the negative bias voltage increases the kinetic energy of electrons in the plasma sheath region. The intensity ratios of H_α/Ar and H_β/Ar provide an approximate measure of atomic hydrogen concentration in the plasma. The observed results indicated that the atomic hydrogen concentration had a maximum between $P = 26$ and 52 Torr. Under the conditions used in this work, the maximum occurred at $P = 38$ Torr, where the atomic hydrogen concentration was roughly 25% higher than when $P = 26$ and 52 Torr (V_b was −100 to −150 V). For $P = 38$ Torr, the ratio of H_β/H_α intensities increased by 5% with V_b, and saturated at $V_b = -150$ V. Thus, during the bias treatment, the atomic hydrogen concentration and the electron temperature increased near the substrate surface.

The positional dependence of the atomic hydrogen concentration and the field strength distribution above a Si substrate surface was investigated in Ref. [234] using plasma emission lines, H_α and H_β, generated in an ASTeX reactor. The gas used was hydrogen, and $P = 15$ Torr, $P_m = 600$ W, and $T_s = 830\,°C$. The applied bias voltage V_b was −200 V. Figure 10.19 shows the intensity of the H_α line as a function of the distance from the substrate surface. The substrate was either a bare Si or a diamond-coated Si with a diameter of 30 mm. For the bare Si substrate, the maximum was located at 1.95 mm above the substrate surface, while it was only 0.8 mm for the diamond-coated Si. The peak intensity of H_α for the diamond-coated Si was approximately 20 times higher than that of the bare Si, and similarly the bias current was one order of magnitude higher in the former case. One of the reasons for this is presumably because a significantly high flux of electrons was emitted from diamond [224] due to bias voltage to create excited hydrogen atoms. The field strength was measured by the Stark splittings of H_α and H_β lines, and the results are shown in Figure 10.20. Although the electric field at the diamond-coated Si surface (0.35 V/μm) is not so high for field emission at room temperature (usually 5–10 V/μm), the high T_s (=830 °C) could enhance thermionic emission of electrons. Finally, from a band shape analysis of the H_α line, the kinetic energy of hydrogen atoms was determined to be several tens of eV.

Optical emission from the plasma center was studied in Ref. [231]. The plasma was generated by a NIRIM-type reactor using CH_4/H_2 as the reaction gas. The substrate temperature was measured by a thermocouple on the backside of the

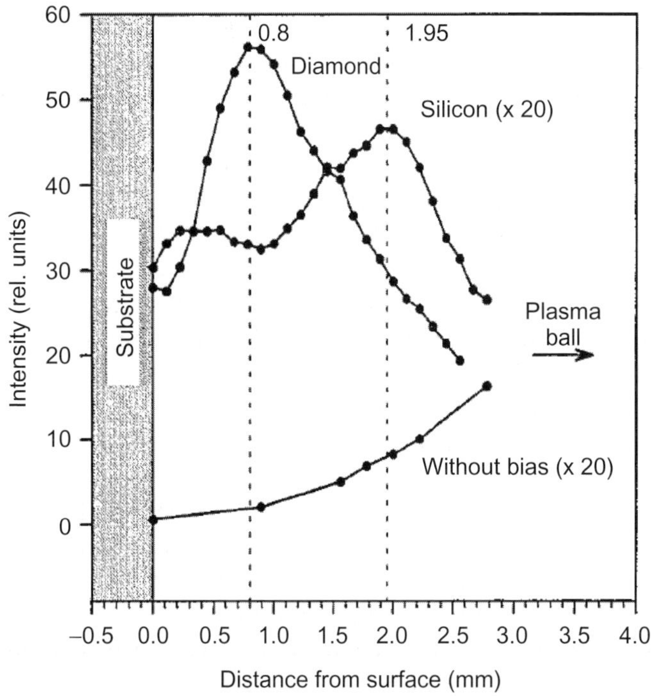

Figure 10.19. Intensity of the H_α Balmer line measured in the plasma (i) without bias voltage and for (ii) a bare Si substrate, and (iii) a diamond-coated Si substrate with $V_b = -200$ V. Each intensity value was an integration over the spectral profile to take into account the broadening [234].

Si substrate. Figure 10.21 (a)–(c) shows the observed spectra with and without the bias voltage under the conditions listed in Table 10.1. Note that in Table 10.1, experiment (a) used standard diamond CVD conditions. The assignments for the emission lines are given in Table 10.2. It should be mentioned that a 15-min BEN treatment under the conditions (c), followed by diamond CVD under the conditions (a), leads to an initial stage of HOD film growth. Note that in the plasma emission measurements, the Ar line at 750.4 nm was used as a reference of actinometer.

The intensities of the optical emission lines changed as a function of the bias voltage, as shown in Figure 10.22. The emissions from CH (387.5 and 430.9 nm), CH^+, C_2, and H_γ increased with the bias voltage V_b. Unlike Ref. [229] ($CH_4/H_2/Ar = 4/496/30$ sccm, $P = 38$ Torr, $T_s = 865\,^\circ C$, and $V_b \geq -180$ V), the ratio of H_α and H_β intensities was unchanged, while the ratio of H_γ and H_α intensities increased, which led to the same conclusion as Ref. [229] that the electron temperature increases with V_b.

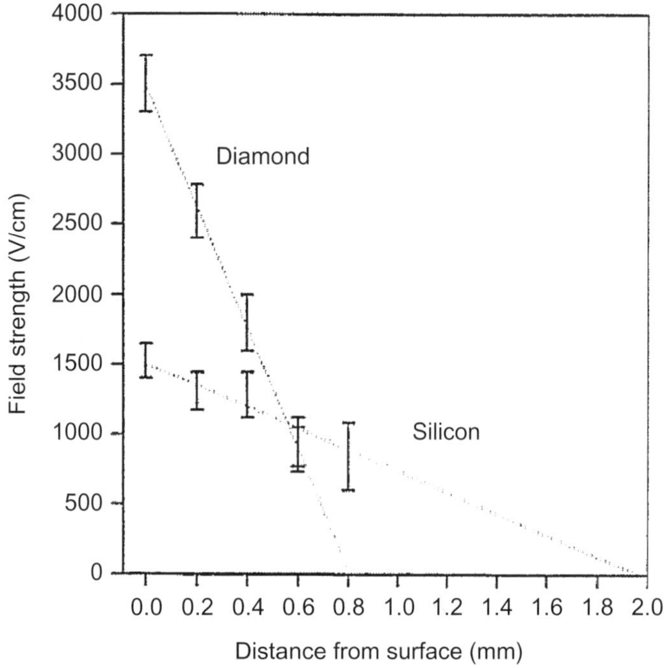

Figure 10.20. Field strength distribution in the cathode plasma sheath above a bare and a diamond-coated Si substrates for $V_b = -200$ V. The field strength was evaluated from the Stark splitting (broadening) of the H_α and H_β lines [234].

The concentrations of atomic hydrogen and hydrocarbon radicals increased with increasing the negative bias voltage, when the ASTeX bell-jar type MPCVD system was used [226, 227]. Figure 10.23 shows a typical optical emission spectrum from the secondary plasma, where $V_b = -250$ V. For other conditions, see Table H.3. The emission intensities from hydrogen and carbonaceous species nearly saturated at $V_b = -200$ to -270 V, and a HOD film could grow under these conditions. It was thus concluded that HOD films can be obtained by the combination of the high dose of carbonaceous species and the increased hydrogen etching effects.

In Ref. [224], optical emission was studied for the plasma generated by an ASTeX reactor under conditions of $c = 2\%CH_4/H_2$, $P = 20$ Torr, $T_s = 750$–$800\,°C$, and $V_b = -170$ V. The substrate was a 3-inch Si wafer. For the purpose of actinometric measurements, 10% Ar was added to the source gas. It should however be noted [235] that this Ar concentration might be too high, and could disturb the distribution of plasma species and other conditions. The bias current was higher if

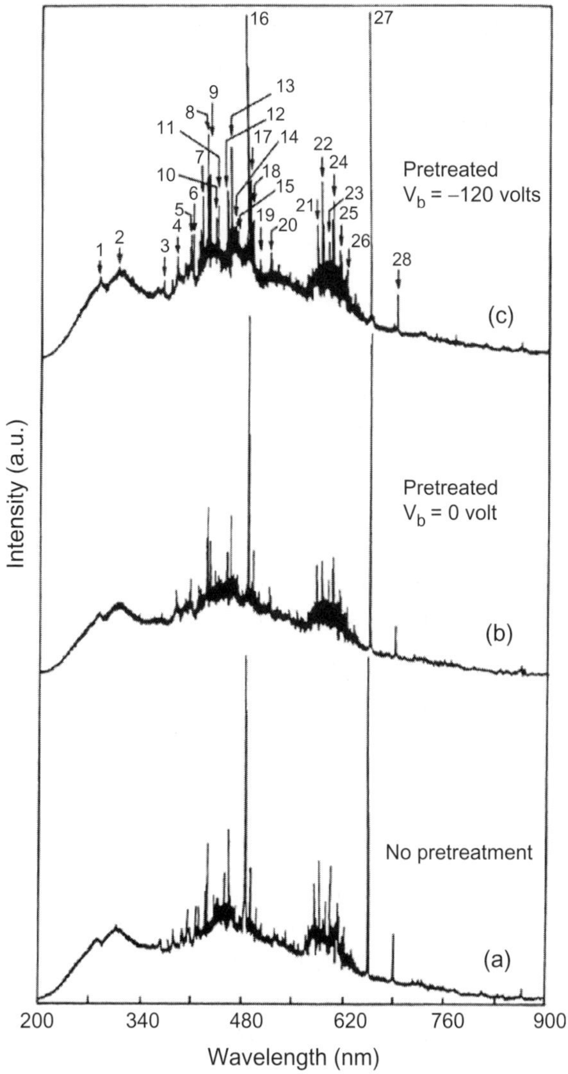

Figure 10.21. Optical emission from the center of the plasma. The plasma conditions are listed in Table 10.1 [231].

Table 10.1. Plasma conditions [231].

Exp.	Source gas	*P* (Torr)	P_m(W)	Bias voltage (V_b)	T_s(°C)
(a)	0.96%CH_4/H_2	40	650	0	800
(b)	5.49%CH_4/H_2	20	650	0	750
(c)	5.49%CH_4/H_2	20	650	−120	750

Table 10.2. Optical emission during the BEN treatment [231].

Peak no.	λ (nm)	Chemical species	Electronic transitions	v' v''
1	282.9	OH	$A^2\Sigma^+$–$X^2\Pi$	1,0
2	308.7	OH	$A^2\Sigma^+$–$X^2\Pi$	0,0
3	369.6	OH^+	$A^3\Pi^+$–$X^3\Sigma^-$	1,1
4	387.5	CH	$B^2\Sigma^-$–$X^2\Pi$	0,0
5	406.6	H_2	...	a
6	410.1	H_δ	$2p^2P^0$–$6d^2D$	...
7	423.9	CH^+	$A^1\Pi$–$X^1\Sigma^+$	0,0
8	430.9	CH	$A^2\Delta$–$X^2\Pi$	0,0
9	434.0	H_γ	$2p^2P^0$–$5d^2D$	...
10	441.0	H_2	$r^3\Pi_g$–$c^3\Pi_u$	0,0
11	444.7	H_2	$v^3\Pi_g$–$c^3\Pi_u$	0,0
12	457.4	H_2	$I^1\Pi_g$–$B^1\Sigma_u^+$	0,0
13	462.9	H_2	$G^1\Sigma_g^+$–$B^1\Sigma_u^+$	0,0
14	466.9	H_2	$K^1\Sigma_g^+$–$B^1\Sigma_u^+$	0,0
15	473.0	H_2	$Q^1\Pi_g$–$B^1\Sigma_u^+$	0,0
16	486.1	H_β	$2p^2P^0$–$4d^2D$	...
17	490.2	CO^+	$A^2\Pi$–$X^2\Sigma^+$	0,0
18	493.2	H_2	–	a
19	501.9	H_2	–	a
20	516.3	C_2	$d^3\Pi_g$–$a^3\Pi_u$	0,0
21	582.1	H_2	$i^3\Pi_g$–$c^3\Pi_u$	0,0
22	588.8	H_2	$h^3\Sigma_g^+$–$c^3\Pi_u$	0,0
23	597.4	H_2	...	a
24	602.0	H_2	$d^3\Pi_u$–$a^3\Sigma_g^+$	0,0
25	613.4	H_2	–	a
26	623.6	H_2	–	a
27	656.2	H_α	$2p^2P^0$–$3d^2D$	–
28	691.3	b	–	–

[a]Transitions with v' and v'' values other than 0,0, respectively.
[b]Unidentified.
(The 691.3 nm band could not be assigned)

the Si wafer was covered with a diamond film due to electron emission from the diamond surface.

In Ref. [235], a plasma emission study was undertaken for CH_4–H_2–Ar plasma in an ASTeX reactor to investigate the effects of V_b, c, and P on H_α (656 nm), H_β (486 nm), and C_2 (516 nm) emission lines. The typical plasma conditions were $P = 30$ Torr, $P_m = 800$ W, $T_s = 600\,°C$, and $V_b \geq -300$ V. The source gas was $CH_4/H_2/Ar = 8/392/11$ sccm, and the substrate was tungsten (W) of 50 mm in diameter and 6 mm in thickness, which had been mechanically abraded with 8-μm diamond powder. An Ar line at 696 nm was used as a calibration line. The observed

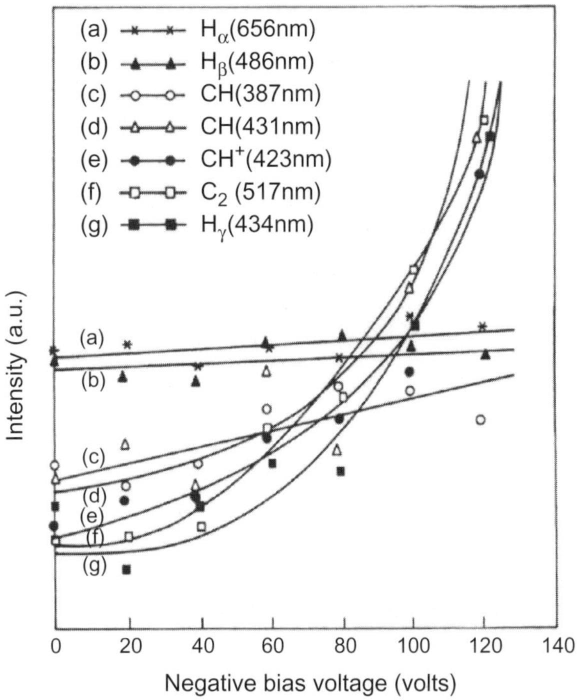

Figure 10.22. Emission intensities of (a) CH (*X–A*), (b) C_2, (c) CH (*X–B*), (d) CH^+ (*X–A*), (e) H_α (*2p–3d*), (f) H_β (*2p–4d*), (g) H_γ (*2p–5d*) lines as a function of V_b. The conditions used were $c = 5.8\%CH_4/H_2$, and $P = 20$ Torr [231].

results at the position approximately 7 mm above the substrate surface are summarized as follows:

(1) The atomic hydrogen density, represented by the H_α intensity, increased monotonically with the bias voltage. The H_α intensity decreased quickly with an increase in the CH_4 concentration up to 4% when no bias voltage was applied, while it was nearly constant when $V_b = -200$ V was applied. The H_α intensity decreased with P from 10 to 20 Torr, but then gradually increased for $P \geq 20$ Torr irrespective of V_b. The H_α intensity was roughly 10% stronger under $V_b = -200$ V than under $V_b = 0$ V.

(2) The ratio of H_β/H_α intensities, which is a measure of the electron temperature, increased with V_b. It was greater with finite bias voltages than when $V_b = 0$ V.

(3) The ratio of C_2 (516 nm) and CH (431 nm) intensities under conditions of $c = 2\%CH_4/H_2$ and $P = 30$ Torr increased with V_b down to -300 V as well as

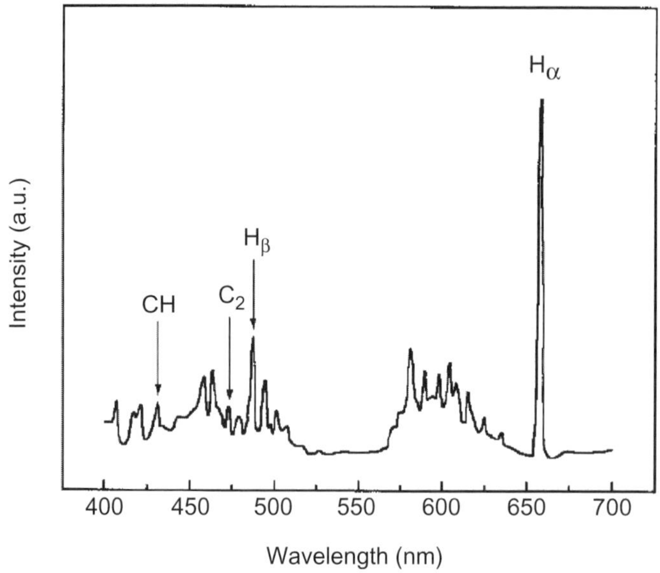

Figure 10.23. Typical emission spectrum from the secondary plasma during BEN ($V_b = -250$ V) [226, 227].

with P up to 50 Torr under a constant $V_b = -200$ V, but decreased with c up to 3%CH_4/H_2 irrespective of V_b.

(4) The positional dependence of the intensity ratios of H_α(656 nm)/Ar(696 nm), H_α(656 nm)/H_β(486 nm), and C_2(516 nm)/CH(431 nm) are shown in Figures 10.24 (a)–(c). In all cases, $P = 30$ Torr and $c = 2\%CH_4/H_2$. These data indicated that the density of hydrogen atoms was higher in the vicinity of the substrate under $V_b = -200$ V, and the electron temperature was also higher near the substrate surface.

From Figure 10.24 (c) and other data, the C_2 intensity also was higher in the vicinity of the substrate surface particularly in the presence of the bias voltage. On the other hand, for the position 16 mm above the substrate surface, the ratio of H_β/H_α intensities increased with V_b down to -300 V, and was approximately 25% greater when $V_b = -200$ V. Thus, the increased fluxes of C_2 and atomic hydrogen due to the negative bias voltage were the important factors for BEN.

For a detailed study of plasma emission under conditions that HOD films could be grown, see Ref. [236]. For studies of plasma emission spectroscopy without bias voltage, the readers can refer to Refs. [237–240].

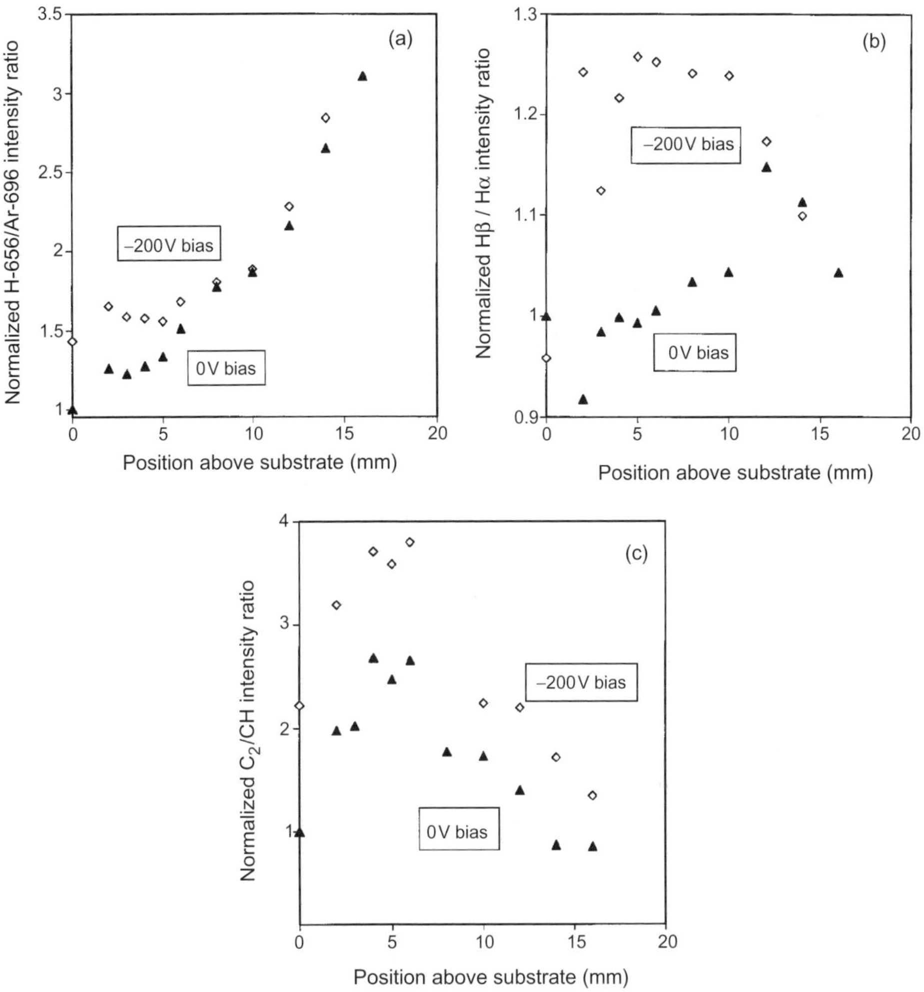

Figure 10.24. Emission intensity as a function of the position above the substrate: (a) H_{α}(656 nm)/Ar(696 nm), (b) H_{α}(656 nm)/H_{β}(486 nm), and (c) C_2(516 nm)/CH(431 nm) [235].

10.2.4. Refractory metals

A BEN study on refractory metals (H_f, Ti, Ta, N_b, and W) in addition to Cu and Si has been done by Walter et al. [241] using an ASTeX reactor under the conditions given in Table H.1. See also Refs. [242, 243]. Figure 10.25 shows the nucleation density as a function of the biasing time for the refractory metals, Si, and Cu. It is seen that Si has the highest nucleation rate and density, while Cu has the

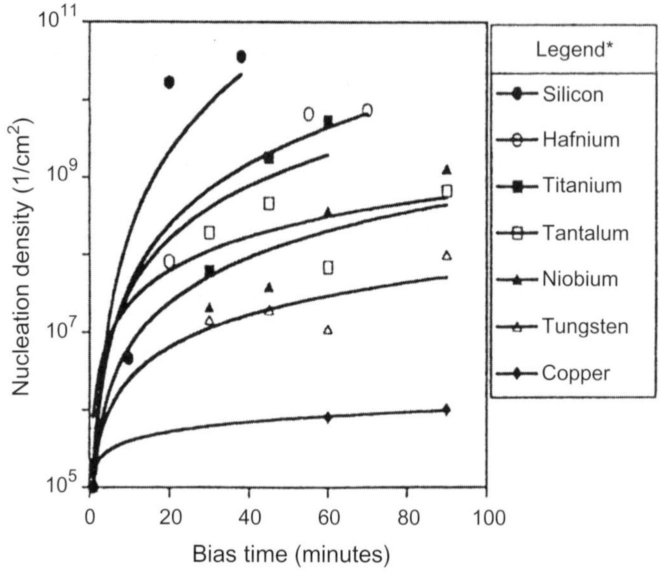

Figure 10.25. Nucleation density as a function of biasing time for various refractory metals compared with Si and Cu [241].

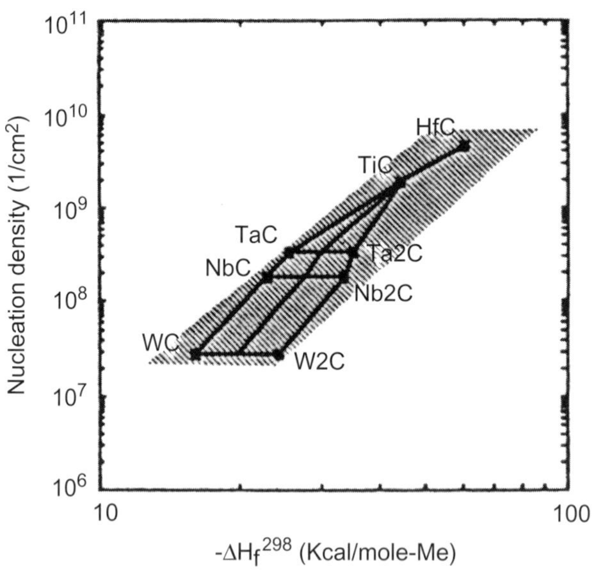

Figure 10.26. Correlation of the diamond nucleation density after 60 min of BEN with the heat of formation of refractory metal carbides [241].

lowest. The nucleation density of diamond on refractory metals after a 60-min bias treatment has a positive correlation with the heat of carbide formation, as seen in Figure 10.26. For instance, H_f has the highest nucleation density of about $10^{10}/cm^2$ because the heat of carbide formation is the highest. It is of interest that although the heats of carbide formation are similar between W and Si, W has a lower nucleation density of about $10^7/cm^2$ than Si ($\sim 10^{10}/cm^2$). This is related to the fact that since SiC has an atomically close-packed crystal structure, while WC has the sodium chloride (B1) or the nickel arsenide (B8) structure, where C atoms can interstitially position in the W lattice, the diamond nucleation is more efficient for SiC. The strong correlation between the nucleation density of diamond with the heat of carbide formation of refractory metals suggests that the carbide formation plays an important role in the diamond nucleation. This result is consistent with the fact that the BEN on Cu was relatively ineffective for diamond nucleation, as carbide is not formed. It is also of interest that the induction period before the onset of diamond nucleation was significantly shorter (~5 min) for Si than the refractory metals. This is because the thickness of SiC layer formed during the bias treatment was less than 100 Å for Si, while it was a few microns for the refractory metals.

Chapter 11

HOD Film Growth

Chapter 11

HOD Film Growth

11.1. HISTORICAL BACKGROUND

The growth method of azimuthally oriented diamond particles and films was first established by Stoner and Glass in 1992 [1] on a β-SiC (100) layer deposited on a 1-inch Si(100) wafer. The process conditions are listed in Table H.2. As a result, most diamond particles were oriented, as seen in Figures 11.1(a) and (b), where D[110]// β-SiC [110]. This is very surprising because the lattice constant of β-SiC (4.359 Å) is roughly 20% greater than that of diamond (3.567 Å). This work was followed by a number of detailed works in Glass' group at NCSU, and the related technologies were fully established in Kobe Steel's overseas laboratories in the State of North Carolina, USA. and the University of Sully, UK, using ASTeX reactors and Si(100) substrates. The HOD film technology was then transferred to Kobe Steel, Japan, and Tachibana et al. were successful in growing HOD films using a NIRIM-type reactor around 1993. The technology includes (i) Si substrate pretreatment such as surface carburization, (ii) optimization of BEN, (iii) ⟨100⟩-oriented columnar growth, and (iv) lateral growth. Unfortunately, these works have not been published. Concurrently, two groups in Fraunhofer Institute, Germany, Koidl's group in Freiburg, and Klages' group in Braunschweig, succeeded in making well-oriented HOD films. In Japan, Kawarada's group has been most active in the research of HOD film growth. At present, there are a number of groups around the world that can synthesize HOD films, and the technology has been progressively sophisticated. Most remarkable success in recent years is the growth of HOD film with a very smooth surface [6, 7]. The flat surface of HOD film (Figure 1.1 (b)) is advantageous over rough surfaces of randomly oriented polycrystalline diamond films (Figure 1.1 (a)) for device fabrication and film over-deposition. Moreover, due to the lower density of grain boundaries in HOD films than in randomly oriented polycrystalline diamond films, the electrical properties of HOD films are significantly better. For instance, hole mobilities of $280\,cm^2/V \cdot s$ [244] and $309\,cm^2/V \cdot s$ [245] have been achieved for HOD films. It is thus expected that HOD films will be used for electronic sensors, actuators, and devices in the coming years, and (100)-oriented single crystal diamond films are likely to be synthesized in the not so remote future.

11.2. HOD FILM GROWTH ON β-SiC (100) LAYER

In the first paper by Stoner and Glass [1], a 4–5 μm thick β-SiC(100) layer deposited on a 1-inch Si(100) wafer [246, 247] was used as the substrate. It was polished with a 0.1-μm diamond paste to smoothen the surface, and oxidized in oxygen gas at 1200 °C to a thickness of approximately 0.1 μm to remove the surface damage created by polishing. The oxide was then removed in dilute fluoric acid (HF). A BEN treatment and diamond growth were done using a 1.5-kW ASTeX reactor under conditions listed in Table H.2. The growth rate of diamond was only 0.05 μm/h, and the diamond CVD was continued for 50 h. As a result, oriented cubo-octahedral diamond crystals, shown in Figure 11.1, was grown in the center of the substrate, while a continuous film was formed near the edge of the substrate, where (100) faces of diamond grains tend to be oriented in the same direction. In the former case, approximately 50% of the diamond crystals were oriented in such a way as D{100} // β-SiC{100} and D[110] // β-SiC[110] within misorientation angles of 3°. As a result of various experiments, a complete removal of surface oxide on both Si and SiC as well as the biasing time were found to be the most important factors for HOD film growth.

Regarding the BEN mechanism to synthesize HOD films [225] (i) at $V_b = -250$ V, the bias current I_b was higher when the diamond-coated substrate holder was used (= 100 mA) than when uncoated or alumina-coated holders were used (~20 mA), suggesting that electrons are emitted from diamond during BEN, (ii) the BEN started from the edge of the substrate and proceeded toward the center, and

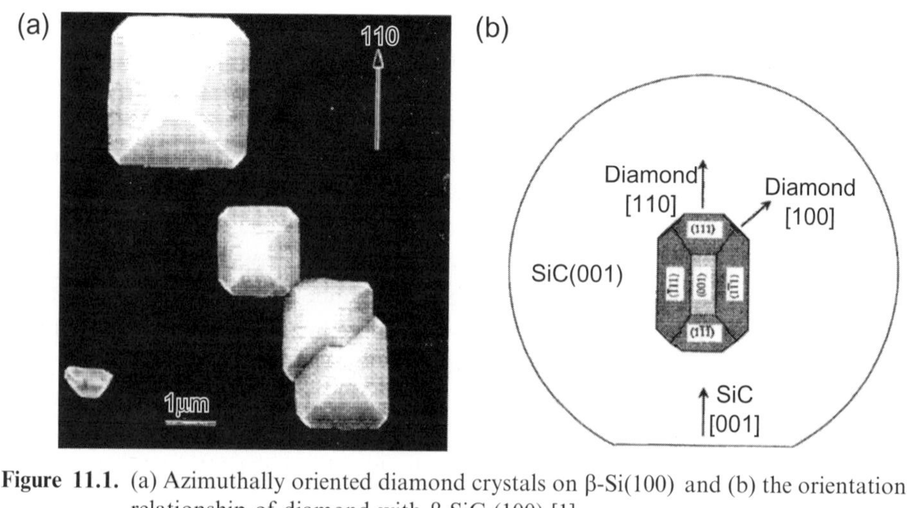

Figure 11.1. (a) Azimuthally oriented diamond crystals on β-Si(100) and (b) the orientational relationship of diamond with β-SiC (100) [1].

(iii) the color of the plasma near the substrate surface turned red by applying V_b, presumably due to an increase in the density of atomic hydrogen. These observations are consistent with the results of BEN research described in the preceding section.

In Ref. [248], HOD films were grown on a 4-μm thick β-SiC(100) layers that had been heteroepitaxially grown on Si(100). When the diamond film thickness was 1 μm, diffraction spots from diamond (220) twins were observed in the XPF diagram, but they became faint when the film thickness became 30 μm. In the (100)XPF diagram, the FWHMs of the diffraction peaks were 11°–12° for azimuthal direction and 12°–13° for polar direction. These values are significantly larger than those of recent results, ≤7°, and were attributed to a low density of oriented diamond nuclei. To examine the effects of growth conditions on HOD film morphology, BEN treatments were done directly on two Si(100) substrates, which were then used for diamond CVD to make 10-μm thick films under conditions that prefer either ⟨100⟩- or ⟨110⟩-oriented growth. Consequently, the former process resulted in an HOD film, while the latter process resulted in a film that contained a considerable number of misoriented diamond grains. These facts showed that to synthesize well-oriented HOD films, both BEN and oriented growth conditions must be finely tuned to create a high density of oriented diamond nuclei, and the ⟨100⟩-oriented growth conditions should be maintained during the diamond CVD.

Using the above procedure, Kawarada's group has undertaken a series of detailed experiments on HOD films. In Refs. [249, 250], fairly coalesced diamond films with a thickness of only less than 6 μm were grown for 8 h on β-SiC (100) [251] (which itself had been heteroepitaxially grown on Si), as seen in Figure 11.2.

Before this work, to obtain a smooth surface consisting of aligned (100) faces, it was necessary to grow diamond films to more than 50-μm thicknesses. In this work, however, a lateral growth of diamond was undertaken using ⟨111⟩-growth conditions in the final step, and a smooth surface was achieved. More concretely, the growth procedure used is as follows:

(i) The BEN process was undertaken for the substrate which is negatively biased (−50 V) with respect to a counter electrode for 5 min in the process gas of $c = 2\%CH_4/H_2$, $P = 50$ Torr, and $T_s = 900\,°C$.
(ii) An oriented growth of diamond was undertaken using $c = 5\%CO/H_2$, $P = 50$ Torr, and $T_s = 840\,°C$ for 2 h. As a result, the film surface consisted of azimuthally aligned diamond grains with a pyramid shape, consisting of (111) faces, as seen in Figure 11.3. In this case, the α-parameter, determined by the CVD conditions, was close to 3. Since the growth rate in the ⟨100⟩-direction was higher than other directions in this step, the misoriented diamond particles formed in thc first step by BEN were totally taken over by the ⟨100⟩-oriented diamond grains.

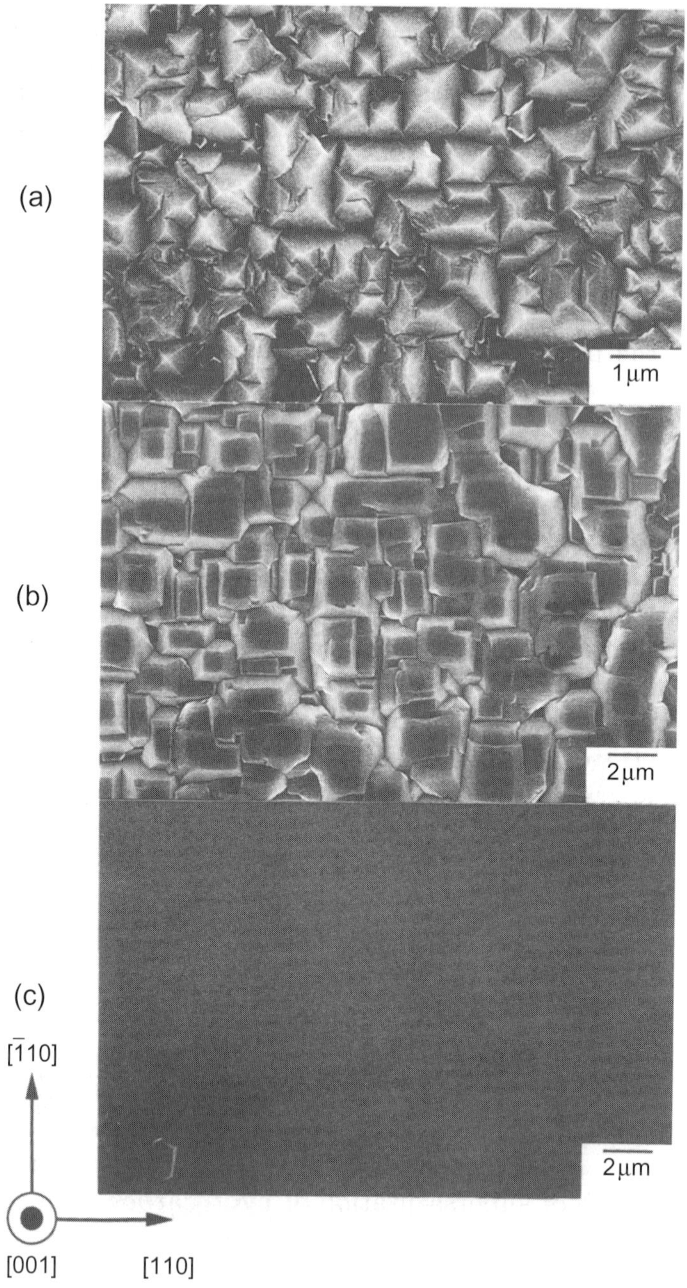

Figure 11.2. SEM images of HOD films grown on β-SiC (100) after 5-min BEN, followed by (a) ⟨100⟩ growth for 2 h, (b) ⟨100⟩ growth for 2 h, followed by ⟨100⟩ growth for 1 h, and (c) ⟨100⟩ growth for 2 h, followed by ⟨111⟩ growth for 4 h [249, 250].

(iii) Finally, a lateral growth of diamond was undertaken using $c = 10\%CO/H_2$, $P = 50$ Torr, and $T_s = 840\,°C$. In this case, the α-parameter defined by the CVD conditions was close to 1, in which case, the growth of (111) faces were faster than other faces. As a consequence, the (100) faces were expanded to fill up the gaps between adjacent diamond (100) grains. This is shown in Figure 11.4 [250].

It was thus demonstrated that the lateral growth after ⟨100⟩-oriented growth is very effective to obtain a smooth surface. Although the lateral growth method had been employed before this work by some groups, this is the first paper that explicitly described the detailed conditions and results. Another interesting observation in this work [250] was that after step (i), mutually perpendicular, white stripes were formed at the substrate surface (Figure 11.5 (b)), and the diamond particles were nucleated along the stripe directions after 3 min (Figure 11.5 (c)). Finally, the substrate was covered with oriented diamond crystals (Figure 11.5 (d)). A more detailed investigation indicated that diamonds actually nucleated in the valleys between the protrusion areas (Figure 11.6 (a)), first grew along the grooves (Figure 11.6 (b)), and then grew towards the protrusions (Figure 11.6 (c)).

A similar study was done in Refs. [6, 252] using the process conditions given in Table H.2. In the AFM image of the smooth surface [252], wrinkles were

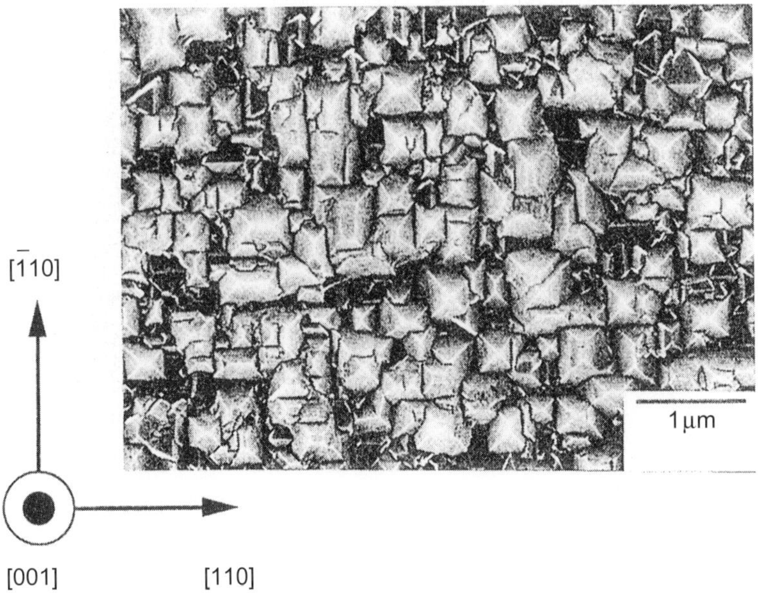

Figure 11.3. Pyramid shape morphology after step (ii) [250].

observed. They were considered to be caused by the internal strain of the diamond film that arises from the low-angle, coherent grain boundaries. In Ref. [6], a HOD film of 300-μm thickness with a perfectly smooth surface was synthesized. The FWHM of the X-ray rocking curve was as small as 0.6°. So far, this is the best data on the orientational alignment, the surface smoothness, and the thickness of HOD film.

The density of azimuthally oriented diamond grains depended on the process conditions such as P, T_s, V_b, and the biasing time. In Ref. [250], using a NIRIM-type reactor, the optimized conditions were found to be $c = 2\%CH_4/H_2$, $P = 50$ Torr, $T_s = 900\,°C$, and $V_b = -50$ V. The input microwave power was 700 W, and the biasing time was 5 min. As a result, 50% of the diamond nuclei were azimuthally oriented. The surface flatness was achieved by changing the growth conditions from the ⟨100⟩ fast growth to the ⟨111⟩ fast growth, as shown in Table 11.1. Regarding the crystal orientations, the pyramid crystals, shown in Figure 11.3, were rotated within ±5° in the plane and within 10° from the direction normal to the substrate surface [250]. As is described later, this originated from the lattice mismatch between β-SiC and diamond. A comprehensive review on the research in Kawarada's group is given in Ref. [137], where CVD processes for homo- and heteroepitaxial growth, surface conducting layer due to hydrogen termination, metal/diamond interface, and device applications are described in great detail.

In Refs. [253–255], BEN experiments were done by ECR plasma CVD under conditions shown in Table H.2. The substrate was a heteroepitaxial β-SiC(100) layer of 0.5-μm thickness deposited on Si [256] by low pressure CVD (LPCVD). Like in

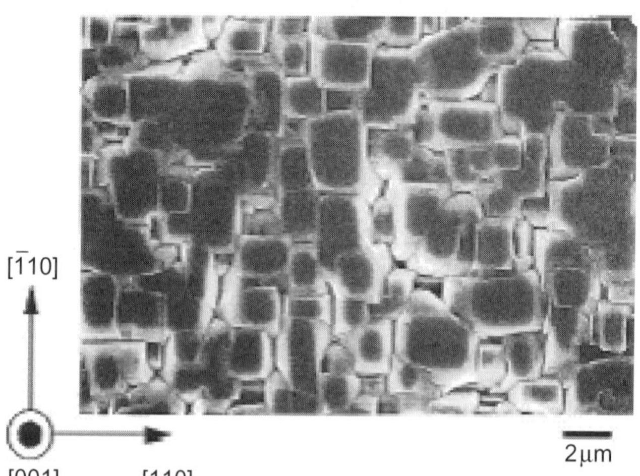

Figure 11.4. Film surface after step (iii). The lateral growth was undertaken for 1 h [249].

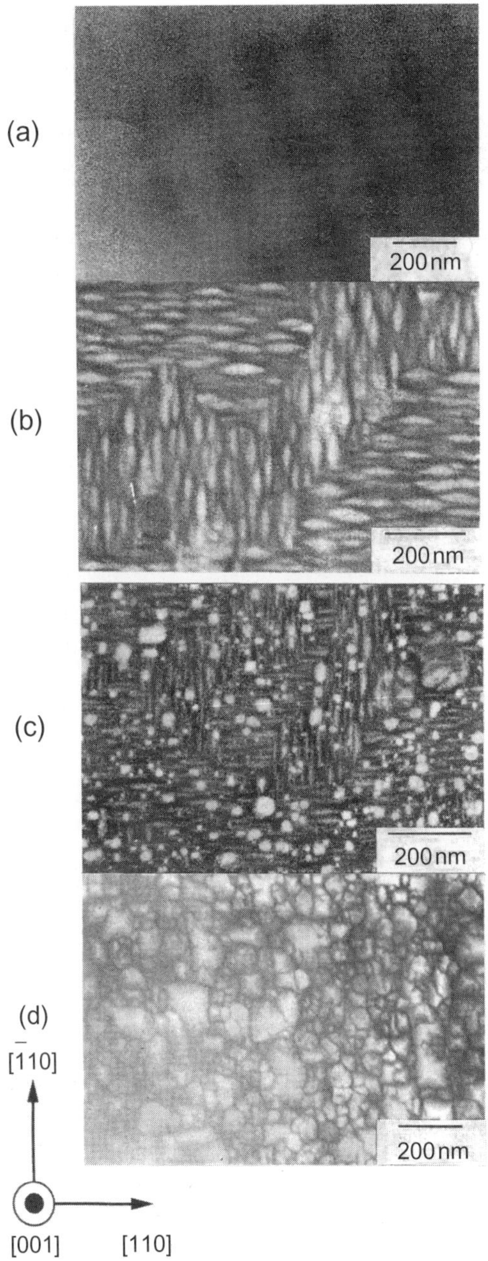

Figure 11.5. SEM images of β-SiC(001) surface after the optimized BEN treatment for (a) 0 min, (b) 2 min, (c) 5 min, and (d) 10 min. [$\bar{1}$10] or [110]-directed bright stripes are β-SiC rectangular protrusions. Diamond particles are observed in (c) and (d) [250].

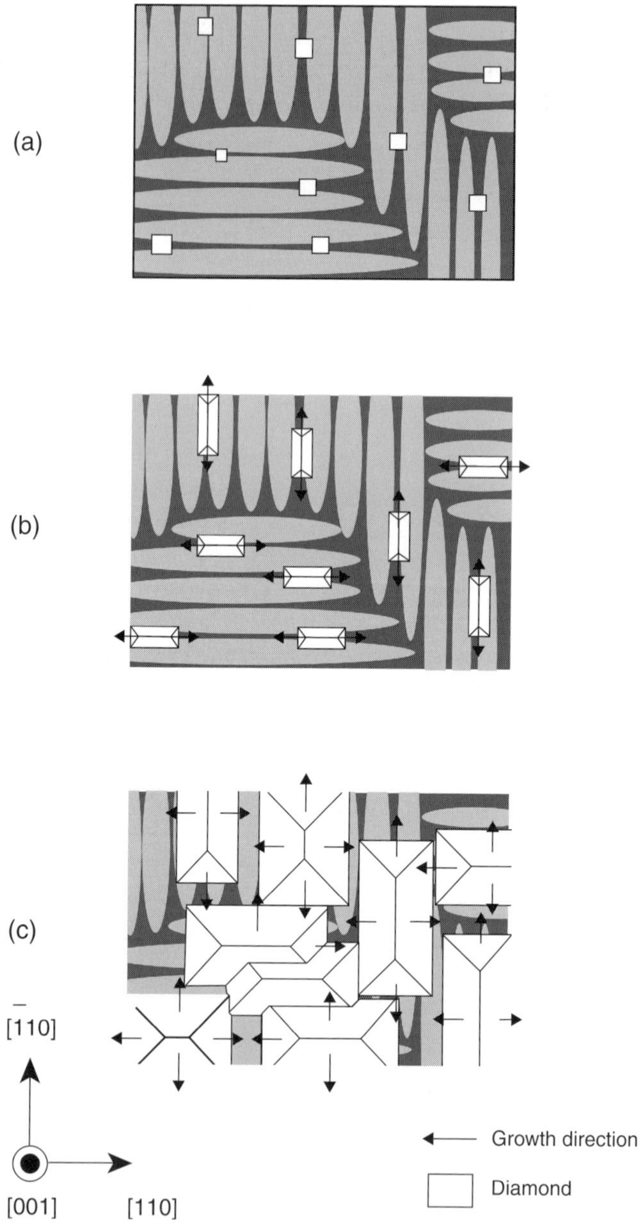

Figure 11.6. Schematic diagram showing a two-step Volmer-Weber growth process of diamond on β-SiC(100): (a) diamond nucleation in grooves (dark areas) between β-SiC protrusions (bright areas) in 5 min, (b) the first growth step where diamonds grow along the grooves, and (c) the second growth step where diamonds grow in the vertical directions in 10 min [250].

Refs. [249, 250], a stripe structure was formed at the substrate surface immediately after the BEN treatment under $P = 0.3$ Torr, $T_s = 750\,°C$, and $V_b = -30$ V, and the SiC(100) surface became very rough. By the end of the BEN treatment, about 30% of diamond particles were heteroepitaxially oriented with respect to SiC(100) over a 40-mm square area. The nucleation density was not influenced by the CH_4 concentration, and remained to be about 10^7–$10^8/cm^2$ for $CH_4/H_2 \leq 50\%$. Unlike the standard BEN treatment at $P = 10$–30 Torr, it took about 60 min to achieve BEN of diamond, presumably because the hydrocarbon ion flux was lower due to low P. The nucleation density and the ratio of the oriented nuclei at $T_s = 400\,°C$ were 10^7–$10^8/cm^2$ and only $\leq 10\%$, respectively. The nucleation density at $T_s = 400\,°C$ was similar to that at $T_s = 700$–$800\,°C$, indicating that the nucleation density was quite insensitive to T_s for ECR plasma.

In Ref. [257], $Si_{1-x}C_x$ alloy films with $x < 0.1$ were deposited on Si by molecular beam epitaxy (MBE) to use them for the substrates of heteroepitaxial diamond films. It was expected that when $x = 4.33\%$, a perfect lattice match of $Si_{1-x}C_x$: D = 2:3 occurs and the degree of orientational alignment could be improved. An HOD film, grown to a thickness of 20 μm using the BEN process, successfully resulted in a ⟨100⟩-oriented film with (100) faces at the film surface, but the FWHM of the (111) XPF was 6°, the same value as when the direct nucleation of diamond was done on Si using BEN. The results of Raman spectroscopy and XRD of the diamond films were not dependent on the x value. It was thus confirmed that the orientational characteristics of the HOD films had no significant dependence on the C content of the $Si_{1-x}C_x$ layers. This work can be compared with that of Ref. [258], where $Si_{1-x}C_x$ layers with $x = 1.4$ and 3.5% were deposited on Si(100) by MBE, which were then followed by a BEN process of $V_b = -200$ V and $T_s = 865\,°C$ for 14 min. The source gas ($CH_4 + H_2$) was at first 100%H_2, and CH_4 was continuously increased to 0.7% during the BEN process. The diamond CVD was then carried out under conditions of (i) $c = 1.5\%CH_4/H_2$, $T_s = 795\,°C$ for 1.5 h and

Table 11.1. Process conditions for HOD film synthesis [250].

	Bias treatment	⟨001⟩ Fast growth	⟨111⟩ Fast growth
Reaction gas (sccm)	$CH_4 = 4$ $H_2 = 196$	$CH_4 = 10$ $CO = 10$ $H_2 = 180$	$CO = 20$ $H_2 = 180$
Reaction pressure (Torr)	20–80	35	35
Microwave power (W)	500–1000	500	500
Bias voltage (V)	−50 to −100	–	–
Substrate temperature (°C)	840–1000	~840	~840

(ii) $c = 2.5\% CH_4/H_2$, $T_s = 825\,°C$ for 2.5 h. The input microwave power P_m was fixed at 2.5 kW, and the substrate temperature was controlled by changing the gas pressure (a higher P results in a higher T_s). A TEM observation indicated that small crystalline β-SiC particles precipitated during the diamond CVD. Furthermore, for the case of $x = 1.4\%$, the carbon concentration x decreased to 0.7 % during diamond CVD. There were diamond particles in contact with $Si_{1-x}C_x$ with a tilt angle of only 4°, and they were heteroepitaxially nucleated on β-SiC(100). Thus, it was inferred that the same phenomenon as that seen in Ref. [257] occurred in these experiments, and it was likely that the deviation from the prescribed stoichiometry of $Si_{1-x}C_x$ was significantly altered during the BEN treatment and diamond CVD. These two works indicated that $Si_{1-x}C_x$ alloy films are not practically useful as the substrates for HOD films.

The atomic structure of HOD film surface was investigated using electron micro-beam diffraction in Ref. [259]. The substrate used was a 500-nm thick β-SiC(100) that had been heteroepitaxially grown on Si(100) and was tilted about the [110]-axis by 4° from the exact (100) orientation. The thickness of the HOD film was ~20 μm. According to the RHEED pattern, using the electron micro-beam (the spatial resolution as observed by SEM was ~0.1 μm), the length of the surface dimer rows was ~1.5 nm, which was significantly shorter than that of the homoepitaxial layer, ~7 to 10 nm.

11.3. THREE-STEP PROCESS: GROWTH OF HOD FILMS ON CARBURIZED Si(100)

In an attempt to maximize the number of oriented grains, a growth of HOD films on Si(100) was reported in Ref. [260]. The method consisted of the three steps: (1) carburization of Si surface, (2) BEN, and (3) oriented growth of diamond. Table 11.2 shows typical process conditions using ASTeX reactors that were placed horizontally or vertically. In the first step, the Si surface was converted to β-SiC(100) which was epitaxial to Si(100). If step (1) was not used, a biasing time of about 30 min was necessary to achieve a sufficiently high nucleation density of diamond, while the biasing time of only 3 to 15 min was sufficient if step (1) was used. The number density of azimuthally oriented diamond nuclei was higher on hetero-epitaxial β-SiC(100) layer, described in Section 11.2, than on carburized Si.

In Ref. [261], a process optimization for oriented diamond nuclei was done using a software for statistical experimental design [262] (design for experimental method). The process parameters for BEN and the oriented growth in the three-step process are listed in Table H.3. The substrate used was Si(100), which had been carburized for 3 h under the following conditions: $P = 20$ Torr, $P_m = 1000$ W, $T_s = 900\,°C$, and

$c = 2\%CH_4/H_2$. Consequently, almost all diamond grains were azimuthally oriented in the film grown for 20 h, and the square or rectangular (100) faces were nearly parallel to the substrate, as seen in Figure 11.7. The FWHM of the Raman line at $1333\,cm^{-1}$ was as small as $3.5\,cm^{-1}$, and there was virtually no band due to non-diamond carbon around $1500\,cm^{-1}$. Since the optimized conditions for carburization, BEN, and oriented growth depend strongly on the CVD reactor used, it would be very useful to identify the optimized process conditions for each CVD reactor by the design for experimental method.

The distribution of azimuthal and tilt angles of (100) faces in HOD films are usually studied by XPF measurements (see Section 5.1) [4, 263]. In Ref. [5], a 30-μm thick HOD film was synthesized, which had (100) faces with 5–10 μm edge lengths at the film surface. The FWHMs of XPF for (100) and (022) diffractions were 8° and 12°, respectively. The FWHM of Raman band at $1332\,cm^{-1}$ was about $8\,cm^{-1}$.

The control of the tilt and azimuthal angles of (100) faces is important to improve the quality of HOD films. In Ref. [264], the tilt $\langle \chi \rangle$ and azimuthal $\langle \phi \rangle$ misorientation angles of the (100) faces were measured by both the X-ray precession method and the X-ray rocking curve measurements for HOD films with thicknesses of 1, 4, 20, and 100 μm, which were made by the three-step process (see Figure 5.3). The results are shown in Figure 11.8.

For the three film thicknesses, $\langle \chi \rangle$ and $\langle \phi \rangle$ have their minima when the thickness was 20 μm. The FWHM of X-ray rocking curve for the 100-μm thick HOD film was 3.1°. The fact that $\langle \chi \rangle$ and $\langle \phi \rangle$ increased between the film thickness of 20 to 100 μm is apparently in contradiction with the van der Drift's theory [77], in which the orientational order should become better as the film thickness increases. The authors of Ref. [264] speculated that this was due to (i) the increase in the

Table 11.2. Process conditions for the three-step process using horizontal and vertical setups of the CVD reactors [260].

	Carburization		Bias		Growth
Parameters	Horizontal reactor	Vertical reactor	Horizontal reactor	Vertical reactor	Both reactors
Bias voltage	0	0	−250 VDC	−170 VDC	0
Current	0	0	~100 mA	~200 mA	0
Time	1 h	1 h	3 min	9 min	>5 h
CH_4/H_2	2%	2%	5%	2%	0.5%
Pressure	25 Torr	15 Torr	15 Torr	15 Torr	25 Torr
Microwave power	1200 W	800 W	900 W	800 W	900 W
Heater set point	–	700 °C	–	600 °C	–/550 °C
Substrate temperature	~900 °C	~800 °C	~750 °C	~700 °C	600–700 °C
Substrate position	Immersed	Immersed	Immersed	Immersed	Remote

(a)

(b)

Figure 11.7. (a) Planar and (b) cross-sectional SEM images of a HOD film after 40 h of CVD growth [261].

intrinsic defect-induced stress, and (ii) the change in the α-parameter from 2.7 to 2.5 for the 20 and 100-μm thick films, respectively. In contradiction to these results, for HOD films made by the *two-step* process and with the film thickness up to 40 μm, the tilt and azimuthal angles of (100) faces decreased with the film thickness, as described later in Section 11.4. The result of Ref. [264] might lead to a pessimistic conclusion that a single crystal diamond film/plate cannot be made even though the HOD film growth technology is further elaborated. However, works by Kawarada et al. (three-step process) [6] and Jiang et al. (two-step process) [265] showed that this is not the case: Jiang et al. [265] showed that an addition of 10 ppm boron (B) to the source gas was very effective to increase the areas of (100) faces by coalescence between adjacent (100) faces (the misorientation angles determined from the FWHM of the X-ray rocking curve was only 2.1°) by enhancing the step-flow mode. This is presumably because B atoms tend to increase the growth rate of (111) faces, enhancing the lateral growth of (100) faces. Kawarada et al. [6] demonstrated that when β-SiC(100) was used as the substrate, the step-flow growth started to occur when the film thickness was over 100 μm, where the coalescence between adjacent diamond grains were fully developed. In this case, $\Delta\chi = 1.8°$ and $\Delta\phi = 0.3°$ for the {111} reflection at $\chi = 54.7°$ in the (100)XPF measurement. One may notice how

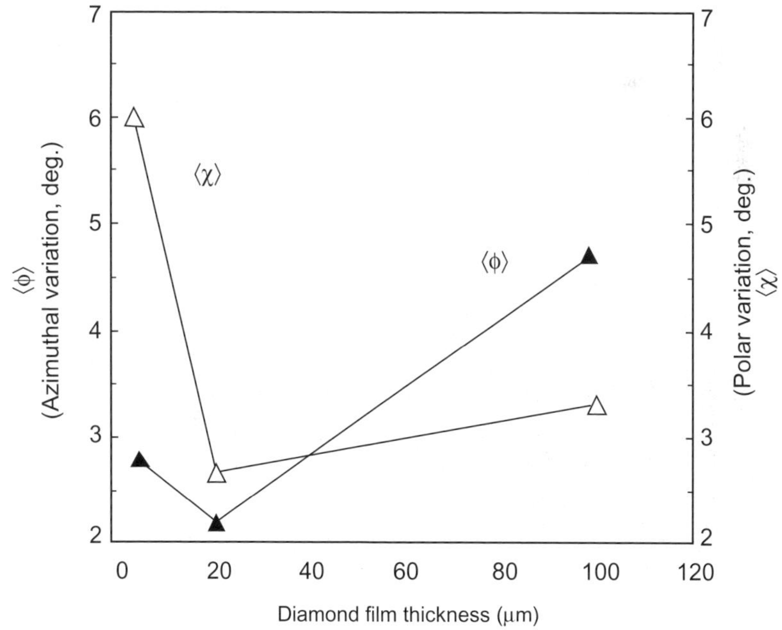

Figure 11.8. Polar ⟨χ⟩ and azimuthal ⟨ϕ⟩ misorientation angles as a function of HOD film thickness [264].

small these values are, compared with the corresponding values presented so far. It was indeed shown that a full coalescence of (100) faces was able to occur: for a 20-μm thick HOD film, the FWHM of XPF for the {111} reflection was 0.92° in the best-coalesced area and less than 1.5° in other areas. The value was as small as 0.62° for a 200-μm thick HOD film.

The carburization method of Si surface was elaborated in Ref. [266] using reaction gases of C_2H_2 and CH_4. For the processing conditions, see Table 11.3. As seen in Figure 11.9, the mesh structure at the Si substrate surface was different when C_2H_2 and CH_4 were used. As a result of BEN and diamond growth, the best fraction of oriented particles was 60%, when C_2H_2 was used. The average length of the mesh span decreased with decreasing the bias voltage, and the average length were 12.7 and 12.1 nm for C_2H_2 and CH_4, respectively, when $V_b = -110$ V. To make HOD films, a lateral growth was undertaken, and a fairly smooth film surface with faint grain boundaries was achieved.

In the three-step process, the thickness of the SiC layer was monitored as a function of time in the carburization and BEN steps, and its thickness reached a maximum at 3–5 min of the BEN treatment [267]. A mesh structure was also observed. The hillock (protrusion) structure was extended along [110] or [$1\bar{1}0$], and hence the boundaries between the hillock structures were considered to be anti-phase boundaries between the anti-phase domains of β-SiC.

In Ref. [22], HOD films with various thicknesses were synthesized using the same BEN and growth conditions except for applying DC bias only in the BEN step.

Table 11.3. Process conditions using C_2H_2 and CH_4 [266].

	Carburization	β-SiC formation (step 1)	Diamond nucleation (step 2)
(a) Experimental conditions for C_2H_4			
Total pressure (kPa)	2.7	2.7	2.7
Substrate temperature (°C)	790	790	790
H_2 flow rate (ccm)	99	99	99
C_2H_4 flow rate (ccm)	0.5	0.5	0.5
Bias voltage (V)	0	−20 to −140	0 to −150
Treatment period (min)	30	3–30	1–20
(b) Experimental conditions for CH_4			
Total pressure (kPa)	2.7	2.7	2.7
Substrate temperature (°C)	790	790	790
H_2 flow rate (ccm)	98	98	98
CH_4 flow rate (ccm)	2.0	2.0	2.0
Bias voltage (V)	0	−50 to −140	0 to −150
Treatment period (min)	30	3–20	1–7

1 Torr ≡ 133.3 Pa

There existed two kinds of (100) faces that behaved differently: the largest (100) faces rapidly increased their areas linearly with the film thickness, and the α-parameter was evaluated to be $\alpha = 2.5$. By contrast, the average (100) faces gradually increased their areas almost linearly until the film thickness reached 6 μm, but the areas were unchanged after this stage. In this case, the α-parameter was evaluated to be $\alpha = 2.75$ (fluctuation of α was ± 0.25). The cause of the difference in the behaviors of the (100)

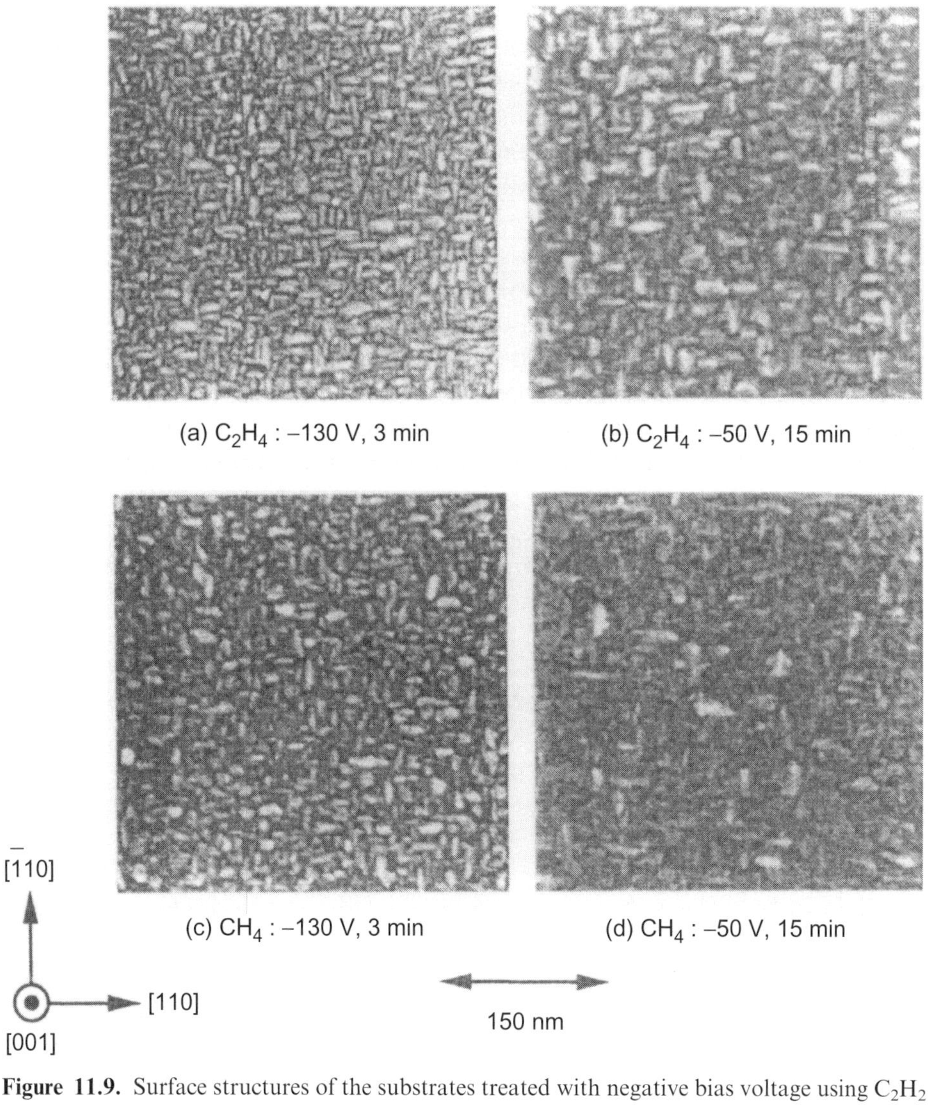

Figure 11.9. Surface structures of the substrates treated with negative bias voltage using C_2H_2 and CH_4 as the source gas [266].

faces and the α-values has not yet been made clear. Nonetheless, this result shows that the mode of the film growth turned from competitive to columnar when the film thickness becomes ~6 μm under given growth conditions. It thus follows that the α-parameter must be changed at this point to further increase the (100) areas. The FWHMs for polar and azimuthal angles of (100) faces decreased monotonically from approximately 20° to 10° as the film thickness increased from about 1 μm to 30 μm. The thermal conductivity increased with the film thickness, and saturated at a value of 7.4 W/cm · K, when the film thickness was 20 μm. From these results, the film thickness of ~6 μm seemed to be a transient point for film morphology and defect density, and film properties evolved much more slowly after this point. A detailed consideration of grain boundaries is also described in this paper [22].

The three-step process for HOD films also was studied in Ref. [268], and a 4-μm thick film was synthesized after a 24-h growth. In Ref. [228], it was shown that the bias current in the BEN process was significantly higher for carburized specimen than that without carburization, as seen in Figure 11.10. Also, the carburization greatly improved the (100) film morphology and orientation.

An HOD film growth was carried out on 1-inch Si(110) substrates [269] using the three-step process. The pretreatment and growth conditions are listed in Table H.3. In this work, (i) there was no apparent grain boundaries in the central area of a diamond film, (ii) an array of misfit dislocations were seen in the periodicity of four D{111} on every three Si{111}, and the spacing between the misfit dislocations was

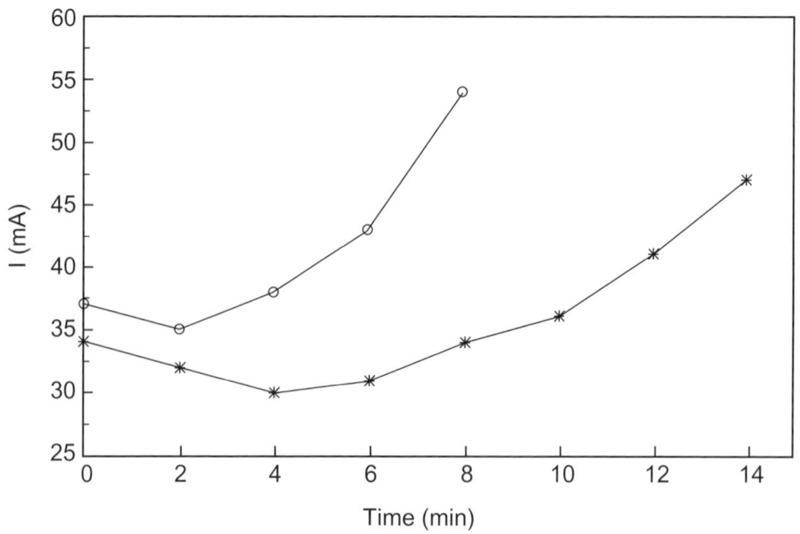

Figure 11.10. Bias current as a function of time for carburized (○) and pristine (∗) Si substrates [228].

0.62 nm, consistent with theoretical prediction, and (iii) there was no interlayer between Si and diamond. It is of interest that in addition to cubic diamond, 2H polytype of diamond (hexagonal diamond, space group *P63/mmc*, lattice parameter $a = 0.252$ nm and $c = 0.412$ nm) [270] was also present, in which the {0001} plane of 2H-diamond had an orientational relationship with Si{110} in such a way that D{0001} // Si{111} and D[11$\bar{2}$0] // Si[110]. By contrast, the cubic diamond had a relationship with Si in such a way that D{111} // Si{111} and D[110] // Si[110].

Finally, it should be noted that the R45 structure, proposed by Verwoerd [271–273] was experimentally ruled out by Ref. [274], but (100) crystals with 45° rotation are often observed [275]. Thus, further investigations are needed to conclude this issue.

11.4. TWO-STEP PROCESS: GROWTH OF HOD FILMS DIRECTLY ON Si(100)

In a series of works, Jiang et al. [263, 276–279] established the two-step process to synthesize HOD films using only (i) BEN and (ii) oriented growth. Usually, 2-inch Si(100) wafers were used as the substrate that had been cleaned by hydrogen plasma to remove a native oxide layer on Si. The process conditions are listed in Table H.4. There was however a considerable non-uniformity in morphology and thickness across the wafer [278]. The nucleation density was $5 \times 10^{10}/\text{cm}^2$, as high as that in the three-step process. Although there was a distribution of film quality measured by Raman spectroscopy across the specimen, the smallest FWHM of XPF was only 4.6° [279]. More recent work indicated that the FWHM was less than 2°, a significant progress from their previous works [265, 280].

The HOD film growth without the carburization step had also been done in Ref. [84]. The (100)XPF of an HOD film, synthesized under the growth conditions expressed by $\alpha \approx 2.8$, had a single peak in the center due to the {100} diffraction, whose FWHM was 7–12°. This means that τ_{100} (see Section 5.5) had a distribution about the film surface normal by this degree. On the other hand, the (220)XPF had four equivalent diffraction peaks of {110} that were 45° off-centered. For a review of works by Klages' group up to 1996, see Refs. [279, 281].

In Ref. [275], both tilt $\Delta\chi$ and azimuthal rotation angles $\Delta\phi$ were measured as a function of film thickness, and the results of XPF measurements are shown in Figure 11.11, where the four peaks correspond to {111} diffractions. For the process conditions of HOD film growth, see Table H.4. The nucleation density was $5 \times 10^9/\text{cm}^2$ that was controlled by the biasing time of BEN. Both angles decreased with the film thickness, but converged to finite values: 3.5° for the tilt angle $\Delta\chi$ and 6° for the rotation angle $\Delta\phi$, as seen in Figure 11.12 [275]. In this figure, $\Delta\phi_{\text{rot}}$

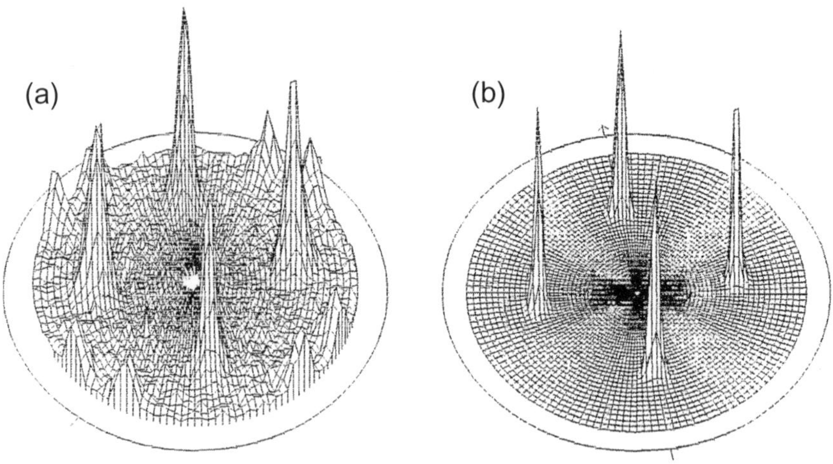

Figure 11.11. (111) XPFs of (a) 0.6-μm thick and (b) 37-μm thick diamond films. The peaks correspond to {111} diffractions [275].

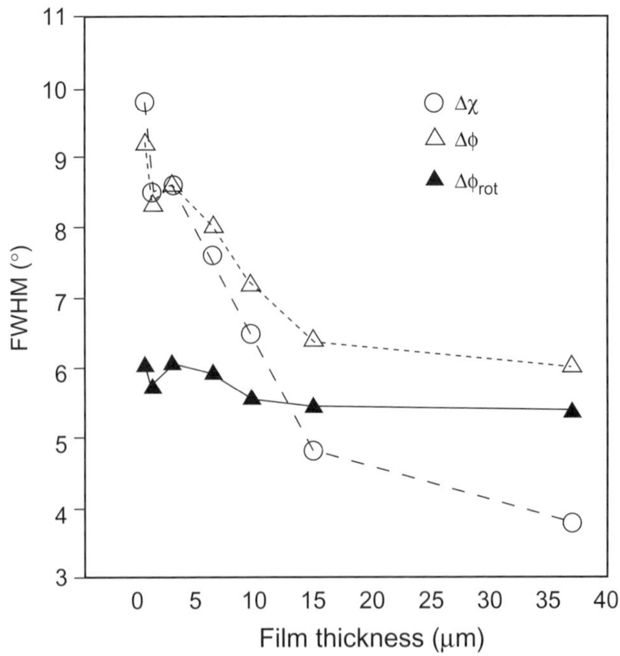

Figure 11.12. Thickness dependence of the FWHMs for (111)XPF. $\Delta\chi$ and $\Delta\phi$ are the tilt and azimuthal rotation angles, respectively, and $\Delta\phi_{rot}$ is the FWHM of the deduced rotational misorientation angle [275].

was calculated to extract pure rotational angle by $\triangle\phi_{rot} = \sqrt{\Delta\phi^2 + (0.71\Delta\chi)^2}$. A similar result was also shown in Figure 11.13 [282]. These data points were obtained from (111)XPF measurements. It is of interest that in the synthesized HOD film, there were diamond grains with (100) faces that were oriented 45°-off from the common direction. As already mentioned in Section 11.3, this might corresponds to the R45 epitaxy proposed in Refs. [271, 272]. For review works of the Schreck and Stritzker group, see Refs. [283, 284]. It is noted [282] that twinnings play an important role in the diamond nucleation of HOD film growth, because twin boundaries are frequently observed in the center of oriented crystallites, and the density of grain boundaries in HOD films cannot be decreased even though the film becomes very thick. If this is the case, in order to try to make a quasi-single crystal diamond film using the HOD film growth method, the initial diamond nuclei must be perfectly oriented epitaxially with respect to Si(100). Whether a single-crystal diamond film can be synthesized by the two- or three-step process is an important issue that remains to be investigated.

In Ref. [285], HOD films were grown on Si(100) by the two-step process. After a 15-h growth, 80% of the grains were oriented, and the mean tilt angle was 8°. The diamond film was characterized by X-ray photoelectron diffraction (XPD) to see the surface atomic structure at the nucleation stage [286]. The incident X-ray was Mg K_α and the kinetic energy of electrons was 964 eV. Hence, the sampling depth was 30 Å. The stereographic projection pattern of a HDO film after a 1-h growth was fuzzy. After an 8-min bias treatment, when the incomplete diamond layer thickness was 50–100 nm, the nuclei were already ⟨100⟩-oriented and also azimuthally oriented.

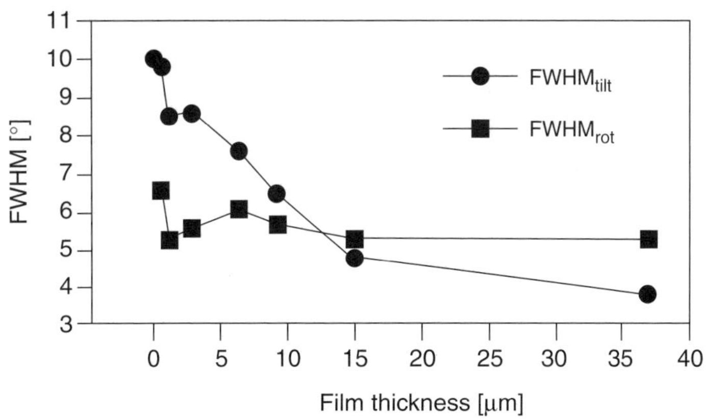

Figure 11.13. FWHMs of tilt and azimuthal (rot) angles as a function of HOD film thickness. The data were obtained by (111)XPF [282].

In a renewed study of the interface [274], the native oxide on Si was removed by hydrogen plasma treatment ($T_s = 780\,^\circ C$ and $P = 15$ Torr), and the Si surface roughness was significantly reduced, if $V_b = -225$ V was applied during the hydrogen plasma treatment. The BEN conditions used were $c = 2\%CH_4/H_2$, $P = 15$ Torr, $V_b = -225$ V, and $T_s = 780\,^\circ C$. As a result, only imperfectly oriented SiC nanocrystallites and heteroepitaxially oriented diamond crystallites were observed. The BEN step was followed by diamond growth using conditions of $c = 1\%CH_4/H_2$, $P = 30$ Torr, and $T_s = 845\,^\circ C$ for 10 min. Consequently, a formation of β-SiC on the Si surface was observed, but the β-SiC was not continuous and homogeneous. XPD measurements indicated that diamond crystallites were grown with a preferential orientation with respect to Si(100). Note that by the BEN treatment, the substrate surface had parallel stripe structures, as already found in Ref. [250]. In the early stage of BEN, only 30–40% of diamond crystallites were epitaxially oriented, but after a 15-h growth, 80% of the crystallites were epitaxially oriented. The mean tilt angle from ⟨100⟩ was 5°. This was in contrast to the fact that if the Si substrate was heated in CH_4/H_2 plasma to carburize the Si substrate surface, only 60% of diamond nuclei were azimuthally oriented, and the mean tilt angle of diamonds was more than 10°. The substrate temperature in the first minutes of diamond growth was essential to achieve better-oriented diamond films, and the BEN step was necessary to generate oriented diamond nuclei [287].

According to Refs. [265, 280] there was a strong coalescence between adjacent diamond grains in the film. Some grain boundaries disappeared or changed directions as the film growth proceeded, resulting in a single crystalline top layer that was extending over the area that was formerly covered by several grains. In Figure 11.14, it is indeed seen that a grain boundary with 2° misorientation is terminated and disappeared.

Effects of Si substrate orientation, (100), (111), and (110), on oriented diamond growth have been studied in Ref. [288] (see also Section 11.16). Continuous diamond films with (100) and (111) orientations could be synthesized on Si(100) and Si(111) substrates, respectively, using the BEN process. The (110)-oriented continuous film was more random in orientation than (100) and (111), but discrete crystallites with (110) orientation were observed. In Ref. [289], (100)- and (111)-oriented growth was achieved on Si(100) and Si(111), respectively. In the former case, D{100} // Si{100} and D[110] // Si[110], while in the latter, D{111} // Si{111} and D[$\bar{1}$10] // Si[$\bar{1}$10]. Note, however, that both (100)- and (111)-oriented growths were possible on Si(100) substrates. The FWHMs of polar and azimuthal angles in XPF measurements were 14° and 11–12°, respectively, for the (100)-oriented film on Si(100). On the other hand, they were 8.5° and 11°, respectively, for the (111)-oriented film. In this case, a twinning occurred between diamond grains that were oriented in such a way as D{111} // Si{111} and those that were 60° rotated from this orientation. Similarly in Ref. [290], the volume fractions of both (100)- and (111)-oriented diamond crystallites grown on Si(100) and Si(111), respectively, were determined by the XPF analysis. For the film synthesis, a BEN treatment was followed by a diamond growth to the film thicknesses of ~2 μm. The volume fractions of diamond crystallites that were textured in the film were 54 and 38% on Si(100) and Si(111), respectively. The twinning was more pronounced for (111) than (100), also according to the XPF analysis.

In Ref. [291], an HOD film growth using BEN was carried out on the reverse surface of an *as received* Si wafer to know the effect of surface roughness. The Si reverse surface was *as cut* so that it was rough and undulating. Although the synthesized HOD film was undulating during the early stage of the film growth along the substrate surface, the diamond film surface became as smooth as a HOD film grown on a polished Si surface by extending the growth time. Thus, the mechanism of BEN was unaffected by the condition of the Si substrate.

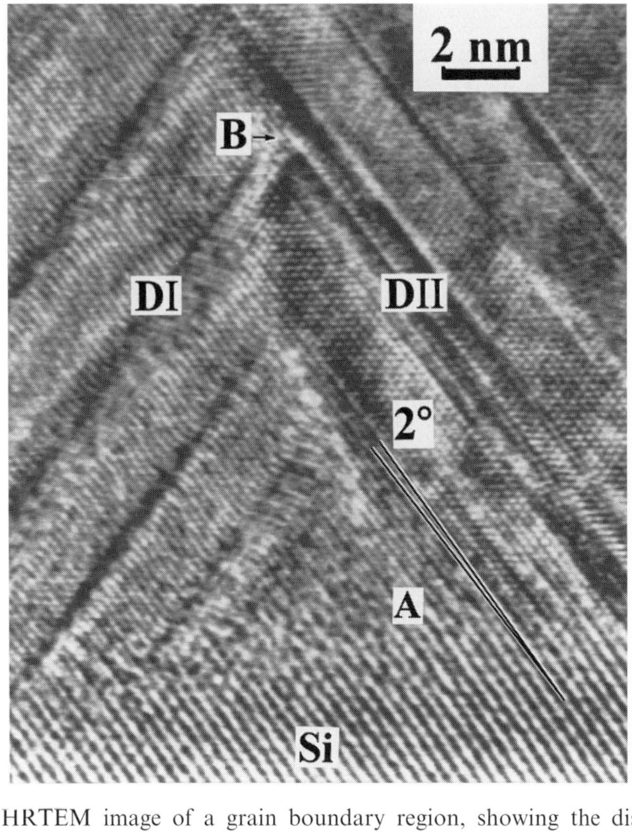

Figure 11.14. HRTEM image of a grain boundary region, showing the disappearance of a 2° low-angle grain boundary [265, 280].

However, the nucleation density on the reverse Si surface was higher than that of a polished, pristine Si substrate, implying that diamond nucleation preferentially initiated at microedges and high spots at the rough Si surface. Inclusion of Si and N impurities in the film was not detected by CL, unlike in the case of HOD films on pristine Si surfaces. This work seems to suggest that the nucleation and orientation of HOD films can be further improved by using properly roughened or textured Si wafers as the substrates.

In Refs. [292, 293], the BEN treatment followed by the (100) growth was undertaken to identify the most influential parameter for oriented nucleation in the two-step process. The Si(100) substrate was placed on a graphite holder. The film characterization was done using the (220) pole density maxima in the (220)XPF diagram, and the FWHM was selected as an orientational parameter.

Figure 11.15 shows one of the specimens made [293]. Figure 11.15 (a) is the film surface morphology observed by SEM, and Figure 11.15 (b) is the (111)XPF. The film thickness was 10 μm. Note that since the two-step process was used for the HOD film synthesis, no carburization step was included. The difference in the substrate temperature, measured both by a thermocouple and a pyrometer, was only within 10 °C. The azimuthal orientation of diamond nuclei was strongly dependent on the biasing time, the gas pressure P, and the bias voltage V_b, but its dependence on the substrate temperature T_s was surprisingly weak. In particular, the optimum biasing time became shorter as $|V_b|$ was higher: it was only 20 s for $V_b = -300$ V. The effect of substrate roughness (using both top and reverse surfaces of a Si wafer, just like the work done in Ref. [291]) was small. The same result was obtained for

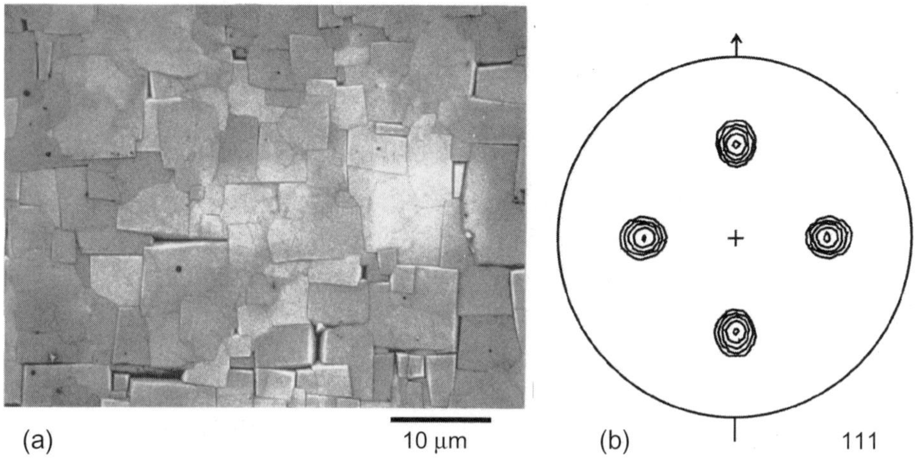

Figure 11.15. (a) SEM image and (b) (111) XPF of an HOD film. The four peaks in the (111)XPF correspond to the {111} diffractions [293].

a 4°-off Si(100) substrate. A prolonged biasing resulted in a loss of epitaxial orientation. Under conditions of $V_b = -200$ V, 10-min biasing time, $c = 8.8\%CH_4/H_2$, $P = 34$ Torr, $P_m = 1300$ W, and $T_s = 790$ °C, an optimum orientation was achieved, where the FWHM of 4.6°. It was hence concluded that the ion bombardment due to the bias voltage was most influential to the nuclei orientations. Figure 11.16 shows the optimum biasing time against the bias voltage. It is of interest that when $V_b = -200$ V, the biasing current increased only slowly with time up to 9 min, but started to increase very rapidly after this point (not shown in the figure. See Figure 10.14). Thus, the optimum biasing time could be determined as the time when the biasing current started to increase rapidly.

A further elaboration of the two-step process was done by depositing a B-doped layer on an undoped (100) HOD film [265]. The source gas was a mixture of CH_4 and trimethyl borane [$B(CH_4)_3$], whose concentrations were 0.5% and 10 ppm, respectively. As a result, a significant lateral growth of B-doped diamond film occurred, which (i) reduced the density of microtwins and (ii) enhanced a step-flow growth. The former result (i) is consistent with those of Refs. [4, 294].

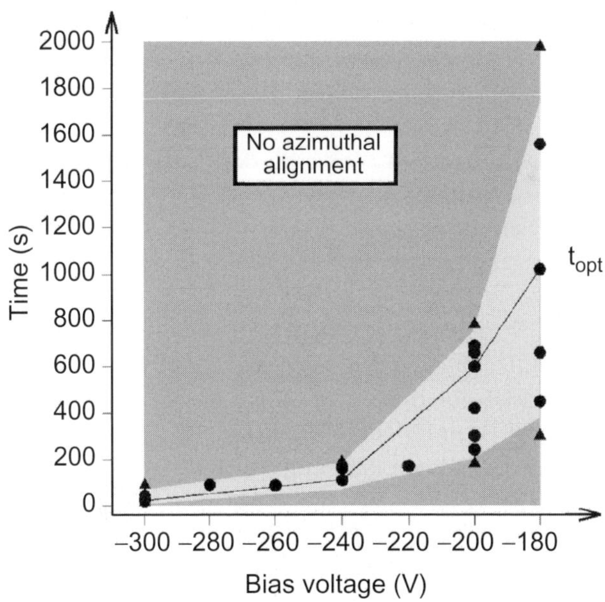

Figure 11.16. Optimum biasing time for different bias voltage. The circles mark heteroepitaxial films, while the triangles correspond to specimens without significant azimuthal alignment. The biasing times t_{opt} of the HOD films with minimum FWHM for the azimuthal tilt angle are connected by straight lines. The vertical extension of the bright area defines the width of the BEN process time window [293].

It is empirically known that B-doped diamond films tend to have (111) faces, and hence the coalescence of (100) faces proceeds because the side (111) faces grow faster than the top (100) faces. As suggested in Section 11.3, the work of Ref. [265] also shows that the growth mode switches from the Volmer-Weber island growth to the step-flow growth mode, if the film surface becomes fairly flat due to the coalescence. An interesting consequence was because of the lateral growth, the grain boundaries between adjacent diamond crystals either disappeared or change directions from {100} to an orientation between {001} and {111}, and some grain boundaries changed their directions to be parallel to the film surface. The result of Ref. [265] can hold for HOD films in general. More generally, it suggests that an addition of B, N, or other impurity atoms in the source gas will be effective to achieve the transition from the Volmer-Weber island growth to the step-flow growth. It should however be noted that the presence of oxygen is detrimental in the BEN treatment step, because SiO_2 is formed at the Si surface, which disturbs the formation of oriented nuclei of diamond [295].

11.5. AFM STUDIES OF FILM SURFACES

AFM images of Si(100) surface during BEN were studied in Ref. [296] by interrupting the BEN step at 2.5-min intervals. After 5 min, island structures with a height of several nanometer were observed. After 20 min, larger islands of several hundred nanometer were observed, where crystal edges began to appear on them. It is known that there are three growth modes of films: (i) the Volmer-Weber mode of three-dimensional island growth, (ii) the Frank van der Merwe mode of layer-by-layer growth, and (iii) the Stranski–Krastanov mode of three-dimensional island growth on one or a few layers. From the AFM study, the initial stage of diamond nucleation was identified as the Volmer-Weber mode. A successive growth of diamond under unbiased conditions for 22 h resulted in an HOD film with an epitaxial relationship of D{100} // Si{100} and D[110] // Si[110].

The orientational order of diamond grains in HOD films, synthesized by the two-step process, was studied using XPF and AFM measurements [297]. The growth conditions were selected so that $\alpha = 2.5$. In the χ-scans of (111)XPF, the width of the (100) diffraction band ($\chi = 54.73^\circ$) of the HOD films with the thicknesses of 0.6 and 37 μm were 9.8 to 4°, respectively. This means that the tilt angles of (100) faces are more narrowly distributed as the film thickness increases. According to a two-dimensional simulation that started from two crystallites, an untilted crystallite A and a crystallite B oriented in the fastest growth direction, the facet area of crystallite A was significantly larger in the intermediate film thickness than that of crystallite B, but it became smaller as the film thickness further increased

[298]. Thus, experimental data on orientational degree should be interpreted with caution as they depend on the film thickness. This also applies to the results of Figure 11.8 [264], Figure 11.12 [275], and Figure 11.13 [282]. Based on the AFM measurements for about 1000 crystallites at the film surface (the film thickness was 11.1 μm), the tilt angle of the maximum mean surface area of (100) faces was 1°, and in terms of the number of (100) faces, they were tilted by ~4°.

Diamond nucleation by the two-step process on Si(100) was also studied by AFM and STM in Ref. [299] in addition to X-ray rocking curve analysis. The nucleation was governed by the Volmer-Weber island growth mode, as stated above, but the growth on existing diamond was governed by the layer-by-layer growth (the Frank van der Merwe mode). Hence, the (100) faces of diamond nuclei were often convex, consisting of a large number of rectangular layers. The heights of the terraces between the adjacent layers were 1–50 nm. Note that a similar but macroscopic structure was also observed in Ref. [23]. The average tilt angle of the (100) faces, measured by AFM, was ± 5.2°. In the neighborhood of each nucleus, there was a depletion region of nucleation, and the average distance between nuclei was in the range of 30–40 nm. The observed STM image was consistent with the 2 × 1 surface reconstruction on the (100) face of the diamond. In more detailed examinations [277] using AFM and RHEED, (i) the formation of diamond nuclei needed 6.5 min, while a SiC layer was formed in 2 min and (ii) the maximum of the nearest neighbor distribution of the nuclei was 10 nm longer compared with random distribution. The result (i) indicates that a depletion area is formed around a nucleus, meaning that the surface diffusion of adatoms (or chemically active hydrocarbons) is responsible for the nucleation. The ion beam bombardment is likely to enhance the surface diffusion of adatoms. The result (ii) indicates that the formation of SiC does not influence the diamond nucleation. In the initial phase of diamond nucleation, unfacetted islands were formed on the substrate surface, which were identified as crystalline diamonds by RHEED. The islands underwent a uniform growth at the beginning of diamond CVD. A similar but macroscopic process has also been observed in Ref. [23].

AFM images of Si substrate surface during BEN were observed in Ref. [300] using a bell-jar type MPCVD system described in Ref. [301]. At the beginning of the BEN step, there was a delay time when the nucleation density remained low. The nucleation density then increased *exponentially* with time to reach a saturation value [2]. The film surface as a result of BEN consisted of well-separated hillocks. A theoretical analysis showed that the diffusion length of hydrocarbon ions, $C_xH_y^+$, was small: each ion species can diffuse less than the average spacing between diamond nuclei across the surface. Furthermore, since incident ions have an energy of 100 eV due to biasing, they penetrate into the surface instead of migrating along the surface.

It was thus assumed to be hydrocarbon radicals that diffuse over the surface [301]. In the exponential regime mentioned above, the nucleation rate was proportional to both the incident flux and the two temperature-dependent terms, i.e. the residence time and the diffusion attempt frequency, according to a theoretical consideration. The etching of sp^2 carbon atoms by atomic hydrogen was considered to be unlikely as the primary cause of leaving sp^3-bonded carbon clusters, because this would lead to a maximum nucleation density when the etching is the fastest, which was not observed experimentally. Furthermore, Raman spectra indicated that the fraction of sp^2-bonded carbon was maximum under the optimum biasing conditions. Taking into account the fact that the nucleation density of diamond on a highly sp^3-bonded tetrahedral amorphous carbon (*ta*-C) was very low, it follows that the existence of sp^3-bonded carbon is only a necessary condition for diamond nucleation. It was thus speculated that the subplantation of carbon ions in the carbon film created a biaxial compressive stress (see Figure 4 of Ref. [302]), and graphite layers are aligned with their c-axes parallel to both the stress direction and the substrate surface. Such oriented nanocrystalline graphite was considered to be the nucleation sites of oriented diamond, because it is known that diamond can grow continuously from the edge of graphite, as described in Section 9.3. It seems, however, that this speculation is not the case because experiments of Refs. [303–305] showed no sign on the presence of oriented nanocrystalline graphite.

11.6. INTERFACE STRUCTURES

The interface between an HOD film and a β-SiC layer formed on Si was investigated by TEM [225, 306, 307]. The experimental conditions used in Ref. [225] are listed in Table H.2. The {111} lattice planes of diamond were tilted by ~5° from those of β-SiC, and there was one dislocation for every seven lattice planes, as shown in Figure 11.17. The observed result was interpreted using the diagram shown in Figure 11.18. A closer examination using an interface theory led to a conclusion that a large number of misfit dislocations were responsible to the tilt, as shown in Figure 11.19. Namely, there was *one edge-type misfit dislocation for every seven lattice planes* of β-SiC{111}. Their theory, however, predicted that in order to completely relieve the strain, there should be *one misfit dislocation for every five lattice planes*, and hence it was assumed that there still remained a strain at the HOD/β-SiC interface. It should be noted that there existed azimuthal rotations of HOD grains by several degrees with respect to the β-SiC[110]. A theory [306, 307] predicted that D{114}/β-SiC{221} had the lowest interfacial energy, and was the best combination for heteroepitaxial growth of diamond.

In Ref. [195], HOD films were synthesized using the three-step process. In the TEM images, there was a location where diamond and Si were in direct contact.

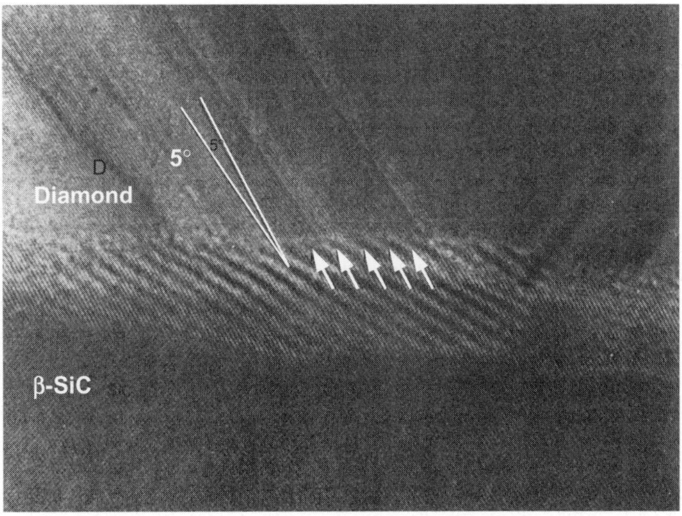

Figure 11.17. HRTEM micrograph of diamond/β-SiC interface. The arrows indicate the position of misfit dislocations [306].

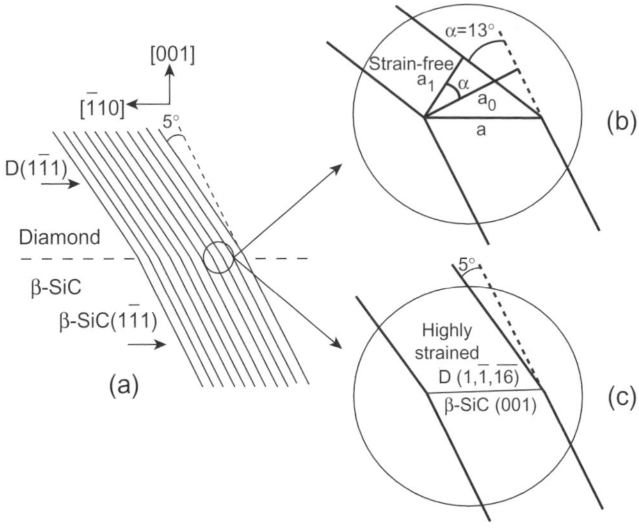

Figure 11.18. Schematic diagrams of (a) the tilted interface shown in Figure 11.17, (b) a strain-free situation which requires a 13°-tilt of the diamond lattice, and (c) a highly strained case which exposes diamond {1, $\bar{1}$, $\overline{16}$} planes at the interface [306].

Both experimental observation and theoretical consideration led to a conclusion that the ratio of D{111} planes and Si{111} planes should be 3-to-2, and the diamond planes should be tilted by 9.5°, as shown in Figure 11.20. This was later confirmed by other research groups.

In Refs. [303–305], the interface structure was also investigated by cross-sectional HRTEM. The diamond films were grown by the three-step process, and the conditions are listed in Table H.3. Consequently, an HOD film was grown in the center of the Si(100) substrate. In the carburization step, there was an *a*-C film of 250-nm thickness on Si, in which β-SiC, diamond, and graphite were embedded. A closer examination indicated that there existed an interlayer of 1.5- to 2-μm thickness between Si and the *a*-C layer, which was identified as *a*-SiC [305]. Since the bias voltage is not usually applied uniformly across the Si wafer, the distribution of these materials depended on the location on the Si substrate. Near the edge of the

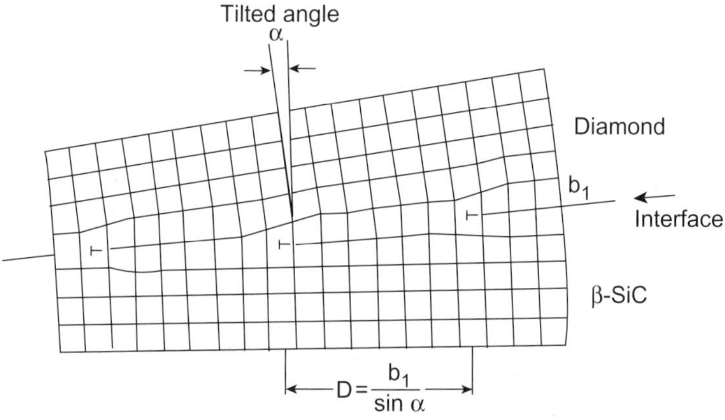

Figure 11.19. Schematic diagram of a tilted interface caused by an array of edge-type dislocations parallel to the interface [306].

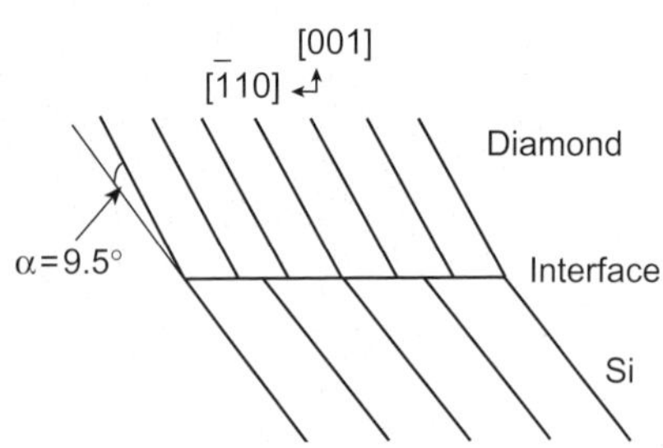

Figure 11.20. Schematic diagram of the D(100)/Si(100) interface for a 3-to-2 lattice matching. The 1.5% lattice mismatch results in a tilt of the {111} planes of diamond by 9.5° with respect to Si{111} [195].

Si substrate, there were β-SiC and diamond, both with a size of 150 nm. The total volume fraction of diamond and SiC was below 5%. Diamond and SiC were not in contact with Si, and approximately 20 nm away from the interface. In the position under the center of the plasma, there were graphite (0001) platelets with a size of 50 nm × 6 nm. They were aligned parallel to the substrate, but randomly oriented in other areas. This experimental result rejects the speculation stated in Ref. [300] (see Section 11.5) that graphite particles are assumed to be oriented perpendicularly to the Si surface to be nucleation sites for diamonds. In this position, no diamond and SiC were observed [303–305]. After the BEN treatment for 10 min, there existed a β-SiC layer with a thickness of 10 nm between Si and diamond crystals of 0.1 μm in size, and the β-SiC layer was epitaxially oriented with respect to Si(100). Diamonds were directly grown on the β-SiC layer, and the selected area electron diffraction pattern indicated that D{111} // β-SiC{111}. Graphite and *a*-C were not observed after the BEN step. The {111} planes of diamond, β-SiC, and Si were all in parallel [304]. After the diamond growth step, an HOD film was formed near the center of the plasma. At the interface between diamond and Si, there were randomly oriented β-SiC particles of nanometer size. The mean tilt angle of diamond (100) faces was evaluated to be 3.5° [305].

In Ref. [308], HOD films were grown using the two-step process, and investigated by HRTEM. The TEM images showed that diamond was in direct contact with Si without interlayer phases (see Figure 11.21), and appeared to grow epitaxially on Si despite the fact that the lattice mismatch is as large as 52%. The diamonds had a tendency to nucleate on the hillocks of Si, and the diamond tilt angles were small

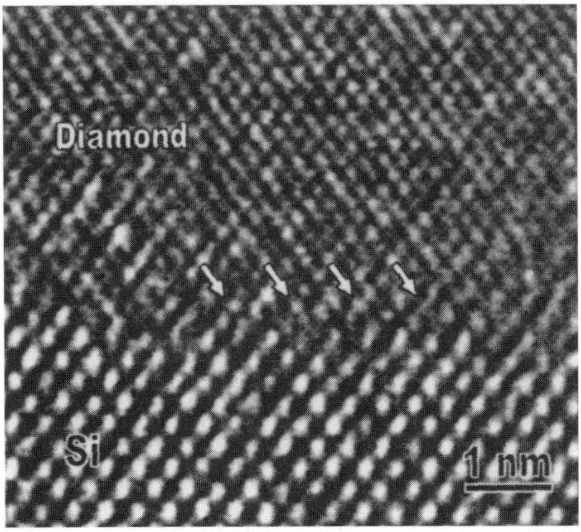

Figure 11.21. A lattice image of the interface between diamond and Si. The micrograph was taken along the [110] axis of diamond and Si crystals [308].

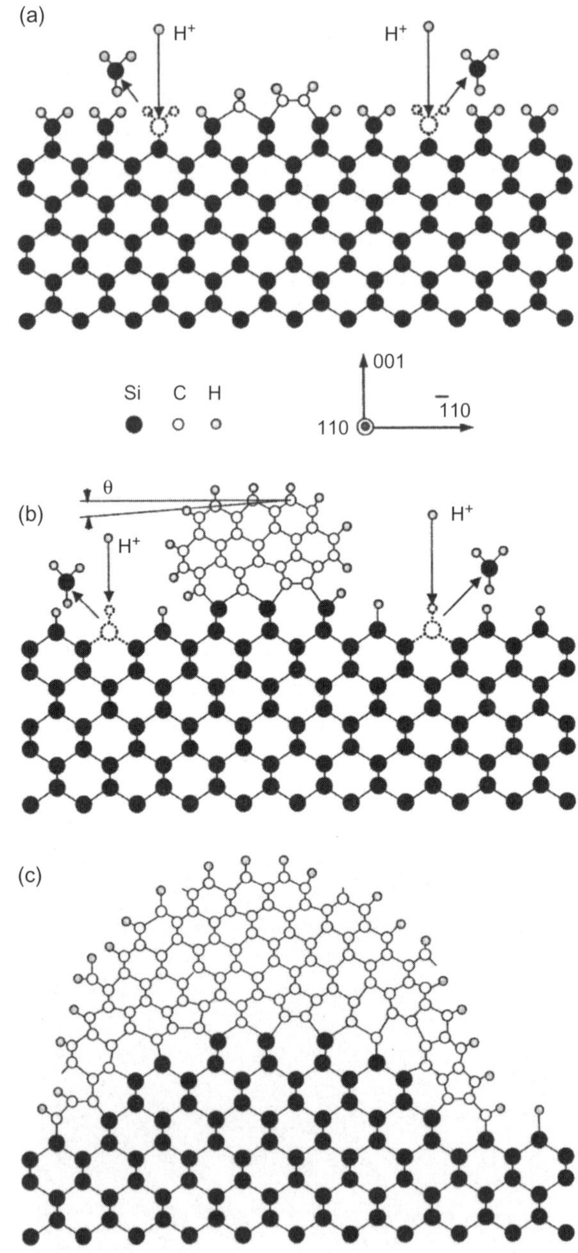

Figure 11.22. Schematic sequence of diamond nucleation on an Si hillock: (a) initially deposited carbon configuration, and H^+ ion etching, (b) etching of Si surface and formation of tilted diamond nucleus, and (c) growth of the nucleus, and formation of a Si hillock due to H^+ ion etching [308].

(~1°), if the Si hillocks were relatively flat. In case that Si hillocks were rough, the tilt angles of diamonds were large, e.g. ~6°. This suggests that diamond could nucleate on Si hillocks in such a way as shown in Figure 11.22.

In Ref. [273], Verwoerd showed by electronic energy computation that the 3-to-2 registry between diamond and Si was stable. The interface of HOD film deposited by the two-step process and the Si substrate was investigated in Ref. [309]. The FWHM of the tilt angle distribution was 4.6°, greatly improved over their past results of about 11°[263, 299]. As seen in the HRTEM of the [110] projection of the interface shown in Figure 11.23, diamond is in direct contact with Si, and two Si{111} fringes match with three D{111} fringes. Namely, there is clearly the 3-to-2 correspondence between Si and diamond, which leads to a lattice mismatch of only 1.5%. This result shows that Verwoerd's computation is actually the case, and SiC is unnecessary as an interlayer for HOD film growth. Thus, even though SiC layers are observed in some cases, it has no influence on the diamond orientation. The interface between Si and diamond was rough presumably due to the hydrogen plasma treatment of Si prior to diamond nucleation under conditions of $P = 15\,\text{Torr}$, $T_s = 850\,°\text{C}$, $V_b = -150\,\text{V}$, $P_m = 900\,\text{W}$ for 20 min [277].

In Ref. [310], an HRTEM study was carried out for an HOD film of less than 1-μm thickness, also made by the two-step process of Ref. [263]. A majority of diamond grains were ⟨100⟩-oriented, and twins, which started from the interface, were frequently found in the grains. At the interface between diamond and Si, a few

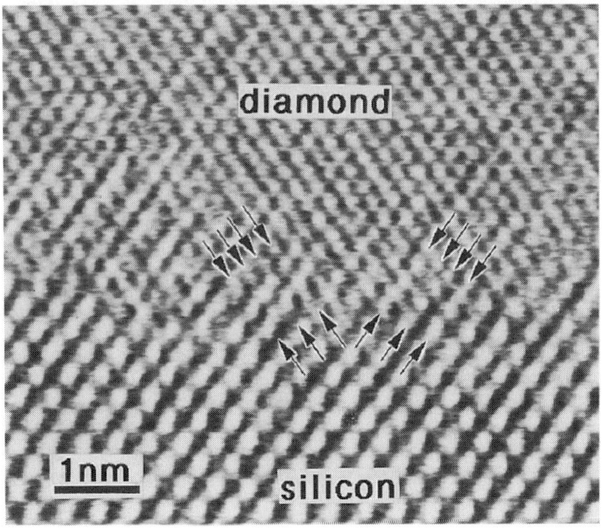

Figure 11.23. HRTEM image vertical to the [110] direction of the diamond/Si interface region [309].

nanometer size β-SiC grains were found, whose basic axes were parallel to those of Si, but independent of those of adjoining diamond grains. Like in Figure 11.23, the 3-to-2 registry was present between diamond and Si, and every third lattice plane of diamond was terminated at the interface due to the misfit dislocation. Diamond grains grew directly on Si, even though the Si surface was considerably rough. The {111} diamond lattice was about 4° tilted from the Si{111} lattice planes, as seen in Figure 11.24. It is of interest that there existed diamond grains that were 70°- and 14°-rotated about ⟨110⟩-axes. The former grains were in twin relationship with ⟨100⟩-oriented diamond grains.

In Ref. [311], diamond films were deposited on Si(100) substrates by the two-step process using a bell-jar type MPCVD reactor [221]. TEM images showed that hillock structures with a density of $7 \times 10^9/cm^2$ were present at the diamond/Si interface. According to the ED pattern analysis, the hillocks contained β-SiC and diamond crystals that were embedded in an amorphous material (carbon). Some β-SiC{220} planes were parallel to Si{220} planes, indicating that some β-SiC particles were in epitaxial relationship with Si. A cross-sectional TEM of the specimen showed that the hillocks of less than 100 nm in height and 50 nm in diameter were formed after the BEN treatment, as seen in Figure 11.25.

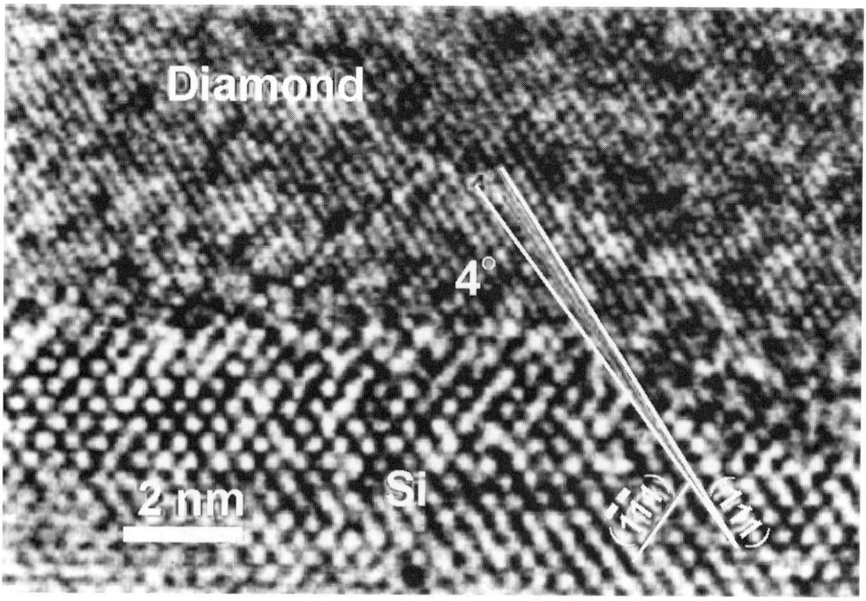

Figure 11.24. A $[1\bar{1}0]$ lattice-fringe image of a diamond grain that is tilted by 4° with respect to Si. Such small angular deviations are compensated for by unequal numbers of $\{\bar{1}\bar{1}1\}$ and {111} terminating planes [310].

An HRTEM image of the hillock showed that there existed a peak of Si substrate underneath, as seen in Figure 11.26. In between the hillocks, the Si surface was covered with an amorphous (carbon) layer. In the hillocks above the Si tip, there were β-SiC particles of a few nanometers in size. These results are in good agreement with Yugo's work [207], which was described previously but done after this work. It should be mentioned that the Si substrate was etched by roughly 10 nm during the BEN treatment for 17 min. It was thus inferred that diamond nucleation occurred from these hillocks, and β-SiC did not play a role of oriented diamond growth.

In Ref. [312], TEM plan views at different depths were observed for an HOD film of 2-μm thickness, and those near the surface were observed for a 25-μm thick HOD film. These results showed that in the HOD film growth, there was a critical thickness that was determined by the growth parameters, the nucleation density, and the density of oriented nuclei. Below the critical thickness, diamond grains contained a high density of dislocations perpendicular to the (100) faces, while above the critical thickness, the dislocation and other defect densities were markedly lower.

The interface structure between a HOD film, made by the two-step process, and Si was observed by TEM in Ref. [313]. (111)XPF measurements indicated that the FWHM was about 9°. In consistent with the results of Tzou et al. [314] there were two areas at the interface: (i) an interfacial region with amorphous structure where the orientations of diamond grains were random, and (ii) an interfacial region

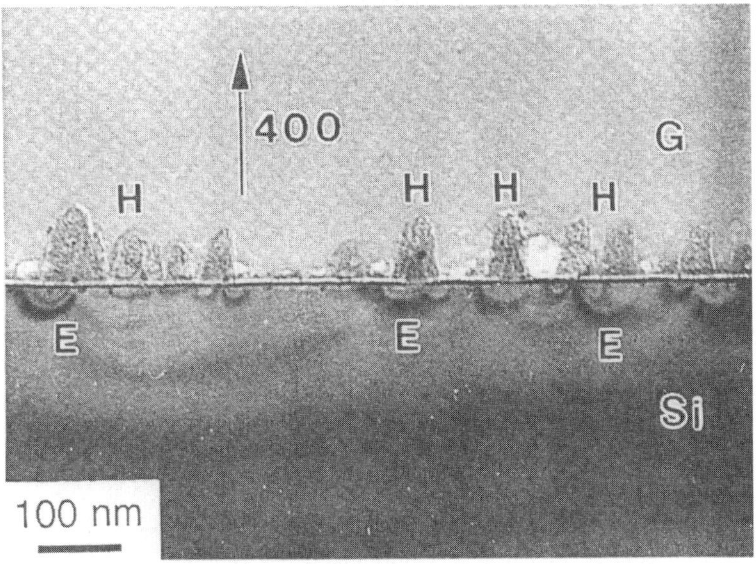

Figure 11.25. Cross-sectional view of a Si specimen after a 17-min BEN treatment. G: glue, H: hillock, E: extinction contours in the Si substrate [311].

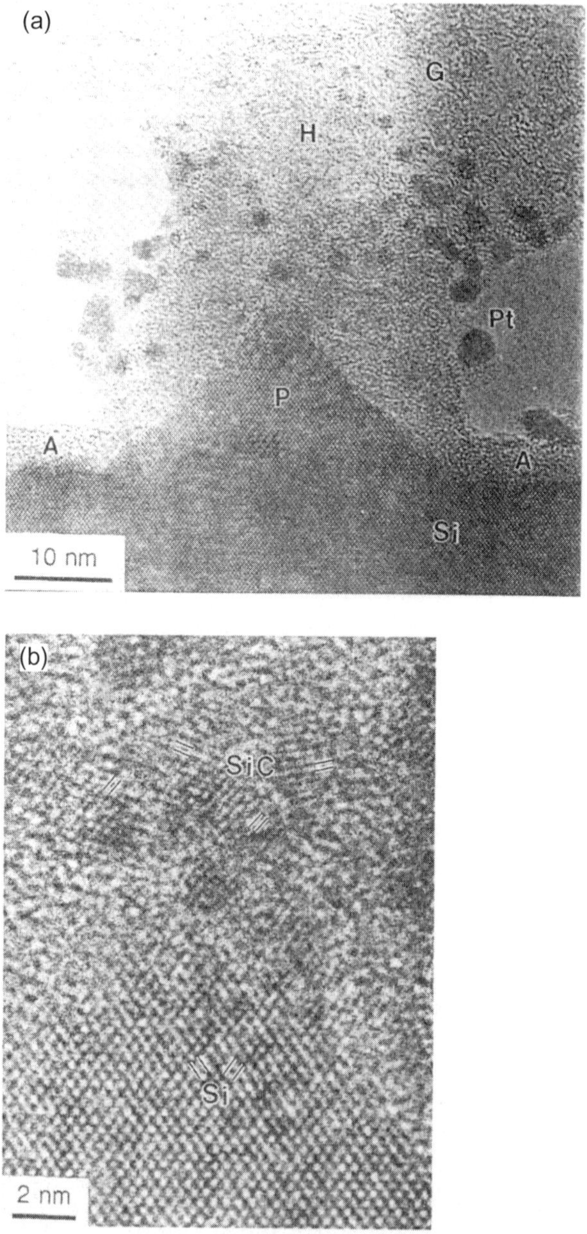

Figure 11.26. (a) HRTEM image of the hillock after a 17-min BEN treatment. H: hillock, P: Si peak, G: glue, A: amorphous layer, and Pt: Pt droplets intended to indicate the original specimen surface and (b) a higher magnification HRTEM of an Si peak tip [311].

where diamond grains were in direct contact with Si, and thus heteroepitaxially oriented. The diamond/Si lattice matching was 3-to-2, and the {111} lattice tilt between diamond and Si was roughly 10°, consistent with the result of Ref. [195]. It was explained that the surface roughness of Si due to hydrogen plasma treatment prior to BEN was the origin of tilted diamond grains, but the hydrogen plasma treatment was necessary for the oriented growth of diamond without carburization step.

In Ref. [315], HOD films were grown using the two-step process based on the BEN technique of Ref. [289], and the interface structure was studied by TEM. As a consequence of the BEN process, (i) the carburization of Si created nanocrystalline, heteroepitaxially oriented β-SiC, (ii) the β-SiC layer formed nanometer size hillocks and ridges that were aligned parallel to the Si[110] direction, (iii) diamonds nucleated on the β-SiC layer that was convex, and (iv) β-SiC existed only under the

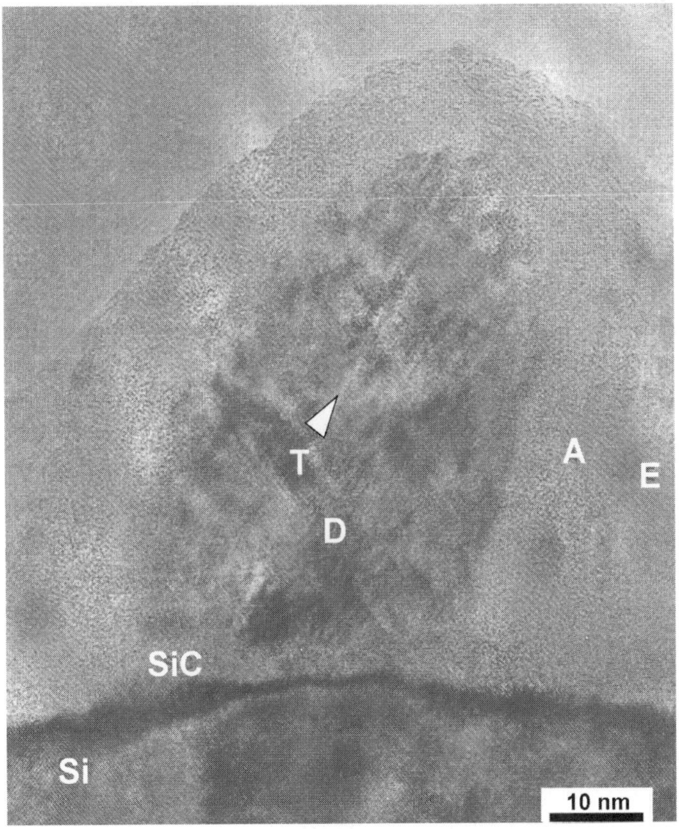

Figure 11.27. HRTEM image of a diamond crystal at the Si surface after a 15-min BEN treatment. D: diamond, T: twin lamella, A: amorphous layer, E: epoxy [315].

diamond grains as β-SiC in other areas were etched out. Figure 11.27 shows an HRTEM image of one of the hillocks formed after a 15-min biasing. One can see diamond (D), twin lamella (T), and amorphous layer (A). The occurrence of tree-like structure, shown in Figure 11.27, is consistent with Yugo's result (Ref. [207], see also Figures 10.2, 10.3, 10.5, and 10.6 as well as Figure 11.25). The fact that ridges of β-SiC developed with the biasing time along Si[110] is also consistent with Kawarada's result [250]. However, the fact that diamonds nucleated on the tops of β-SiC ridges is in contradiction with Kawarada's result [250] and Maeda's result [267], where the three-step process was used, and diamonds nucleated in the valleys of the β-SiC mesh structure. It is not clear at the present stage whether this difference arose from the two-step versus the three-step processes. It would be of great interest to study whether β-SiC is really etched by hydrogen plasma during the BEN process, because this would lead to a revision of the common understanding that β-SiC actually plays an important role for heteroepitaxial nucleation of diamond in the three-step process.

A planar TEM observation of HOD film (see Figure 11.28) was done in Ref. [316] in the film positions at the surface, in the middle, and at the interface. The films were synthesized by MPCVD using the two-step process, and more than 10 μm thick. Regarding the tilt angles of (100) faces, the maximum misorientation decreased from 14° to 5° as the growth temperature was increased from 880 to 970 °C. At the interface (see Figure 11.29), the average grain size was 0.2 μm, and the grains were randomly oriented. The ED measurements showed that every diamond grain contained internal microtwins. Amazingly, a dramatic change in the microstructure was observed at 1 μm above the diamond/Si interface, where rectangular

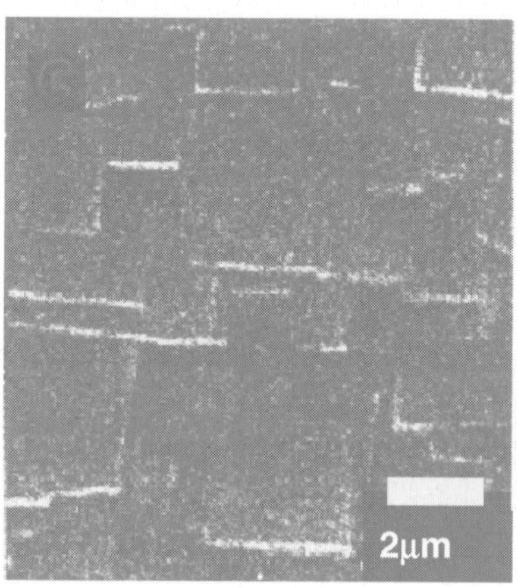

Figure 11.28. SEM image of the HOD film synthesized by the two-step process. The CVD was done for 4 h at $T_s = 970\,°C$ [316].

grains were very well aligned (see Figure 11.30). The azimuthal misorientation angles were less than 10°. A similarly good alignment was also observed at the film surface.

In Ref. [317], diamond growth was done by MPCVD (ASTeX reactor) using BEN on 6H-SiC single crystal substrates by changing the CH_4 concentration between 4–10%, and the epitaxial relationship was investigated by TEM and ED. For related works, see Refs. [49, 318]. The BEN treatment was undertaken for 24 min under conditions of $V_b = -200$ V, $c = 10\%CH_4/H_2$, $P = 20$ Torr, and $T_s = 760$–$790\,°C$. The diamond growth was carried out using $c = 1\%CH_4/H_2$ for 30 min, and a polycrystalline film was deposited. TEM measurements indicated the presence of an

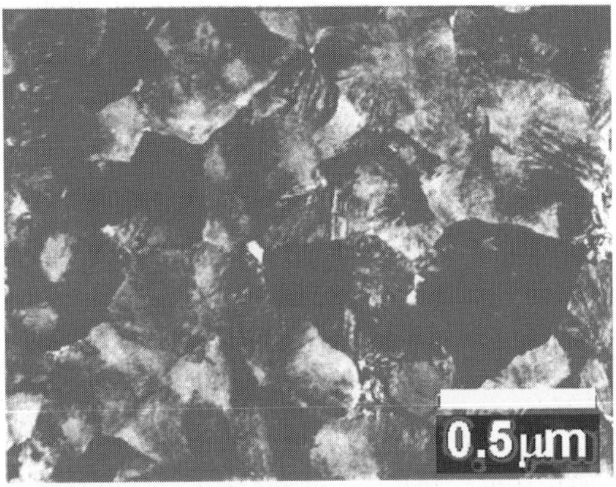

Figure 11.29. Planar TEM image of the diamond/Si interfacial area of the HOD film synthesized by the two-step process. The CVD was done for 6 h at $T_s = 940\,°C$ [316].

Figure 11.30. Planar TEM image of the area just 1 μm above the diamond/Si interface of the HOD film synthesized by the two-step process. The CVD was done for 4 h at $T_s = 970\,°C$ [316].

amorphous interfacial layer of 40 nm thickness, and β-SiC was formed on 6H-SiC with an orientational relationship of β-SiC[110]//6H-SiC[11$\bar{2}$0] and β-SiC{111}// 6H-SiC{0001}. A few nanocrystalline diamond particles were embedded in the amorphous layer. When the BEN treatment was done using $c=5\%CH_4/H_2$, the diamond grains had a columnar structure, and the presence of the interlayer and β-SiC was similar to the above. The {100} crystal planes of diamond were parallel to the basal 6H-SiC{0001} plane. The interface between diamond and 6H-SiC was smooth, but the defect density was high in the diamond grains near the interface. For the BEN treatment using $c=4\%CH_4/H_2$, diamonds were less oriented than in the case of $c=4.5\%CH_4/H_2$.

The same orientational relationship was found in Ref. [45] using HFCVD. As seen in Figure 11.31, triangular (111) faces of diamond are well aligned in parallel to the film surface.

Figure 11.31. (111)-oriented diamond film grown on 6H-SiC by HFCVD [45].

11.7. INTERNAL DEFECTS

The internal structure of diamond grains in HOD films was investigated in Ref. [319] by TEM. Figure 11.32 is one of the observed TEM images. There were dislocations at and near the grain boundaries, but the defect density was extremely small inside the grain away from the grain boundaries. Most grain boundaries were parallel to the [100] direction, but in Figure 11.32, the grain boundaries are parallel to both [100] and [110] directions. At low angle grain boundaries observed in a higher magnification, a series of dislocations were found with a spacing of 20 nm, as shown in Figure 11.33. Most misorientation angles between adjacent diamond grains were less than 2°, but in some cases, they were as much as 6°.

11.8. POST-TREATMENT OF HOD FILMS

In Ref. [320], an HOD film was treated under a negative bias voltage using conditions of $V_b = -50$ to -200 V, $P = 37$ Torr, and $c = 8.9\% CH_4/H_2$. These conditions were also used for the HOD film growth. For $V_b = -50$ V, there was no change in the HOD film morphology. For $V_b = -100$ V, the (100) faces became microcrystalline, consisting of pyramid shape structures of single crystal diamond. This result was considered to arise from defect formation by ion bombardment and a preferential etchings starting from the local defects. For $V_b = -150$ V, the (100) faces were covered with hillocks with a diameter of about 120 nm, and for $V_b = -200$ V, bulb-like structures were grown at the edges of (100) faces.

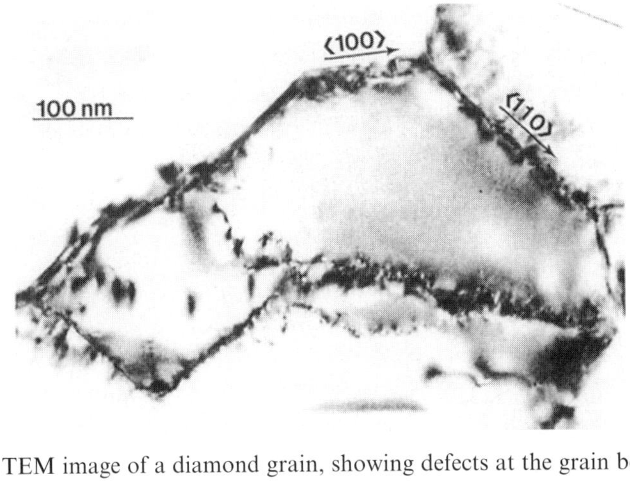

Figure 11.32. TEM image of a diamond grain, showing defects at the grain boundaries [319].

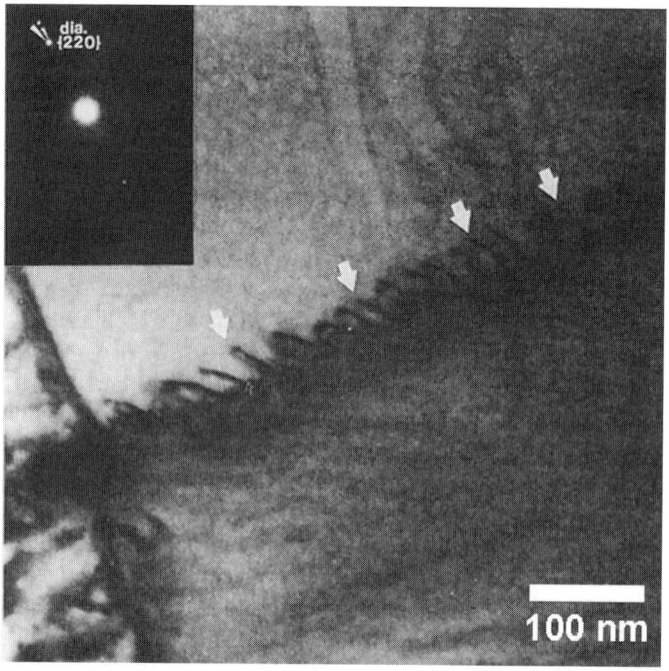

Figure 11.33. TEM image showing widely spaced dislocations at the grain boundary. Arrows indicate every 5th dislocations, where the spacing between each dislocation is approximately 20 nm. The inset shows an ED pattern from the grain boundary [319].

This work should be compared with the results of Refs. [321–323]. In Ref. [321], $V_b = -100$ V was applied during the diamond CVD step, and as a result, the film surface consisted of azimuthally oriented (100) faces. Namely, application of V_b suppressed the growth of non-(100)-oriented grains. For the two-step growth conditions, see Table H.4. Also, Refs. [324, 325] are related in this context.

11.9. UNIFORMITY OF NUCLEATION

A practical issue of BEN is that the nucleation density is significantly non-uniform across the substrate due to a non-uniform generation of secondary plasma. In Ref. [326], formation of a secondary plasma during BEN was studied using an MPCVD reactor similar to the ASTeX-type. The secondary plasma was at first generated along the edge of the substrate and the substrate holder, and then extended toward the center of the substrate with time. The change in the bias current with time was understood by the observed fact that the diamond nucleation extended from the peripheral region of the substrate to its center. A most useful technological finding was that the secondary plasma was uniform over the Si wafer of 13 mm in diameter, when a grounded ring anode of 30 mm in diameter was placed at 10 mm above the substrate, and the bias voltage of −200 V was applied to the substrate. As a result, a uniform BEN was achieved for both disk and square substrates, as seen in Figure 11.34. The plasma shape depended on whether the substrate holder was precoated with diamond film, because diamond can easily emit electrons, which will expand the secondary plasma and facilitate the BEN process at lower bias voltages.

However, as stated in Section 10.2.2 [222], a secondary plasma could be generated in the center of the substrate between the main plasma ball and the substrate holder as well as the outer rim of the substrate holder. The substrate holder has a complicated shape and was precoated with diamond film, and the 3-inch Si(100) substrate had been carburized prior to the BEN treatment. As a result, the diamond nucleation occurred both in the center and the peripheral of the substrate. Thus, the position of the secondary plasma depends on the geometry of the reactor chamber,

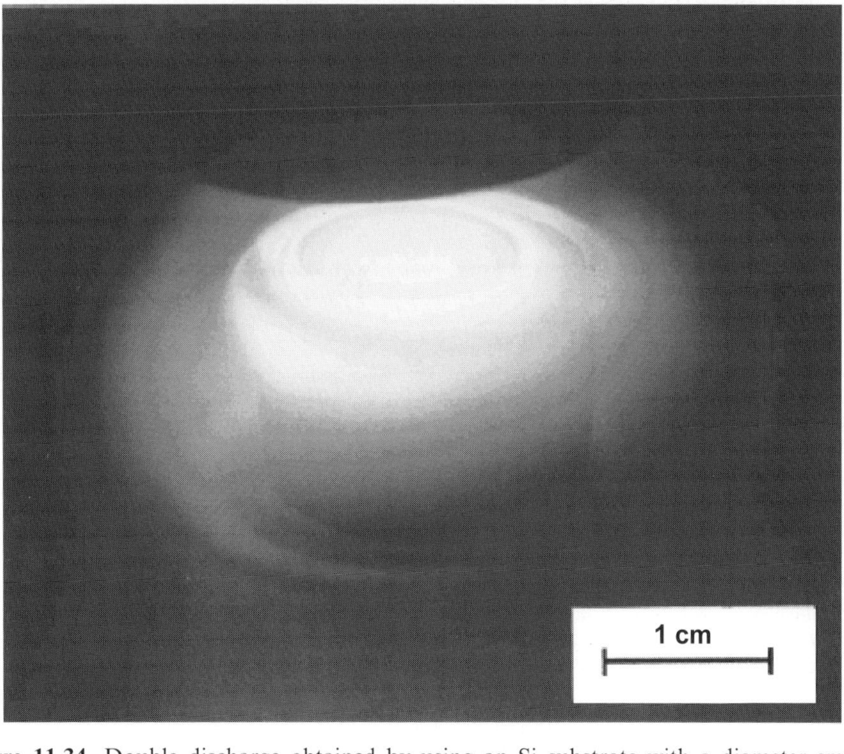

Figure 11.34. Double discharge obtained by using an Si substrate with a diameter smaller than the substrate holder and an additional anode [326].

the position of the anode, the substrate material, the configuration of the substrate holder, and the diamond precoating on the substrate holder.

It should be mentioned here that when the NIRIM-type MPCVD reactor with a biasing capability is employed for the BEN treatment, using a W-wire ring as an anode at the top of the plasma ball (see Figure 3.1), there is no preferential generation of the secondary plasma at the peripheral region of the substrate and the substrate holder, perhaps because the electric field due to the bias voltage is vertical to the substrate surface as the chamber is quartz. Consequently, the nucleation due to BEN is fairly uniform over an Si substrate of at least 1-cm square. Thus, the uniform nucleation by BEN is an advantage of using the NIRIM-type reactor.

Milne et al. [228] synthesized HOD films using the two-step process. The carburization step of Si substrate was not essential, but effective in reducing the period of the BEN step. The (100)-oriented diamond area was formed in a region of a concentric ring that was 25% of the deposited area of a 4-inch wafer.

In Refs. [327, 328], effects of short biasing time of less than 1 min, or even 15 s, on HOD film synthesis were investigated using the two-step process. The process conditions are listed in Table 11.4 [328]. The diamond nucleation density, the homogeneity of nucleation, and the nuclei size distribution were fairly high and homogeneous. The nucleation density was $1\text{–}2 \times 10^9/\text{cm}^2$. After a growth for 16 min, there were multiply twinned (63.8%), single crystal (26.1%), and epitaxially oriented (5.6%) diamond particles on the substrate in addition to undetermined particles (10.1%). The density of epitaxial crystals was approximately $10^8/\text{cm}^2$. A feature of this process is that the bias voltage used, $V_b = -120$ V, is fairly lower compared with the conventional conditions, $V_b = -200$ to -300 V. This is presumably because the gas pressure used, $P \sim 7.5$ Torr, is lower than the conventional values,

Table 11.4. Process conditions used in Ref. [328].

Time	Heating 5 min	Etching 3 min	Bias 30 s	Growth 20 min/20 h
Total pressure (hPa)	13	13	13	13
Microwave power (W)	–	300	300	300
Total flow rate (sccm)	400	400	400	150
Gas composition (vol.%)	100% (H_2)	100% (H_2)	4%CH_4 (in H_2)	1%CH_4 (in H_2)
Bias voltage (V)	0	0	−150	0
Temperature (K)	813	1073	1113	1073
Substrate	Silicon (100) square of $10 \times 10\,\text{mm}^2$			
Substrate-holder	Molybdenum			
Electrode	Graphite $\phi = 30$ mm			

Note that 1 Torr = 133.3 Pa and 1 hPa = 100 Pa.

$P = 20$–30 Torr. It seems, however, that the degree of azimuthal alignment of (100) faces in the diamond film grown for 3 h is significantly lower than that made by using the standard conditions [329].

11.10. *IN SITU* MONITORING

The nucleation of diamond and the successive growth can be monitored by laser reflection interferometry (LRI) [330, 331] using a setup shown in Figure 11.35. This method was first invented by K. Nishimura, and used for a NIRIM-type reactor. In this particular example, it is an ASTeX reactor that was used, but the method can be applied to any type of CVD reactors.

Figure 11.36 shows a typical datum obtained by the LRI. During the BEN pretreatment, the reflection intensity decreased gradually, and once the diamond growth started, it exhibited a periodic change in intensity due to the interference between the incident and reflected (at the Si surface through the diamond film) laser lights. Without the pretreatment, the intensity simply exhibited a periodicity with minor structures. From the period, one can estimate the diamond film thickness. Similar data were obtained in growing polycrystalline diamond films, but the amplitude of the period became smaller with the growth time due to the increased surface roughness.

Similarly, *in situ* measurements of laser reflectivity and laser scattering as well as substrate temperature have been done to determine the time to switch from annealing to diamond growth on Ni [174, 176, 177]. Generally speaking, *in situ* measurements of bias current, substrate temperature, optical interference, and optical scattering will bring useful information on the timing from BEN to growth steps.

In Ref. [332], the BEN treatment was carried out using the conditions listed in Table H.3. After the treatment for 30 min, the diamond deposition was not uniform,

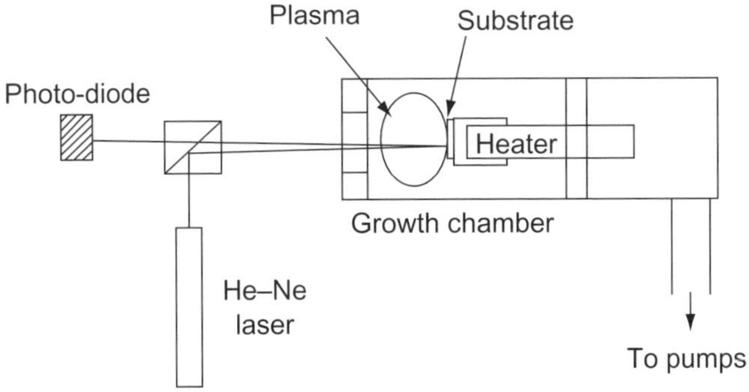

Figure 11.35. Diagram of the MPCVD system and the LRI setup [330, 331].

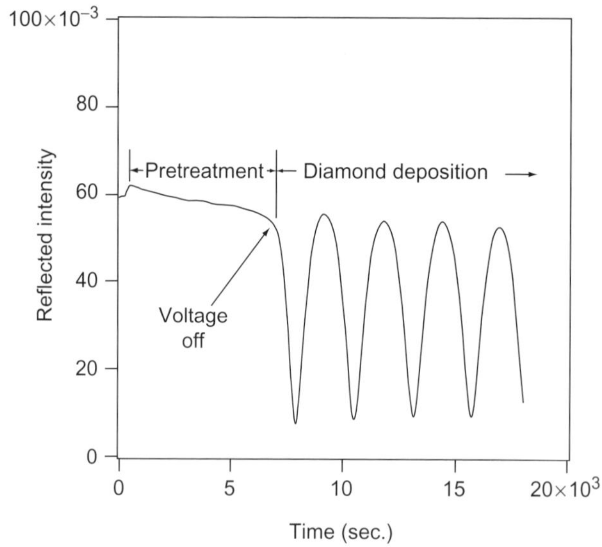

Figure 11.36. Typical datum of LRI showing the interference cycles in the reflected intensity during diamond film growth, and the incubation period at the beginning of the deposited run [330, 331].

and there was a region (at a distance of 4 mm from the center of the substrate) where the optical reflectivity was minimum. This was due to the interference effect between the light scattered from different interfaces present in the nucleating layers and that from the diffuse scattering arising from the surface roughness. Interestingly, the surface features observed by AFM were more regular in this region, and after the diamond CVD to the film thickness of ~2 μm, HOD grains were grown, although the degree of azimuthal alignment was not very good. This experiment suggests that an *in situ* observation of the optical reflectivity is very useful to determine the end point of the BEN step.

11.11. OPTIMIZING BEN CONDITIONS

The influence of the process parameters on BEN was studied in Ref. [331] using an ASTeX reactor. The substrate used was Si(100) of 1-inch in diameter, which was placed on a Mo substrate holder. No carburization was done before the BEN process. The microwave power P_m was 900 W, and the substrate temperature T_s was 725 °C. Laser reflection interferometry (LRI) [330, 331] was used to monitor the diamond nucleation and the subsequent growth rate. The biasing time was determined from the inflection point of the intensity of the LRI. At this point, which was less than 20 min after the plasma ignition, approximately 80% of the substrate was covered with diamond particles. This was followed by diamond CVD for 30 min to facilitate the evaluation of nucleation density. A use of low gas pressure $P < 15$ Torr and low bias voltage $V_b > -150$ V resulted in no nucleation, while high gas pressure $P > 25$ Torr led to a poor uniformity of diamond films. Using the statistical experimental design technique [262], the sweet spot was found to be $P \sim 20$ Torr, $V_b = -320$ V, and $c = 5\%CH_4/H_2$.

In Ref. [333], HOD films were grown on a 1-inch Si(100) wafer using the three-step process, and both the growth parameters and the material properties were investigated in detail. The process parameters were given in Table H.3. The film growth rate was 0.4 μm/h. A well-oriented HOD film was formed in the peripheral areas, more than 4 mm from the center of the 1-inch wafer. A similar result was reported in Ref. [326] that the BEN and the growth of HOD film started from the peripheral region. In the TEM studies of diamond/Si interfaces, the known 2-to-3 registry was observed, but unlike previous studies by Wolter et al. [260], Maeda et al. [267], and Kawarada et al. [250], the carburized surface of Si was not β-SiC but amorphous. Moreover, there was no interlayer between diamond and Si. Regarding the period of BEN, a prolonged carburization of more than 120 min was detrimental for HOD film growth. According to XPF measurements of the HOD films, FWHMs for the tilts about $\langle 110 \rangle$ and $\langle \bar{1}10 \rangle$ axis, and the twist about $\langle 001 \rangle$ axis were 3.66°, 3.72°, and 2.33°, respectively. Unlike the result of Tachibana et al. [334], there was no splitting in the diffraction peaks. In view of orientational degree based on the XPF measurements, these FWHM values are quite small. In terms of the film morphology, however, there is still a large room to improve the flatness and uniformity.

11.12. VARIOUS TECHNIQUES FOR BEN

11.12.1. Cyclic nucleation process

In Refs. [335–337], a cyclic nucleation process was used for the BEN treatment in the three-step process of HOD film synthesis by an ASTeX reactor using Si(100) substrates of 1-cm square. In Ref. [335], the carburization was first done under conditions of $c = 2\%CH_4/H_2$, $P = 20$ Torr, $T_s = 860$ °C, $P_m = 900$ W for 2 h. In the nucleation step, the conditions were $P = 20$ Torr, $T_s = 860$ °C, $P_m = 900$ W, and $V_b = -200$ V. In the growth step, the conditions were $c = 2\%CH_4/H_2$, $P = 25$ Torr, $T_s = 700$ °C, and $P_m = 1000$ W. The cyclic process was used in the nucleation step in such a way that an introduction of CH_4/H_2 for 3 min followed by an introduction of H_2 for 1/2 min was repeated twice. In comparison with the steady state BEN process, the cyclic nucleation process increased the nucleation density by about 6 times ($\sim 6 \times 10^9/cm^2$), reducing the grain size, and increasing the number of (100)-oriented grains. This latter fact was observed by the FWHM of X-ray rocking curve of the (400) diffraction: for the films after a 18-h growth, the FWHM was $\sim 8.6°$ using the cyclic process, better than that of the conventional steady process ($\sim 9.6°$). Furthermore, the uniformity of diamond nucleation was remarkably improved. Thus, the cyclic BEN process improved the degree of oriented nuclei, the uniformity, and the nucleation density. It seems that the H_2 plasma treatment step in the cyclic process has a similar effect to that described in Ref. [338].

It seems, however, that the chamber volume was so large compared with the gas flow rate that the gas in the chamber was not completely refreshed by the cyclic change in the source gas. Hence, the effect mainly arose from the change in the average CH_4 concentration rather than a complete refreshing of the source gas in the chamber.

An HOD film treatment using H^+ ion flux, generated by applying a negative voltage to the substrate in hydrogen plasma [324, 339], was proved to be an effective method to remove misoriented diamond grains and achieve a lateral growth. In Ref. [340], the H^+ treatment was undertaken immediately after the BEN. The treatment removed most twinned nuclei, leaving 30% of oriented nuclei on the Si(100) substrate. The diamond growth resulted in a uniform HOD film morphology over the entire area of 2-inch Si(100) substrate. For this process, a 3-kW ASTeX reactor was used, and the process conditions are listed in Table 11.5.

So far, the BEN process was undertaken using a DC bias voltage. It is of great interest to investigate how HOD films are grown by using an alternating current (AC) bias voltage. Such works have been done in Refs. [341–343] using the two-step process. In Ref. [341], an ASTeX reactor was modified so that a 50-Hz

Table 11.5. Process conditions for H^+ ion treatment. Step 1 is the BEN process, Step 2 is the H^+ treatment, and Step 3 is diamond CVD [340].

Step	1	2	3
CH_4 flow rate (sccm)	15	0	13
CH_4 flow rate (sccm)	0	0	2
H_2 flow rate (sccm)	285	500	285
Microwave power (W)	800	1000	1200
Substrate temperature (°C)	740	800	720
Pressure (mbar)	20	30	40
Bias voltage (V)	−150	−140	0
Time (min)	20	30	30–900

1 mbar = 0.76 Torr.

AC voltage was applied to the substrate. The bias voltage was actually applied between the stainless steel chamber and the graphite susceptor, and the effects of AC and DC bias voltages on film morphology were compared. The process conditions are shown in Table 11.6.

The nucleation process was monitored by LRI (see Section 11.10), and the application of bias voltage was terminated if diamond nucleation was detected by LRI. No appreciable diamond nucleation was observed in the AC-BEN treatment, if the voltage was less than 125 V_{RMS} (±175 V peak to peak). The optimum biasing time for the AC-BEN treatment was 45 min, much longer than the DC-BEN treatment that needed only 12 min. In terms of the number of oriented nuclei, it was more than 50% for the AC-BEN treatment, while it was only less than 10% for the DC-BEN treatment. It was hence inferred that concurrent processes of forming an epitaxial SiC layer and the diamond nucleation, induced by the AC-BEN treatment, were responsible for the increase in the number of oriented diamond nuclei.

In Ref. [342], a 200-V, 60-Hz AC bias voltage was applied to the substrate for 20–40 min. In the ED pattern of the specimen with a 20-min bias application, the presence of β-SiC crystallites were identified, where β-SiC crystallites were heteroepitaxially oriented with slight (±5°) rotations about the ⟨100⟩ axis. Among the biasing times studied, 25 min was the best to form heteroepitaxially oriented β-SiC, and hence best-aligned HOD films. In Ref. [343], two series of the process conditions were employed, as shown in Table 11.7. The AC bias was 200 V and 60 Hz. The orientation of (100) faces in the HOD films was very sensitive to the duration of BEN, and the best results were obtained when the biasing time was 25 min in Series 1 experiments, and 20 min in Series 2 experiments, where 90% of the diamond grains were azimuthally oriented. TEM and TED observations indicated the presence of heteroepitaxially oriented β-SiC, suggesting that β-SiC is important for diamond heteroepitaxy. It is so far not elucidated, however, how AC-BEN is different from

Table 11.6. Process conditions for AC and DC biasing [341].

Parameters	BEN AC source	BEN (common)	BEN DC source	Diamond deposition
V_b (V)	±210		−210	0
I_b (mA)		10–20		0
c (%)		2		1.5
P (Torr)		15–20		30
P_m (W)		700		700
T_s (°C)		850		750–800
Biasing time (min)	45		12	

Table 11.7. Process conditions for AC-BEN [343].

Parameters	BEN	Growth
(*Series 1*)		
CH_4/H_2 (%)	4	2
AC (60 Hz)		
Biasing time (min)	20, 25, 30, 40	1200
(*Series 2*)		
CH_4/H_2 (%)	8	2
AC (60 Hz)		
Biasing time (min)	15, 20, 25, 40	1200
P_m (kW)	1.0	1.5
P (Torr)	20	30
T_s (°C)	845	820

Table 11.8. Process conditions for synthesizing (100)-oriented diamond films [344, 345].

Process step	1	2	3	4	5
CH_4 concentration (%)	0	1.0	1.3	1.7	1.7
N_2 concentration (ppm)	5	50	60	60	0.5
Substrate temperature (°C)	810	860–880	750	830	830
Bias voltage (V)	0	−(0–230) pulsed	0	0	0
α-value	–	–	2.8–3.0	2.3–2.5	–
Time	5 min	17 min	20 h	40 h	45 min

the standard DC-BEN, and thus a further study will be necessary to establish the advantage of AC-BEN.

11.12.2. Repetitive pulse bias for nucleation

In Refs. [344, 345], repetitive pulse bias was used for BEN (RP-BEN). The whole process to grow an HOD film consists of 5 steps, as shown in Table 11.8. The reactor was a custom-made MPCVD system with a maximum output of 6 kW. The pulse voltage was −220 V, and the optimum pulse was 5 Hz. As a result, oriented nucleation with a nucleation density of $8 \times 10^8/\text{cm}^2$ was achieved over the substrate area up to 30 cm^2. An HOD film was grown over a 20-cm^2 area. For a 60-μm thick HOD film, the FWHM of (111)XPF peaks was 2.8° for the tilt angle and 5.5° for the rotation angle. The tilt angle was large. Since this is related to the control of α-parameter during the film growth (see Table 11.8), not only the BEN step but also the film growth step must be properly controlled and optimized to obtain high quality HOD films.

11.13. LATERAL GROWTH

In the (100) HOD film synthesis, (100)-oriented growth is performed after the BEN step under conditions of $\alpha \sim 3$ (see Section 5.5) so that the film surface consists of pyramid shape diamond grains, most of which are azimuthally (100)-oriented. To form (100) faces, the (111) faces of the pyramids must be grown faster than (100) faces using CVD conditions of $\alpha \sim 1$. Consequently, the (111) faces of the pyramids disappear, and the film surface is filled by (100) faces, resulting in an HOD film. More practically, since it is difficult to set up conditions of $\alpha = 1$ and 3, conditions of $\alpha \sim 1.5$ and ~ 2.5 are used in the actual experiments. As stated in Section 5.5, growth parameters of c, T_s, and P must be controlled to attain the these α-parameter values. Addition of B, N, and O also influences the α-parameter, as stated before.

11.14. SUPPRESSION OF SECONDARY NUCLEATION

In Ref. [86], HOD films were grown using the three-step process under conditions listed in Table H.3. In the growth step, however, grain boundaries and steps between adjacent (100) faces were still visible. A further extension of the growth time led to an occurence of secondary nucleation. To solve this problem, the source gas was changed to $4\%CH_4 + 1\%CO_2$, diluted with H_2, and a higher substrate temperature, $T_s = 900\,^{\circ}C$, was used. After a growth for 10 h, the surface became fairly continuous, markedly reducing the grain boundaries and steps to make a better-coalesced surface. The film thickness was 20–30 μm. Thus, the time needed to achieve a significantly coalesced film was greatly shortened as compared with the past work [84]. Just like the planarization described in Section 11.13, a control of the α-parameter is also necessary to avoid the secondary nucleation.

11.15. EFFECTS OF ADDITIVES

It is well known that nitrogen can be easily incorporated in diamond, and usually, it must be very careful of the air leak in the CVD chamber to prevent nitrogen contamination of diamond films. It is also known that addition of too much nitrogen to the source gas resulted in microcrystalline diamond films rather than well-faceted films. However, Koidl's group found [90] using a NIRIM-type reactor that an addition of small amount of N_2 gas to the source gas of $c = 1$ $2\%CH_4/H_2$ resulted in pronounced ⟨100⟩-textured films, while the film became microcrystalline without N_2 addition. More specifically, an addition of 40–200 ppm of N_2 to the source gas of 0.5–2%CH_4/H_2 led to smaller (100) tilt angles and stabilized the (100) orientational growth. An example is an addition of 60-ppm N_2 in 1.5%CH_4/H_2, which resulted in a (100)-faceted, ⟨100⟩-oriented diamond film. By contrast, the film became microcrystalline without the addition of N_2. Moreover, the nitrogen addition resulted in a better film quality as evidenced by Raman spectroscopy. The relationship between the tilt angles and the N_2 concentration is depicted in Figure 5.21.

Effects of N_2 addition to the source gas on film morphology were studied in Refs. [92, 93, 346] for the growth of polycrystalline diamond films. An addition of 200 ppm N_2 in $c = 1\%CH_4/H_2$ resulted in the appearance of (100) faces at the film surface. Also, as stated previously, by using MPCVD, the growth rate increased 5 times by adding 25 ppm N_2 to the source gas [92], and the growth orientation of ⟨114⟩ was preferred [93]. Similarly, it is known that an addition of B_2H_6 to the source gas of CH_4/H_2 tend to form (111)-faceted diamond films, when the CH_4 concentration is low such as 0.5%. However, (100) faces appear at the film surface, when the CH_4 concentration is high such as 2–3%. In both cases of N_2 and B_2H_6, it has not been understood yet why (100) morphology or (111) faces are created under such conditions. Practically, however, these methods are useful to synthesize better quality HOD films.

In Ref. [347], CO_2 was added up to 3000 ppm to the source gas during the BEN treatment. However, the in-plane orientation of the film determined by X-ray diffraction was deteriorated as the CO_2 concentration was increased. Thus, the addition of oxygen in the BEN process gives a negative effect on oriented diamond nucleation.

In Ref. [348], the oriented growth of diamond film was optimized using the two-step process by changing the source gas composition during the growth. The reactor used was an ASTeX-type MPCVD reactor. The conditions for the BEN step were $c = 4\%CH_4/H_2$, $P = 20$ Torr, $P_m = 400$ W, $T_s = 927\,°C$, which was continued for 90 s. This was followed by the first growth step using $c = 0.5\%CH_4/H_2$, $T_s = 850\,°C$ for 30 min. In the growth steps 2 and 3, combinations of CH_4, CO, H_2, and N_2 were used. The best result was obtained using $c = 1\%CH_4/H_2 + 180$ ppm N_2 for 10 h, followed by $c = 10\%CO/H_2$ for 4 h. Optimizing the growth step is really a laborious work, but this work demonstrated that it is rewarding.

11.16. (111)- AND (110)-ORIENTED GROWTH ON Si(111) AND Si(110) SURFACES

In Ref. [288], oriented diamond films were deposited by the two-step process on Si(100), (111), and (110), surfaces by MPCVD using an ASTeX 1.5-kW reactor.

The source gas was $c = 0.5\% CH_4/H_2$. The bias voltage was $V_b = -100$ to -300 V. During the growth step, $P_m = 600$–1000 W, and $T_s = 800$ °C. The diamond growth was continued to make 4- and 100-μm thick diamond films. On the Si(100) surface, a HOD film was formed with diamond (100) faces of approximately 2 μm square. For the 100-μm thick diamond film, the FWHM of the (400) X-ray rocking curve was 5°. The films and the orientational relationship with the Si(100) substrate are illustrated in Figure 11.37. The diamond film grown on Si(111) surface was (111)-oriented, where the [110] edges of the (111) faces were mutually parallel between diamond grains. Secondary nucleation, however, was significant at the film surface, presumably because the growth conditions were not best optimized for the ⟨111⟩-growth. For the case of diamond films grown on Si(111) with a low nucleation density, there were (111)-oriented crystals that were inversely directed. One might think that such (111)-oriented heteroepitaxial films should have a number of twins due to antiphase domains. As is seen later in Section 12.1, however, the antiphase domain structures can disappear at a certain stage for unclarified reasons, and all (111) faces are oriented in the same direction. It is thus possible that a similar situation could take place for the (111)-oriented diamond films on

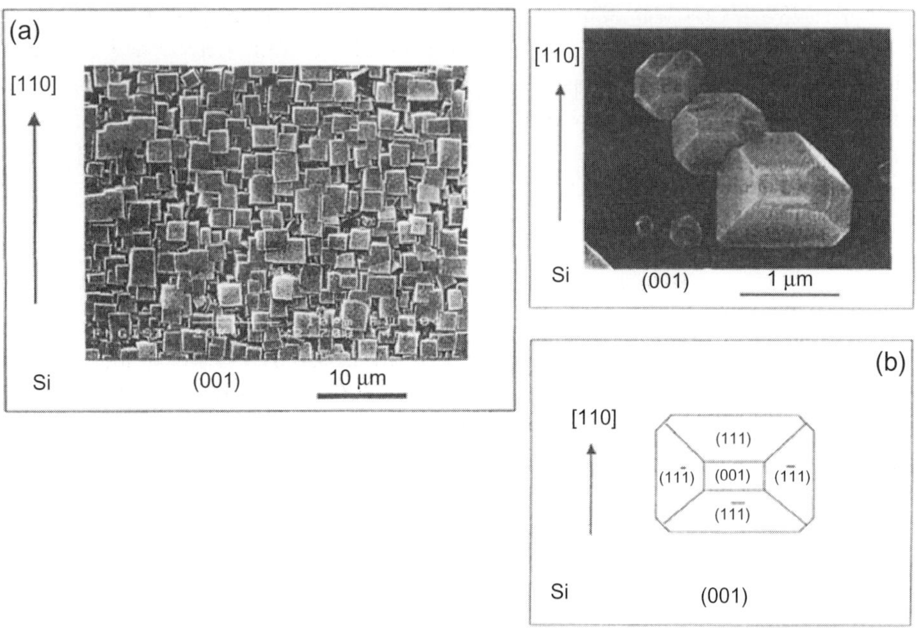

Figure 11.37. SEM image of (a) HOD film and oriented crystals grown on Si(100). The diagram (b) shows the orientational relationship of diamond with Si(100) [288].

Si(111). The FWHM of the (111) X-ray rocking curve for the 4-μm thick diamond film was 7°, which is smaller than that of (100), i.e. 12°, and the orientational relationship with the Si(111) substrate was as follows (see Figure 11.38): D{111} // Si{111} and D$[\bar{1}10]$ // Si$[\bar{1}10]$.

Finally, for the case of diamond growth on Si(110), a (110)-oriented diamond film was grown on the Si(110) substrate with an orientational relationship: D{110} // Si{110}, and D$[\bar{1}10]$ // Si$[\bar{1}10]$ (see Figure 11.39). This study is very encouraging as one can grow (100)-, (111)-, and (110)-oriented diamond films on Si(100), (111), and (110) substrates, respectively, and it seems that a further study is necessary to optimize the growth procedures and conditions.

A similar result was obtained in Ref. [284] that (111)-oriented diamond particles were azimuthally aligned in such a way that the [110] edges of many diamond (111) faces were parallel to the [110] direction of the Si(111) substrate. This article also contains the works of Schreck and Stritzker's group up to 1996.

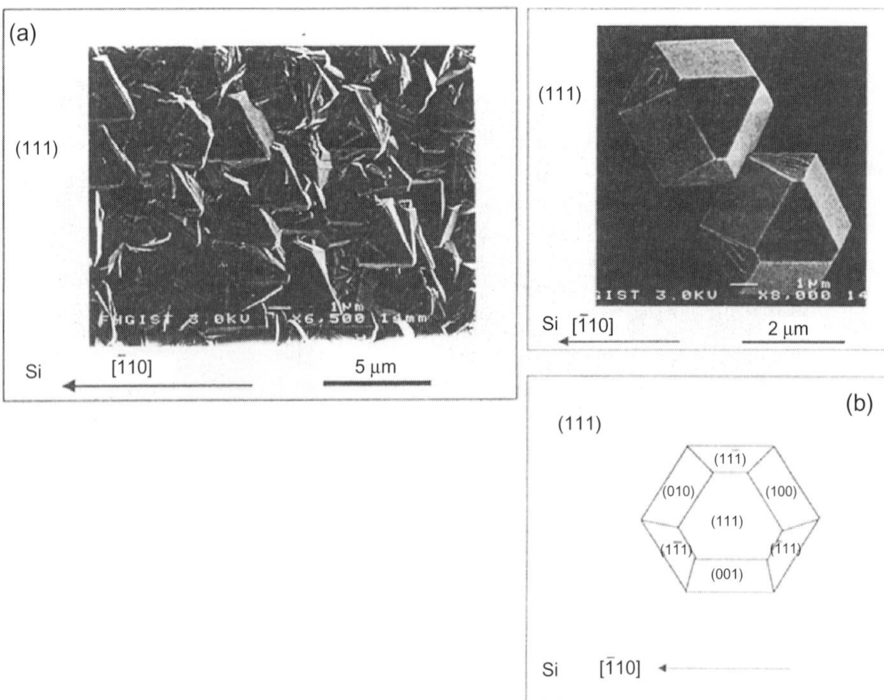

Figure 11.38. SEM image of (a) (111)-oriented film and oriented crystals grown on Si(111). The diagram (b) shows the orientational relationship of diamond with Si(111) [288].

The possibility of (111)-oriented diamond growth on 6H-SiC(0001) has been suggested in Ref. [318]. Indeed, a growth of diamond on α-SiC(0001) was undertaken in Ref. [349]. The substrate was heated in H_2 at 1200 °C for about two minutes, and the diamond growth was done by MPCVD under conditions of $c=50\%CO/H_2$, $P=30$ Torr, and $T_s=900$ °C for 12 h. As a result, 5–20 μm size diamond crystals were grown, in which some crystals were 180° rotated with respect to the others. The orientational relationship of diamond with α-SiC(0001) was as follows: α-SiC{0001}//D{111} and α-SiC$[11\bar{2}0]$//D$[1\bar{1}0]$ So far, only discrete diamond crystals were grown, and it would be of great interest to synthesize a continuous film. See also Section 11.6 on this topic.

In Ref. [267], a BEN treatment was done using the conditions listed in Table H.3, and consequently, pyramidal particles were deposited on Si(111) surface. Such a feature had also been observed for diamond deposition on cBN(111), and hence this is a common feature for the initial stage of (111)-oriented growth.

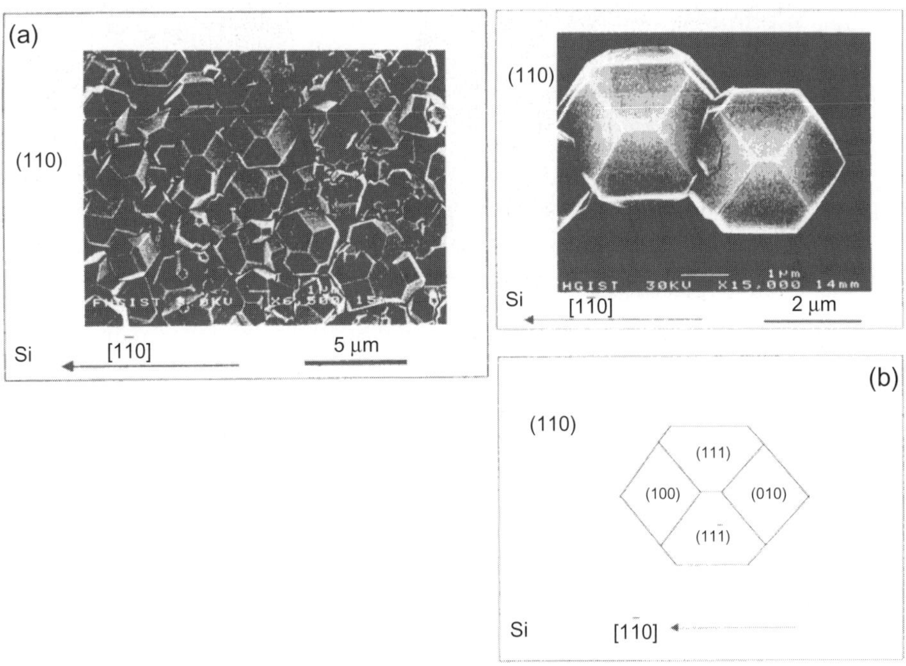

Figure 11.39. SEM image of (a) (110)-oriented film and oriented crystals grown on Si(110). The diagram (b) shows the orientational relationship of diamond with Si(110) [288].

11.17. RECENT PROGRESS

Apart from the work of Refs. [6, 252], one of the best HOD films in recent years is presented in Ref. [338]. The growth method is the two-step process, applying BEN directly on a Si(100) wafer. The process conditions by MPCVD (using an ASTeX reactor) are listed in Table 11.9, and the SEM images of the film are shown in Figure 11.40. An important point of this process technology is to periodically use H^+-ion etching in addition to optimized growth conditions. This is in contrast to the conventional HOD film growth method in which the lateral growth method (see Section 11.13) is used to expand the (100) faces by choosing an appropriate α-parameter. The effects of H^+-ion etching are described by Jiang et al. in Refs. [324, 339, 340]: an H^+-ion flux etches (111) faces faster than (100) faces and hydrocarbon species generated by etching diamond redeposit on the side surfaces of (100) faces, thus increasing the (100) area, and removing misoriented grains.

The same method was transferred to HFCVD, where the process conditions, listed in Table 11.10, were used. The substrate was a 2-inch Si(100). Figure 11.41 shows a SEM image of the film surface grown by HFCVD. The diamond deposition was done for 70 h and the film thickness was 140 μm. It is obvious that the coalescence between adjacent (100) faces developed significantly, and most (100) faces have edge lengths as long as 30 μm. The FWHM of the X-ray rocking curve was 3.9°.

11.18. CHARACTERIZATION OF HOD FILMS

11.18.1. Polarized Raman spectroscopy [350]

The in-plane orientations of both HOD film and Si(100) substrate were investigated by polarized Raman spectroscopy in Ref. [263], and the results are shown in Figure 11.42. Define ϕ as the angle between the polarization of the laser light and

Table 11.9. HOD growth conditions using the two-step process [338].

Process	H_2 concentration (%)	CH_4 concentration (%)	Gas pressure (mbar)	Substrate temp. (°C)	Bias voltage (V)	Power (W)
Plasma etching[a]	100	0	25	850	0	900
Nucleation[a]	99	2	20	850	−150	900
Growth	98.5	1.5	25–30	700–780	−150 to +150	900
Bias-assisted H^+ etching	100	0	30	780	−150	1000

[a]Plasma etching for cleaning the substrates lasts 20 min; the nucleation time is 15 min.
1 mbar = 0.76 Torr.

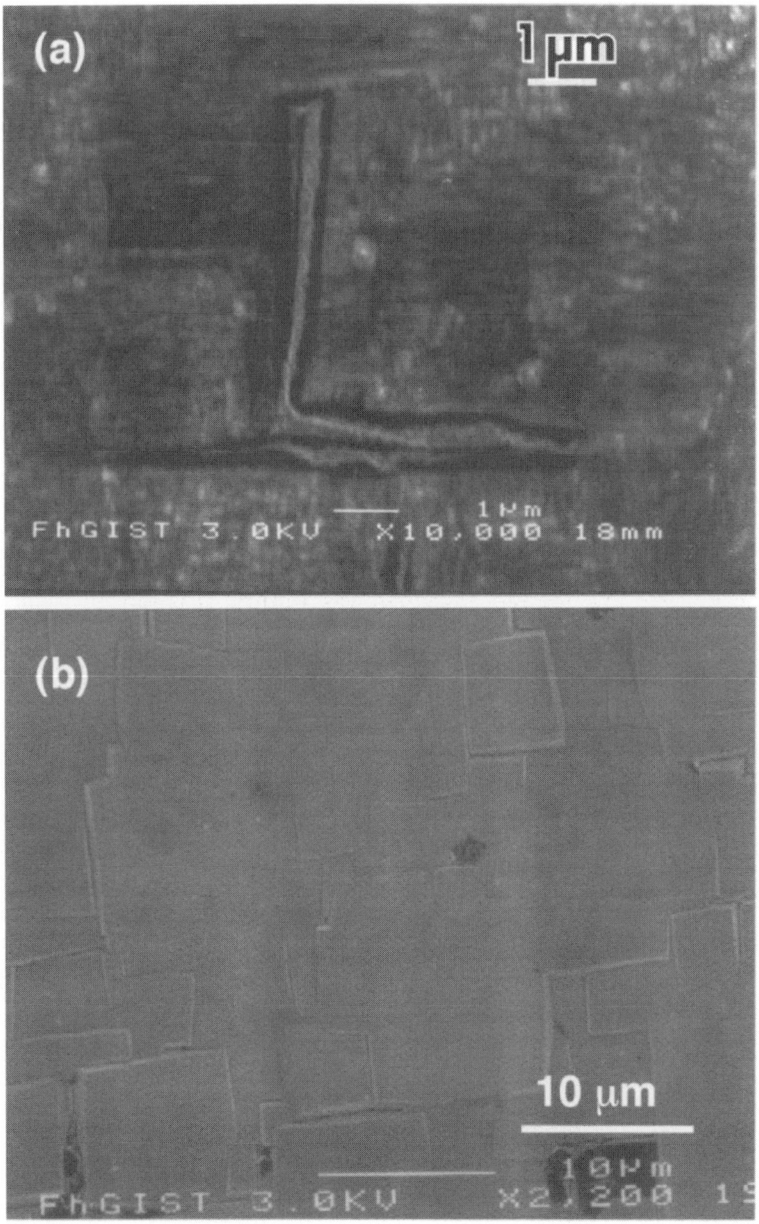

Figure 11.40. SEM images showing (a) (100) diamond film surfaces after H^+ etching and (b) a growth step without biasing the substrate [338].

the azimuthal angle of diamond (100) face, then the angular Raman intensity distribution of the Si substrate was expressed that $I \sim \cos^2(2\phi)$, while that of the HOD film was expressed that $I \sim \cos^2(2\phi) + B$, where the constant B is originated from the minor portion of diamond crystals with no common orientation or texture, and the depolarization effects arises from the scattering by grain boundaries and the background signal.

Polarized Raman spectroscopy of HOD films was also studied in Refs. [228, 351, 352]. The HOD film was grown by the three-step process under conditions listed in Table H.3 [351]. The angular dependence of polarized Raman spectroscopy was measured for both the 1332-cm^{-1} line of the diamond film of 40-μm thickness and the 515-cm^{-1} line of the Si(100) substrate. The results are shown in Figure 11.43.

Table 11.10. HOD growth conditions using the HFCVD system [338].

Filament (Ta) diameter	1 mm
Filament temperature	2100–2200 °C
Bias voltage	−275 to −325 °C
Filament–substrate distance	8–10 mm
Gas pressure	15–40 mbar
Hydrogen flow	1000 sccm
Methane flow	8–12 sccm

1 mbar = 0.76 Torr.

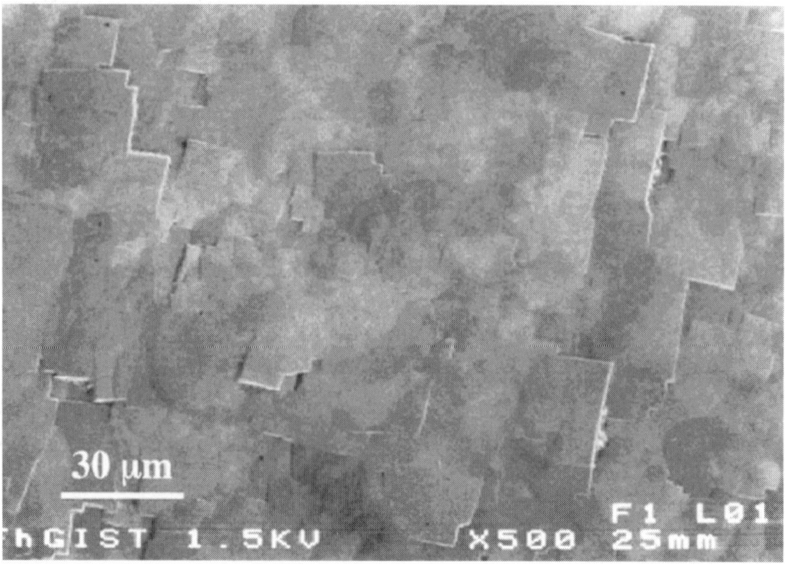

Figure 11.41. SEM image of a 140-μm thick HOD film synthesized by HFCVD [338].

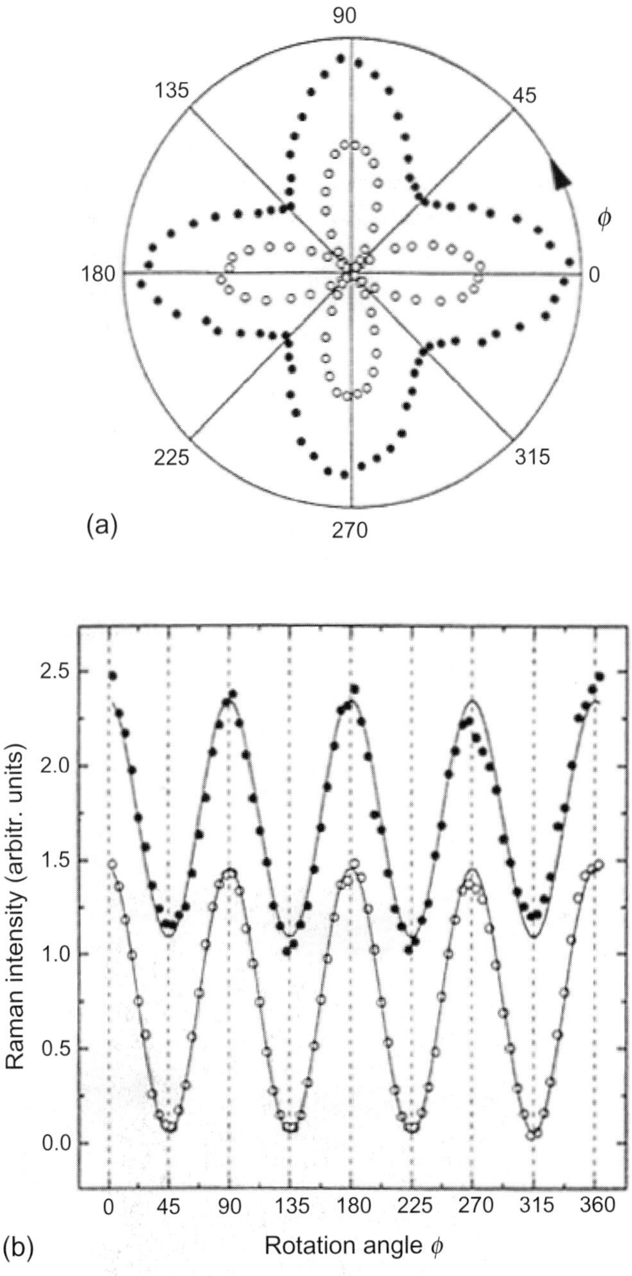

Figure 11.42. (a) Polar plot and (b) linear plot of Raman intensities of diamond (solid circles) and the Si substrate (open circles) as a function of the rotation angle ϕ. The solid curves in (b) are the calculated $\cos^2(2\phi)$ behavior [263].

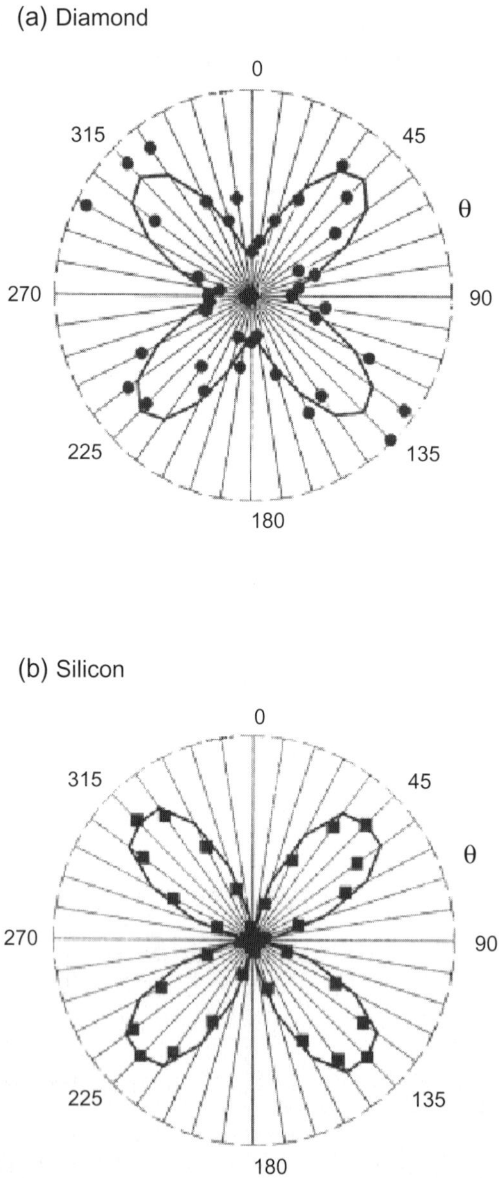

Figure 11.43. Polar plot of the Raman intensity of (a) a single diamond crystallite, and (b) the Si substrate as a function of the rotation angle θ. The solid curve is the calculated $\sin^2(2\theta)$ dependence for the adopted scattering geometry [351].

In both cases, a four-fold symmetry is obvious. There were, however, scattering centers even within the ⟨100⟩-textured and (100)-oriented diamond grains, which caused a significant depolarization in the Raman spectrum of the HOD films. According to P. John [351], there were a few key points in the observed data of Figure 11.43. First, with high optical quality layers, one could observe the Raman signals of both silicon and diamond at the same time at the same position. Thus, the figure shows the alignment of the diamond with the underlying silicon. Second, the diamond crystallites had a range of tilt angles, and therefore the diamond 'clover leaf' was slightly asymmetric because the laser was always focused on a single diamond crystallite as the laser plane of polarization was altered (in fact the sample was rotated). Finally, it was not a confocal arrangement, and the Raman signal came from the entire depth of the diamond. These facts explain why the origin of the diamond signal does not go to zero (as it does in silicon).

11.18.2. Confocal Raman spectroscopy

The stress distribution in HOD films was studied in Ref. [353] by polarized confocal micro-Raman spectroscopy using a 514.5-nm Ar^+ ion laser. The process conditions for the HOD films are listed in Table H.4, and the SEM image is shown in Figure 11.44. This paper is so comprehensive, and includes a number of relevant references so that the readers are recommended to read it to know the details of the experimental and analysis techniques. The results obtained are listed below:

1) The stress distribution in diamond film is basically compressive, and rapidly decreases at 10–15μm away from the diamond/Si interface.

Figure 11.44. SEM image of the HOD film [353].

2) Near the interface, a biaxial stress exists in the diamond film, while near the growth surface, the stress tensor is more complicated.
3) Grain boundaries are the major source of the intrinsic stress.
4) The micro-Raman spectra of several locations in a single grain give significantly different spectra, as shown in Figure 11.45. Some spectra have a splitting presumably due to non-uniform local stress. Also, some have a peak below the reference position (1332 cm^{-1} for type IIa diamond single crystal without stress), meaning that a tensile stress is locally present.
5) There exists a significant variation in the micro-Raman spectra in depth, as shown in Figure 11.46. Each spectrum can be deconvoluted by Lorenzian band shapes, where the peak positions were located at 1332, 1335, and 1341 cm^{-1}.

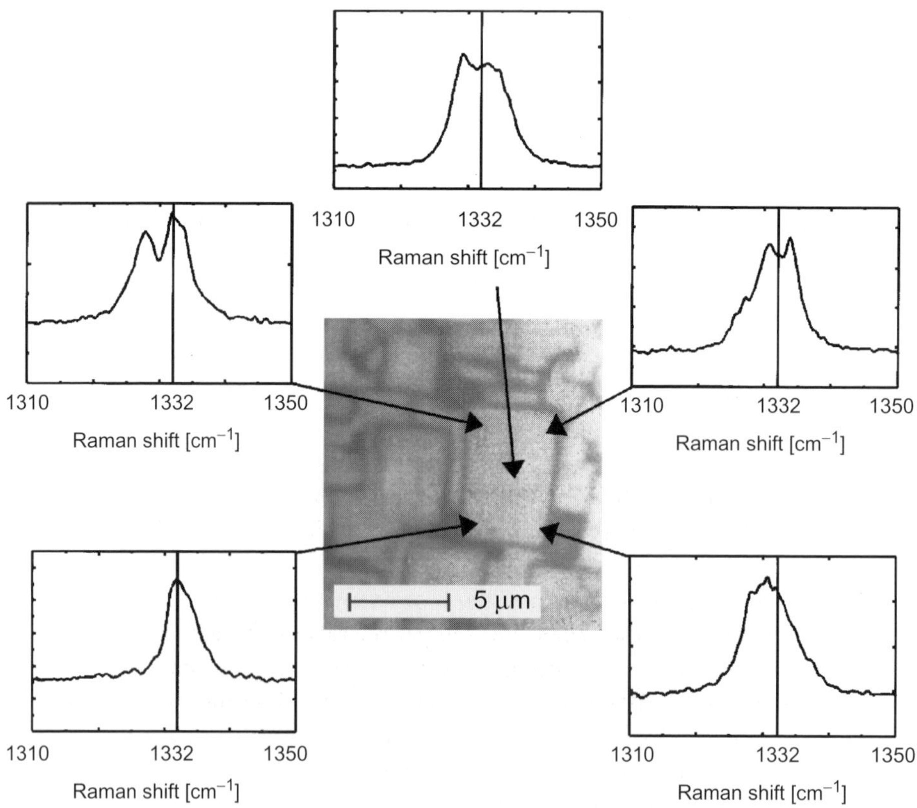

Figure 11.45. Raman spectra of the same diamond grain at the surface of a 10-μm thick HOD film. The spatial and depth resolutions are <1 μm and about 2 μm, respectively. The spectral resolution is <0.2 cm^{-1} [353].

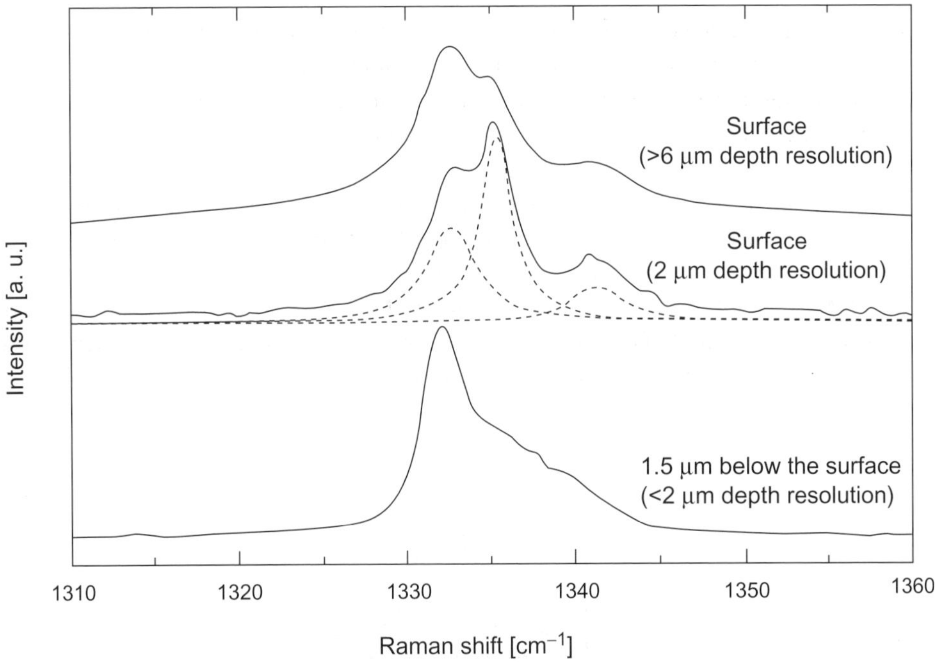

Figure 11.46. Effect of the depth on the micro-Raman spectrum. The dashed curves represent the Lorenzian deconvolution (the FWHMs ~3 cm^{-1}, and the peak positions are 1332.53, 1335.32, and 1341.21 cm^{-1}) [353].

The change in the depth causes the change in the band intensities but not in the band peak positions. The corresponding compressive stresses were evaluated to be 0, 1.7, and 5.3 GPa, respectively.

6) For a 47-μm thick free-standing diamond film, the Raman peak shifted from the interface to the growth surface, as shown in Figure 11.47.
7) There exists a stress variation in the radial direction, as shown in Figure 11.48. The stress at the position 4.5 mm away from the center is approximately 30% greater than that in the center. This is attributed to the difference in the CVD conditions between the center and the peripherals of the substrate.
8) From the Raman band width, it is evaluated that the standard deviation of the stress is ±2 GPa, which is five times greater than the mean stress.
9) According to the theoretical calculation, the intrinsic stress at the diamond–Si interface is tensile, and not compressive. Such a tensile stress may be relaxed by plastic deformation of the Si substrate.

A similar study was further carried out by the same group [354]. In the Si substrate surface layer after diamond CVD, there were a number of dislocation loops, and the Si substrate was indeed subject to a plastic deformation. For a diamond layer of 300-nm thickness, there were numerous twins and stacking faults within the grains. For an HOD film of 12-μm thickness, synthesized by the two-step process, the {111} growth sectors, which were the side walls of (100) faces, disappeared by the coalescence with the adjacent (100) grains. When two {100} growth sectors merged during the diamond growth, low angle-grain boundaries were formed, and then the {111} growth sectors disappeared. According to Raman spectroscopy of a 12-μm thick HOD film, the {100} growth sector of a single grain with a (100) face had only the diamond line at $1333\,cm^{-1}$, but the {111} growth sector at the corners of the grain had both G and D bands (see Figure A.2). These bands originated from grain boundaries associated with the ⟨111⟩ growth. Also, within the same grain that had a (100) face, there was a distribution of internal stress of +0.2, −3.4, and +1.4 GPa.

The standard method of confocal Raman spectroscopy is presented in Ref. [355] using HOD films grown by the three-step process. The process conditions are listed in Table H.3. The film thickness was approximately 0.8 μm. In the Raman spectra of

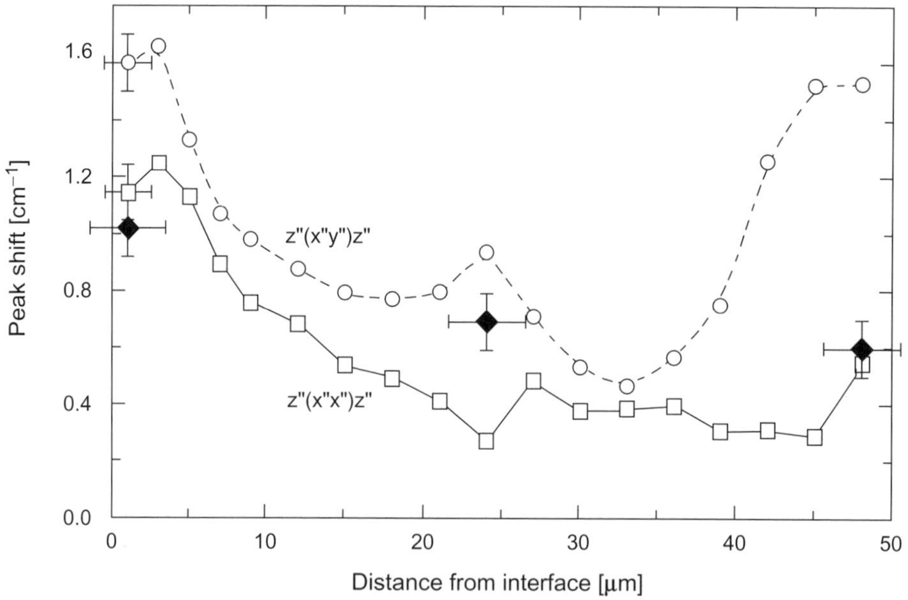

Figure 11.47. Raman shift of diamond peak as a function of the distance from the interface for three measurement geometries. The solid points represent the backscattering measurement data from the (100) growth surface, while open squares and circles represent the backscattering measurement data from the (110) cross section [353].

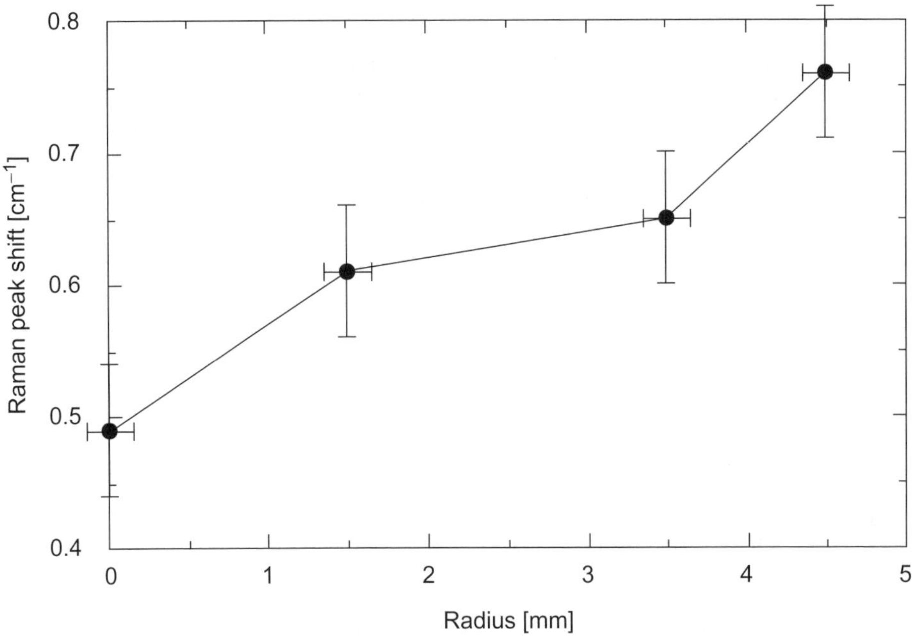

Figure 11.48. Radial stress variation determined by micro-Raman spectroscopy for a circular specimen [353].

the HOD film, there existed a new line at 1210 cm^{-1}, when the focusing locations were 0.6 and 0.8 μm deep from the surface, i.e. the 1210 cm^{-1} band was intense near the diamond/Si interface. The origin of the band was not yet known, but could be attributed to random networks of tetrahedrally coordinated carbons or small diamond crystallites. By contrast, the 1210-cm^{-1} band was not detected in randomly oriented polycrystalline diamond films.

11.18.3. Infrared absorption spectroscopy

A study of FT-IR spectroscopy on HOD films was carried out by Milne's group in Ref. [329]. The TO mode of β-SiC was intense and observed at 800 cm^{-1}. The observed spot was radially scanned across a 100-mm Si(100) wafer to know the SiC thickness variation in the processes of carburization, BEN, and growth. After the carburization, the SiC thickness (actually the FT-IR intensity) was greater as the observed position was away from the wafer center. On the contrary, by the BEN treatment, the intensity decreased toward the center, and significantly increased in the region where HOD film was formed. In the growth step, the spectral intensity dramatically decreased near the center of the wafer. This result indicated that an

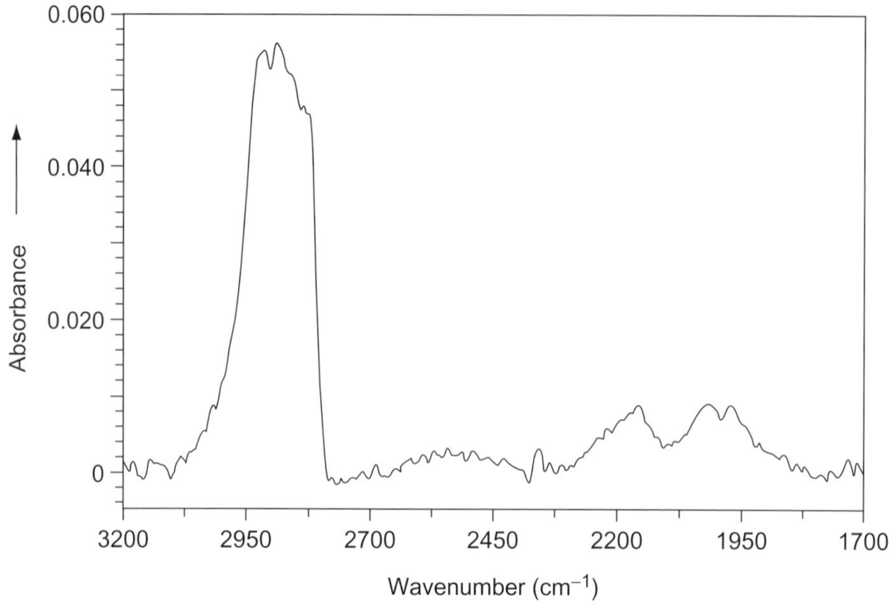

Figure 11.49. FT-IR spectrum of a diamond film in the C–H vibrational region. The band around 1700–2700 cm^{-1} is the two-phonon band [352].

appropriate β-SiC layer in thickness, epitaxy, and quality are needed to later grow a well-oriented HOD film. In the growth step, the β-SiC layer became thinner than in the BEN step, and indeed, the thickness of the SiC layer was estimated to be only a few nanometer. However, the thickness of SiC was greater in the regions where the azimuthal alignment of (100) was lost. The average β-SiC thickness was evaluated to be approximately 5 nm. It is known that in the range between 1700 and 2700 cm^{-1}, there exist a band at 2990 cm^{-1}, due to C-H vibrational modes, as shown in Figure 11.49. Works of Milne's group up to 1996 is very well summarized in Ref. [352].

11.18.4. Cathodoluminescence (CL)

Scanning CL spectra at 77 K of an undoped and a B-doped HOD films, which were selectively overdeposited on an undoped HOD film to 4-μm thickness, were studied in Ref. [356]. For the undoped film, the CL spectrum consisted of "blue band A" at 438 nm and "green band A" at 474 nm. Furthermore, sharp bands due to nitrogen were seen at 531, 552, 618, and 672 nm. On the other hand, the B-doped (5×10^{17}/cm^3) film exhibited a single blue band A at 435 nm. In this case, a weak free-exciton band also was detected at 242 nm. It is of interest that for the

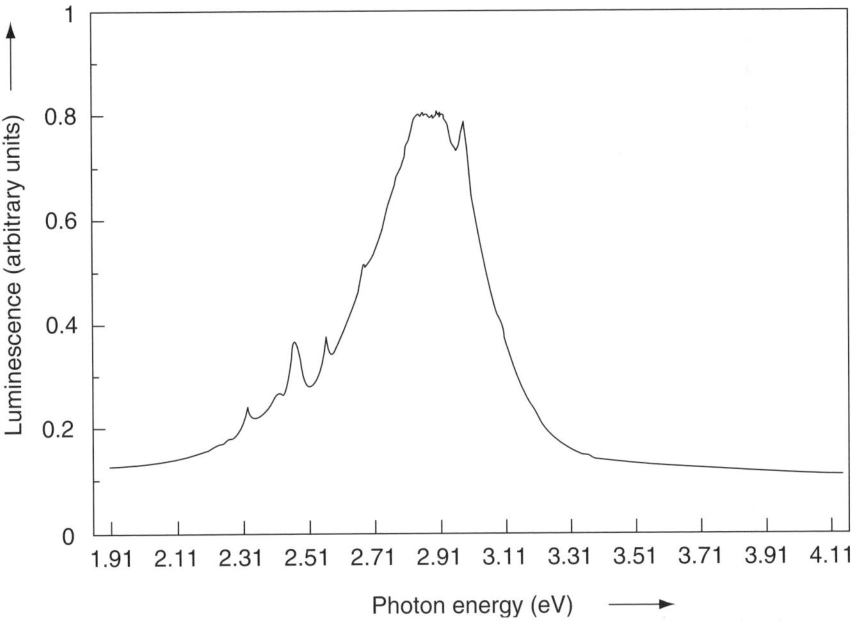

Figure 11.50. CL spectrum of diamond film with a broad "band A" at 2.9 eV [352].

undoped HOD film, there was a clear correspondence between the SEM and the CL images. By contrast, the image correspondence was very poor for the HOD film with a B-doped layer. Since the blue band A of the B-doped HOD film originated from grain boundaries, defects were considered to be the source of the blue band A. A cathodoluminescence spectrum of an HOD film was also measured in Ref. [352], and only band A was observed, as shown in Figure 11.50. The peak position was 2.9 eV. There were no peaks due to Si (1.681 eV) and N (2.156 and 3.188 eV).

11.18.5. Electron Energy Loss Spectroscopy (EELS)

In the energy loss spectrum due to plasmon, it is known that plasmon loss by Si is located at 17 eV, π^* plasmon losses due to graphitic carbon at 6 eV, SiC at 20 eV, and graphitic carbon at 22 eV. In Ref. [357], the plasmon bands were observed by high-resolution photoelectron energy loss spectroscopy (PEELS), and the energy-filtered images of the HOD/Si interfaces were observed. Consequently, a thin SiC layer existed at the entire interface, while *a*-C layer existed only in limited regions. An inclusion of non-diamond carbon in the {111} growth sectors of diamond was responsible for the high defect densities and the high compressive stress.

11.18.6. Thermal conductivity

The thermal conductivities of an HOD film and uniaxially aligned (111)-faceted diamond film were measured in Ref. [358]. The films were synthesized using the AC-BEN method (see Section 11.12), and the thicknesses were approximately 70 μm. The thermal conductivity of the HOD film was $\sim$1120 W/m · K, roughly twice as high as that of the uniaxially aligned film, $\sim$550 W/m · K. Note that the thermal conductivity measurements were done along the film direction. The higher thermal conductivity for the HOD film was ascribed to a lower density of grain boundaries.

11.19. SELECTIVE DEPOSITION

Since HOD films have flat surfaces much smoother than ordinary polycrystalline diamond films, it is expected that HOD films will be used for electronic sensors and devices as well as micro electro-mechanical systems (MEMS) in the future. To this end, a selective deposition of (100)-oriented HOD films was undertaken using the two-step process on a 2-inch Si(100) wafer, and a 0.5-μm thick SiO_2 film was used for a masking film [295]. It was demonstrated that a micropattern of 6-μm thick HOD films was possible, and even a 8-μm wide channel was also formed. Selective deposition using polycrystalline films has been studied before [359], but this work demonstrated that a similar fabrication is possible using HOD films.

11.20. MODEL MECHANISMS

There are a number of mechanisms for oriented nucleation by BEN that have been proposed so far. In Ref. [2] (see Figure 10.9), it was considered from TEM, AES, and XPS that (i) initially, adsorbed carbons on Si form Si–C bonds, and as the biasing continues, an *a*-SiC region increases; (ii) Si is then preferentially etched from the *a*-SiC layer by energetic H^+ ions, and a continued flux of carbonaceous species creates an excess concentration of carbon on the Si surface; (iii) after the Si surface is significantly carburized (the carbide thickness is 9 nm), the mobility of carbon atoms is enhanced by ion bombardment during the biasing; and (iv) then the diamond nucleation starts to occur on SiC. This model, however, did not explain how oriented diamond grains are formed by BEN.

Based on TEM and EELS measurements, Ref. [216] argued that the *a*-C film, produced during the BEN treatment, provided with nucleation sites for diamond. In this work, only small crystalline regions of SiC were present, and otherwise SiC did not play an important role for diamond nucleation. In Ref. [221], the ion energy distribution was measured, and it had a maximum at 70–90 eV. Since this energy is close to the optimum energy for subplantation of carbon into the film, it was inferred that the BEN process creates diamond nucleation by carbon subplantation. In Ref. [360], it was proposed that incident CH_x^+ ions with 70–80 eV stabilize randomly oriented graphitic layers, making some layers oriented normal to the surface, and diamonds heteroepitaxially nucleate on the layer edges. As already described in Section 10.2.2, Kulisch et al. [222] concluded by a Monte Carlo calculation that the average carbon ion energy is 15–20 eV, its maximum being 40–50 eV, so that the subplantation model was not accepted. In Ref. [316], it was argued that no oriented nuclei was necessary for HOD film synthesis. This means that it is not the oriented nuclei but the growth process that creates HOD grains, somewhat similar to the case of diamond growth on Pt(111) that will be described later. However, all other models either need oriented diamond grains or an oriented SiC layer in the BEN step. Thus, the argument of Ref. [316] does not seem to be relevant.

In Ref. [361], HOD films were grown by the three-step process using the conditions listed in Table H.3, and SiC particles were observed after BEN by plan-view TEM and ED. The depth analysis of AES showed that β-SiC was present under diamond. The β-SiC particles were aligned with Si in such a way that β-SiC{100} // Si{100} and β-SiC[100] // Si[100]. From these and other results, the following model was proposed: first, in the initial stage of BEN, the subplantation of carbonaceous ion species ($C_xH_y^+$) forms β-SiC particles that are aligned in registry with the basal Si(100) lattice. At the same time, Si atoms in the SiC particles are etched by the bombardments of atomic hydrogen, and the vacancies are filled with carbonaceous species. As a consequence, sp^3 carbon clusters are formed in the SiC lattice that is epitaxial to the Si lattice, and become the nucleation sites for diamond. Atomic hydrogen also plays a role of removing (i) sp^2 carbon clusters from the surface and (ii) misoriented diamond clusters that are weakly bonded to the substrate. In the HOD films made in Ref. [361], tilt and azimuthal angles of (100)-oriented crystals were 10 and 9°, respectively, and the FWHM of the Raman band of the film at 1332 cm^{-1} was only 3 cm^{-1}.

A novel model of diamond nucleation by BEN was proposed by Stöckel et al. [362] as a result of their series of works [363–365]. In their BEN experiments, a NIRIM-type MPCVD reactor was used. The BEN and diamond growth conditions are listed in Table H.4. When a negative bias voltage was applied to the Si(100) substrate against a W-wire placed above the plasma ball, a secondary plasma was generated between the main plasma and the substrate, as schematically shown in Figure 11.51. The secondary plasma was initially localized on the substrate holder, which had been covered with diamond film by the precedent growth experiments. The plasma front then proceeded toward the center of the substrate, and the BEN experiment was terminated at this stage. The diamond nucleation occurred in the area of the substrate that was within the distance of about 200 μm from the plasma front. At the front of the secondary plasma, cone-like structures were observed, as shown in Figure 11.52. In the successive growth of diamond using the CVD conditions listed in Table H.4, the HOD film was formed only in this area. However, an HOD film was deposited on the whole substrate surface of 1-cm square by reducing the substrate temperature from 900 to 800 °C in 45 s after the plasma sheath started to cover the substrate [364].

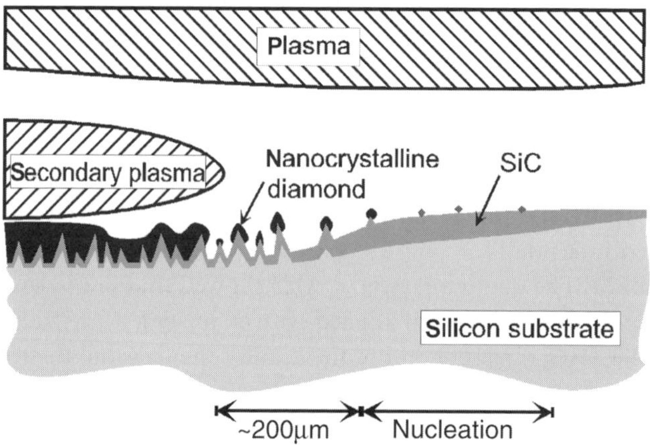

Figure 11.51. The morphology and the chemical state of the Si substrate at the front of the secondary plasma [362].

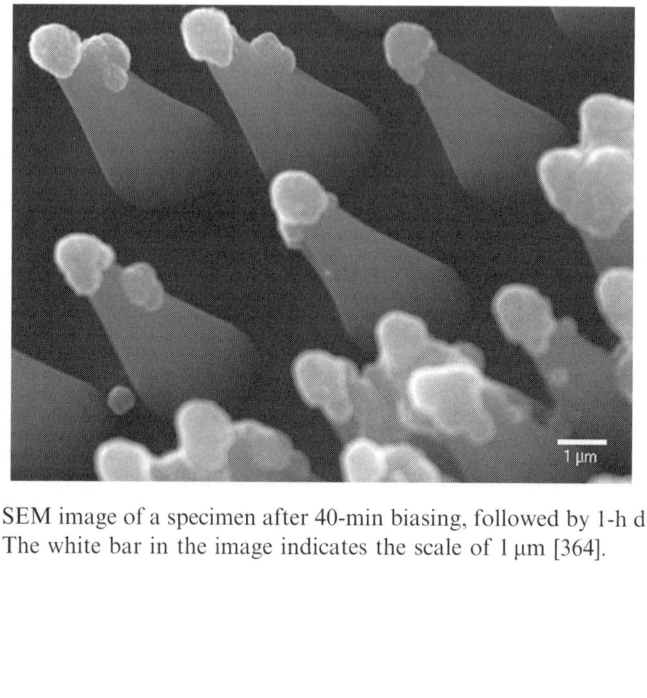

Figure 11.52. SEM image of a specimen after 40-min biasing, followed by 1-h diamond CVD. The white bar in the image indicates the scale of 1 μm [364].

The bias current also exhibited an unusual behavior, as seen in Figure 11.53 [364]. This result should be compared with Figure 10.14 [222]. One may notice the difference between the two results, but it might arise from the difference in BEN conditions, as seen in Tables H.1 and H.4.

The XPS spectra clearly indicated the presence of SiC, as seen in Figures 11.54 [364]. These data should be compared with those of Figure 10.7 [2]. In their XPS energy loss spectra for C(1s) core level, shown in Figure 11.55 [363], there was a known bulk plasmon band at 34 eV and the surface plasmon band at 24 eV. In addition, a band due to SiC was observed at ~20 eV.

It is of interest that in the Raman spectrum [363], there was no evidence on the presence of diamond (1332 cm^{-1}), and the spectrum was rather typical of a carbon material, as seen in Figure 11.56.

The core of the nucleation model proposed in Refs. [362, 363] is an assumption based on the experimental data that epitaxially oriented nucleation sites are formed in the SiC layer of about 10-nm thickness during the bias treatment. These sites are exposed at the SiC surface, while plasma etching of SiC is occurring during both the BEN treatment and the successive diamond growth process. The model of nucleation process is schematically depicted in Figure 11.57:

(a) During the first few minutes of the BEN treatment, nanometer size β-SiC crystallites are formed. Orientationally, they are in registry with the Si substrate, but there are misorientation angles up to 6°. Simultaneously,

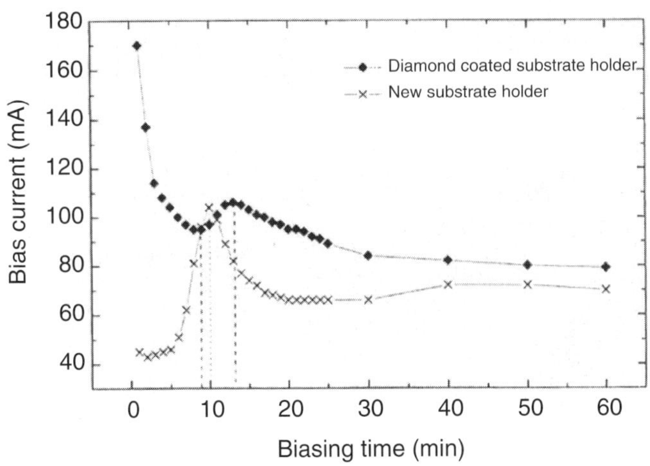

Figure 11.53. Current from the electrode to a new (×) and an already diamond-coated (◆) substrate holders during the BEN treatment as a function of the biasing time [364].

both Si and β-SiC surfaces are etched by H^+ ions and H radicals, but Si is etched more rapidly than β-SiC.

(b) After about 10 min, β-SiC islands are formed with a thickness of about 10 nm. At this point, the excess carbon ion flux due to subplantation of C causes a supersaturation of C within β-SiC crystals to create epitaxially oriented diamond crystals. The β-SiC layer is constantly subject to etching, whose rates were 100 nm/min under the biasing conditions, and 10 nm/h under diamond growth conditions.

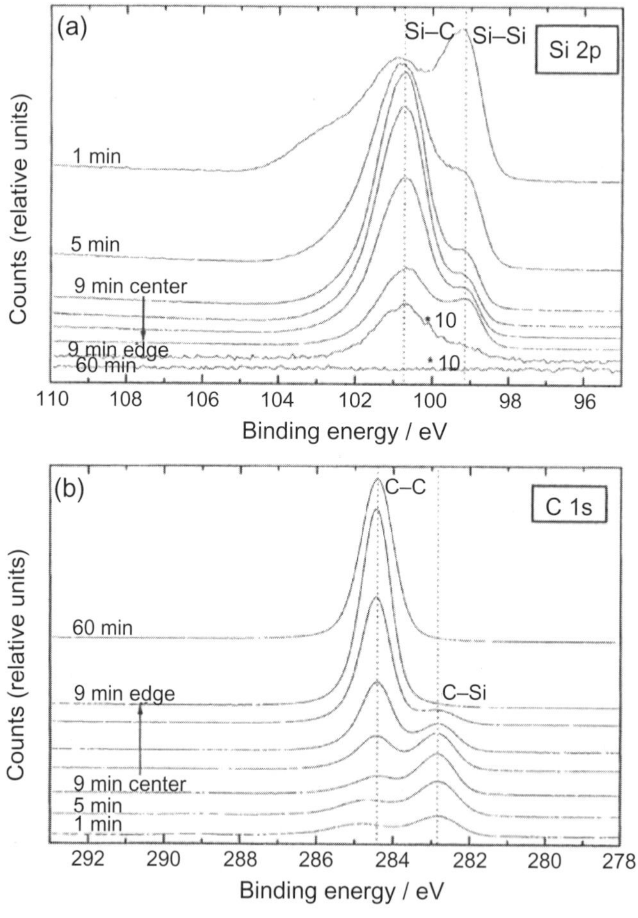

Figure 11.54. XPS spectra of Si(2p) and C(1s) core levels after different BEN treatment times [364]. See also Figure 10.7 [2].

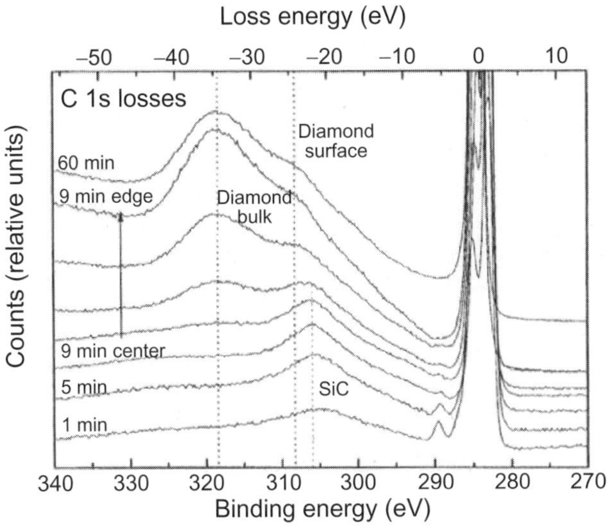

Figure 11.55. XPS energy loss spectra of C(1s) core level. For the specimen area covered by the bright plasma, the energy loss spectrum is typical for diamond (bulk plasmon at 34 eV, and surface plasmon at 24 eV) [364].

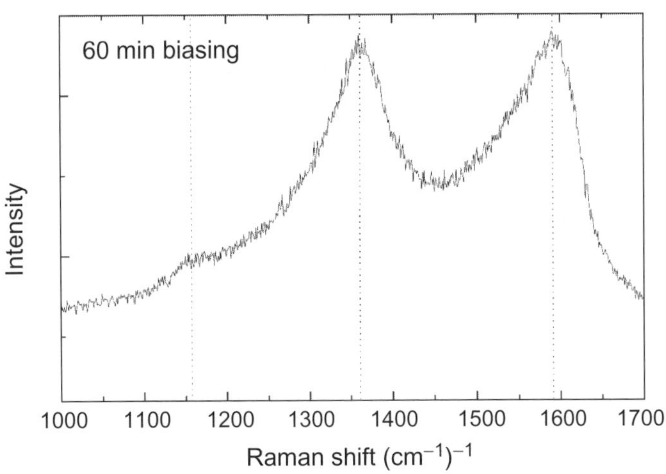

Figure 11.56. Raman spectrum of a specimen treated for 60 min under biasing conditions. No diamond line at 1332 cm^{-1} appears. The bands at 1360 and 1590 cm^{-1} are assigned to disordered graphitic carbon, and the shoulder at 1160 cm^{-1} is due to nanocrystalline diamond [364].

(c) As the BEN treatment continues, the β-SiC islands are etched by the plasma to expose diamond nuclei or nucleation sites to the surface of the β-SiC layer. The diamonds have cone-like hillock structures with the heights of 10 nm up to 100 nm.
(d) Oriented diamond crystals are then grown from the nuclei or nucleation sites, as β-SiC is continued to be etched even under the diamond growth conditions.

In this model, the SiC layers play a role of (i) a diffusion barrier of C and Si, (ii) an accumulation layer of C for supersaturation and forming diamond nuclei or

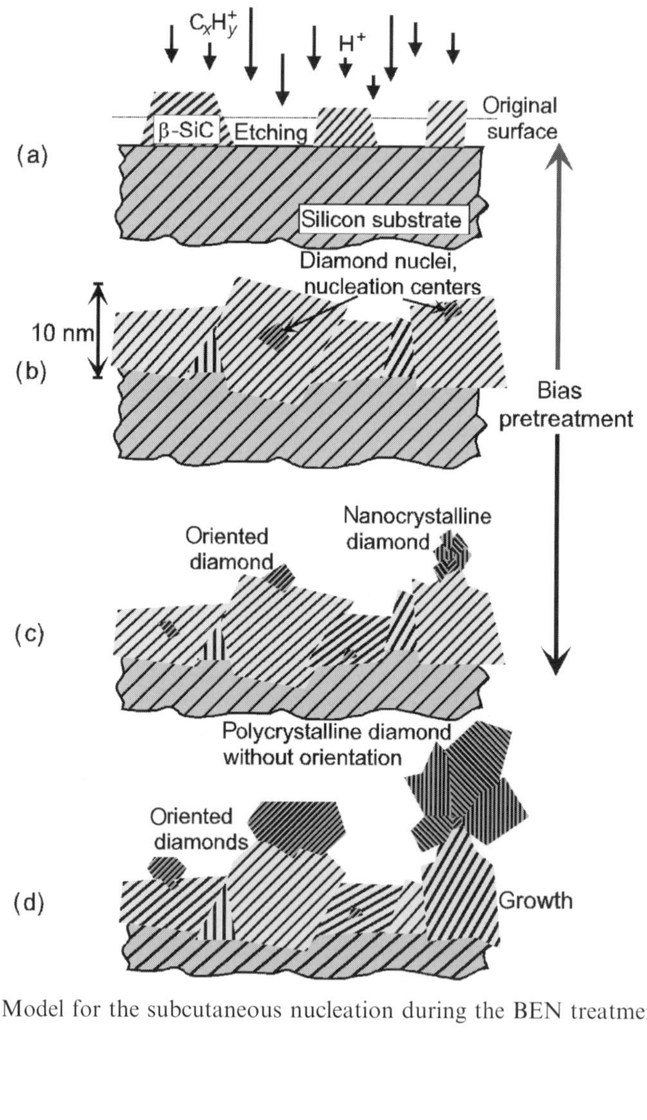

Figure 11.57. Model for the subcutaneous nucleation during the BEN treatment [362, 363].

Table 11.11. Etching rate of different materials under bias and growth conditions [362, 363].

Material	Bias pretreatment	Diamond growth
Diamond	Growth	Growth
Fine crystalline graphite	100 μm/min	100 μm/min
Amorphous carbon	Strong etching	Strong etching
Turbostratic carbon	Strong etching	Strong etching
β-SiC covered Si	–	2.5 nm/h
Silicon substrate surface	≤200 nm/min	≤20 nm/h

nucleation sites, and (iii) a structural transfer layer from the Si(100) surface, although there remains misorientations up to 6°. According to the TEM observation, β-SiC crystallites were found on the substrate surface area further than 200 μm from the secondary plasma front. The thickness of β-SiC within the 200-μm area was about 10 nm, β-SiC(111) was approximately parallel to Si(100), and the diamond crystallites grown on β-SiC was nearly epitaxial to β-SiC. A key to create the model is the etching rate, as shown in Table 11.11 [362, 363].

It is of most intrigue that unlike previous nucleation models, this model assumes a formation of diamond nuclei or nucleation sites inside the β-SiC layer, while the β-SiC layer concurrently plays a role of basal lattice for diamond epitaxial growth like in the precedent models. The second point of intrigue is the fact that the exposure and survival of the diamond nuclei or nucleation sites are made possible by a subtle balance of etching rates of Si, β-SiC, diamond, and other forms of carbon. This is consistent with the fact that HOD films can be formed only when the substrate was pretreated by proper BEN conditions.

11.21. SUMMARY DISCUSSION ON HOD FILMS

Since HOD films have well aligned structure of (100) faces at the film surface, and can further coalesce by lateral growth, HOD films have much better electrical characteristics than randomly oriented polycrystalline diamond films. In the best-oriented HOD films, the FWHM of the X-ray rocking curve was as small as 0.5°. Judging from the fact that the X-ray probes all the diamonds on the Si surface including the small diamonds in the vicinity of the Si substrate, the true FWHM value of X-ray rocking curve at the diamond film surface should be significantly smaller than 0.5°. Furthermore, the film made by Kawarada et al. [6, 7] has no visible grain boundaries, and the surface is very smooth. Thus, there is no doubt that HOD films can be used for a semiconductor material. However, in the normal HOD films of a few 10-μm thickness, there still remain misorientations between (100) faces, and the grain boundaries have not been completely removed. Thus, the issue is firstly whether the small misorientations can be completely removed to make a single crystal, and secondly, whether the area of HOD films can be expanded to several inches in diameter for industrial use. Regarding the first issue, XPF showed a splitting of (220)XPF band by roughly $\pm 2.5^\circ$ [334], indicative of the presence of misorientation. Thus, a new method to remove the misorientations for both tilt and azimuthal angles must be developed. The second issue depends on the reactor size as well as the biasing method. So far, the largest MPCVD reactor can make diamond coating over a 6-inch substrate, but has no biasing capability. For larger area coating, HFCVD might be more suitable, and the biasing would be easier. Unfortunately, HFCVD has problems of poor controllability and stability, and an inclusion of filament material in the deposited diamond. Thus, HFCVD must solve many technological problems. Removing the misorientations between (100) faces is a very challenging problem, but if it is solved, HOD films will be more widely used in industry, particularly for electronic devices.

In terms of science, the real mechanism on the nucleation of oriented diamonds does not seem to have been elucidated. Most importantly, the role of SiC has not been understood yet. This is because SiC is present in the BEN step, but tend to disappear in the diamond CVD step. It certainly plays an important role, but there remains a controversy over what the actual role is. The new method of making better-oriented HOD films, mentioned above, could be found if the role of SiC is made clear.

It is really amazing how extensively the BEN treatment and the HOD film synthesis have been studied by so many groups, as one could see in the review presented so far. It seems that the research is almost terminated as so many works have been done. However, it has been found recently that a significantly coalesced HOD film could be synthesized, in which the average surface region without visible grain boundaries is $\sim$60 μm in size, and the maximum size is 100 μm [366]. Figure 11.58 (a) shows a SEM image of the film surface, and Figure 11.58 (b) is a cross-sectional view of the film. It is seen that the film initially underwent a known columnar growth, but when the film thickness reached 50 μm, large diamond grains suddenly took over the growth. The cause of creating large grains is currently under study. This indicates that there is still room to improve the quality of HOD films and discover a new mechanism of oriented and heteroepitaxial growth.

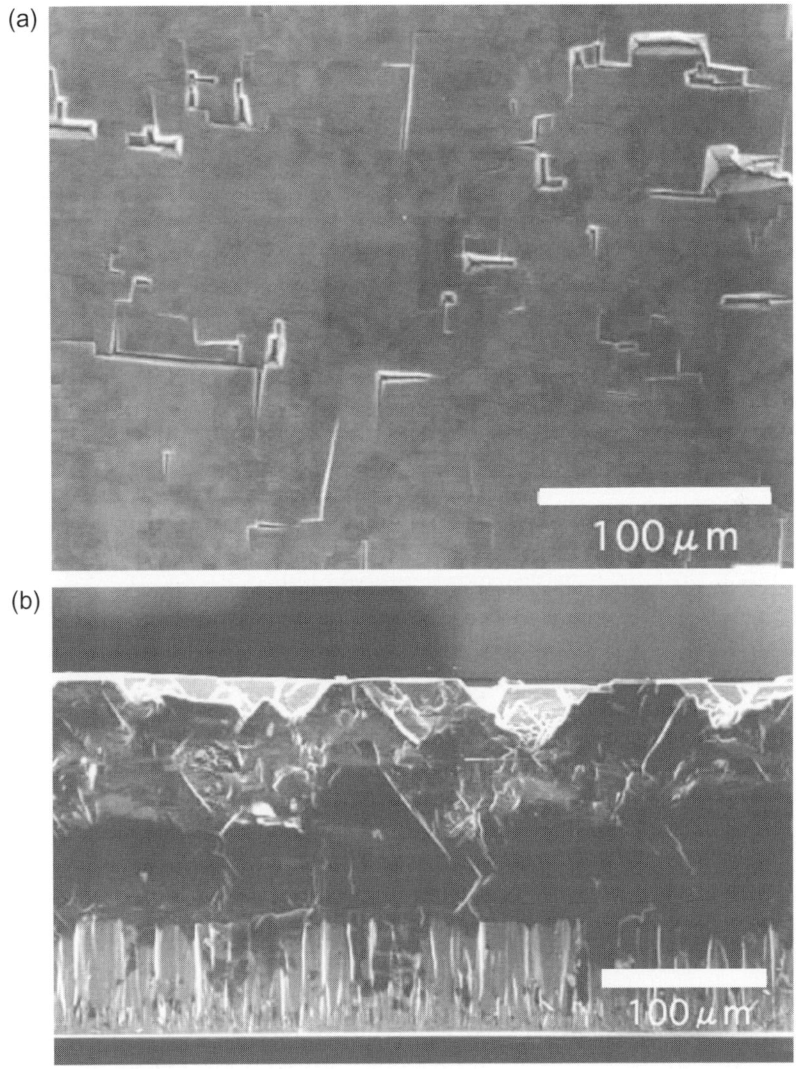

Figure 11.58. SEM image of an HOD film with large coalescence regions: (a) film surface and (b) its cross section [366].

Chapter 12

Oriented Growth on Noble Metals

Chapter 12

Oriented Growth on Noble Metals

It is of interest to investigate whether diamond can be deposited on precious metals such as Pt, Ir, Pd, and Au. The first motivation was to study catalytic properties of Pt under the conditions of diamond growth. Such works have been done by Belton, Schmieg and their co-workers in Refs. [367–369]. It was found that graphitic carbon, adsorbed at the Pt surface, was first converted to hydrocarbons and then to diamond as the CVD continued. Takaya and Sakamoto [370] observed by XPS that there was no chemical shift in the Pt(4f) and C(1s) bands, suggesting that no chemical reaction and compound formation took place between Pt and carbon. In 1992, Shintani found for the first time that a spontaneously coalesced, azimuthally oriented (111) diamond film is grown on single crystal Pt(111), as seen in Figure 12.1 [8]. In fact, the diamond deposition was done on a specially processed Pt foil with fairly large (111) domains at the surface. Successively, the work was repeated using a bulk Pt(111) crystal to confirm the result [9].

Diamond growth on Pt, Pd, and Au foils by HFCVD is described in [371]. For Pt and Pd [371–373], SEM images indicated that the diamond nucleation occurred in the bulk as the diamond particles were embedded in the materials, similar to Figure 12.5. It is inferred that there was a Metal (M)–C–H complex formed in the bulk, which played an important role in diamond nucleation.

The above works stimulated the study of diamond growth on Ir(100), and Sawabe's group showed by DC plasma CVD that a (100)-oriented, flat diamond film was heteroepitaxially grown on Ir(100) that also had been heteroepitaxially deposited on MgO(100) [10]. This was followed by the works of Sawabe's and Morooka's groups [11, 13, 72, 374–376] as well as Schreck and Stritzker's group [12, 377–379], and Kawarada et al. [380, 381]. Furthermore, diamond growth on Pd has been studied by Ikeda et al. [382]. So far, azimuthally (111)- and (100)-oriented diamond films without grain boundaries have been successfully deposited on Pt(111) and Ir(100), respectively.

12.1. Pt(111)

12.1.1. Oriented growth by MPCVD

In what follows, the results obtained in Refs. [383–387] will be summarized.

a. Pretreatment and nucleation

The Pt surface was scratched by the ultrasonic treatment, and the XRD of the specimen indicated that diamond powders were embedded in Pt, although no diamond particles were observed by SEM. The scratched Pt(111) surface was very rough, as seen in Figure 12.2 [383]. A detailed study of the Pt surface by TEM revealed that the Pt surface actually consists of polycrystalline columnar structures of Pt. Their average length was 400–500 nm [388]. The scratches on the Pt surface completely disappeared to be a smooth surface only after a 1-min exposure to hydrogen plasma, and the single crystal quality was improved over *as received* Pt substrates. Moreover, XRD indicated that the embedded diamond particles were preferentially ⟨111⟩-oriented by the hydrogen plasma treatment. The growth of azimuthally oriented diamond film on scratched surface is not characteristic only for Pt: it has also been achieved on scratched surfaces of Ni and Co substrates, as described in Sections 9.2.1 and 9.2.2 [168, 182, 195].

Diamond growth was carried out using a NIRIM-type MPCVD reactor. The optimized conditions were determined to be $c = 0.3\%CH_4/H_2$, $P = 50$ Torr, and $T_s = 875\,°C$. The growth rate of diamond under these conditions was 0.2–0.3 μm/h. The initial stage of diamond nucleation was observed by SEM by interrupting the CVD after (a) 20 min, (b) 30 min, (c) 45 min, (d) 60 min, (e) 75 min, and (f) 105 min from the start, and the same spot on the specimen was observed. The results are shown in Figures 12.3 (a)–(f), respectively. It is clearly seen that "dissolution and recrystallization" of diamonds take place on the Pt surface. For instance, diamond particle A repeatedly undergoes this process: it is not visible in Figure 12.3 (a), and only vaguely seen in Figure 12.3 (b). It then appears on the Pt surface in Figure 12.3 (c), but becomes smaller in Figure 12.3 (d). In Figure 12.3 (e), diamond

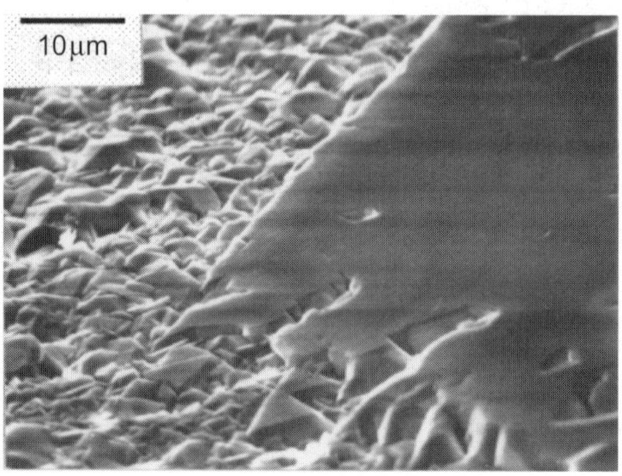

Figure 12.1. A large (111) diamond grain coalesced on a (111) domain of Pt foil [383].

particle A appears to be well-faceted, but is then dissolved in the Pt substrate in Figure 12.3 (f). Similar dynamics have also been observed for other crystals. In addition, positional changes as well as rotations of diamond crystals occurred. Such features have also be observed by Ishikura and co-workers for diamond growth on Cu [389].

Since the melting point of Pt (1772 °C) is much higher than the CVD temperature (875 °C), and the solid solubility of C in Pt is less than 1 atomic % at 875 °C (see Appendix G and the Pt–C binary phase diagram [390] of Appendix F), it is likely that a Pt–C–H ternary system plays an important role for the molten substrate surface. Indeed, Roy and co-workers [391, 392] have reported a recrystallization of diamond from Ni–C and Cu–C solutions in hydrogen plasma due to a metal–C–H complex formation. The SEM images of Figures 12.3 (a)–(f) indicate that the epitaxial growth of diamond arises from the physical motion (positional and orientational changes) and chemical interaction ("dissolution and recrystallization") of small diamond crystals at the Pt surface.

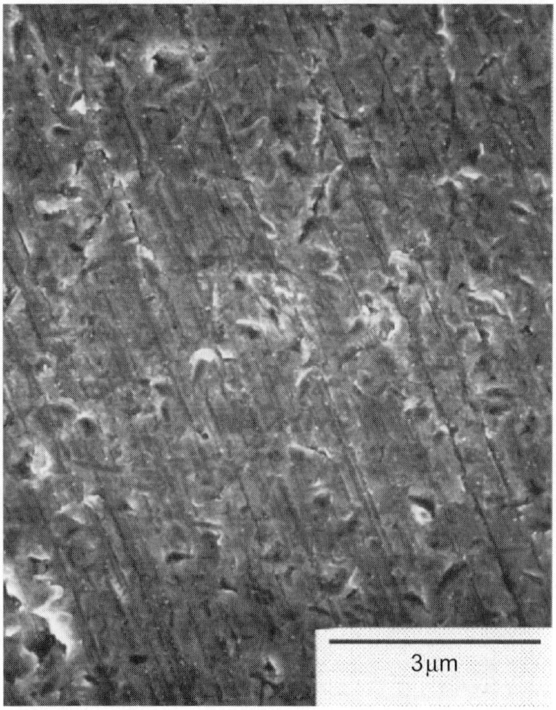

Figure 12.2. Ultrasonically scratched Pt(111) surface [383].

b. Growth

Figures 12.4 (a)–(e) show the time evolution of diamond growth. After 1 hour (Figure 12.4 (a)), both (110)- and (100)-oriented crystals, indicated by A and B, respectively, are clearly visible. After 2 hours (Figure 12.4 (b)), (111)-oriented crystals indicated by C and D grow faster than crystals A and B. After 3 hours (Figure 12.4 (c)), the substrate surface is almost entirely covered with diamond crystals to begin a vertical columnar growth. After 5 hours (Figure 12.4 (d)), crystals C and D become larger than other crystals such as crystal B, and after 10 hours (Figure 12.4 (e)), crystal C grows over crystal B. The (110)-oriented crystal A still survived in Figure 12.4 (e) because of its high growth rate, but (111)-oriented crystals

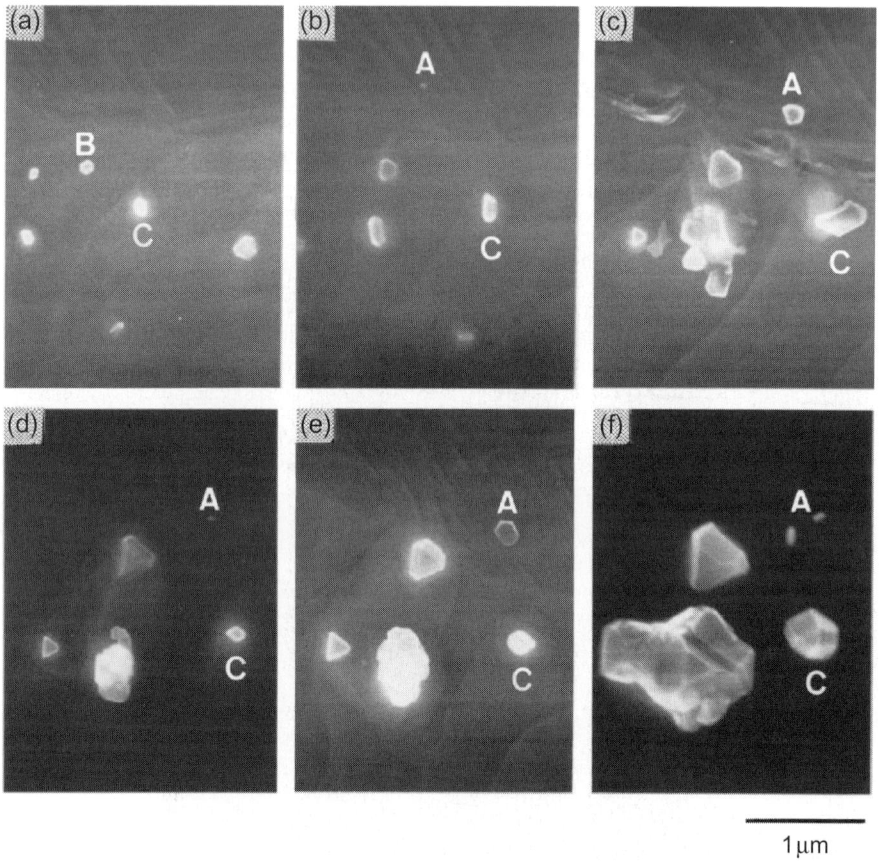

Figure 12.3. SEM images showing the time evolution of diamond particles after (a) 20 min, (b) 30 min, (c) 45 min, (d) 60 min, (e) 75 min, and (f) 105 min from the start [383].

are totally dominant after 30 hours. A remarkable feature in Figure 12.4 (e) is the occurrence of the coalescence between adjacent (111) faces of crystals C and D, implying that those crystals are almost perfectly aligned azimuthally.

The results in Figure 12.4 show that the growth of (111)-oriented diamond films proceeds by the van der Drift texture evolution mechanism [77]. It was observed that even though there were diamond crystals that had (111) faces, they were taken over by azimuthally oriented diamond crystals with (111) faces that were more parallel to the substrate surface. From the diamond crystal shapes in Figure 12.4, the condition of $\alpha \sim 1$ is satisfied, and the diamonds undergo a stable $\langle 111 \rangle$ growth. This result is in contrast to work by Barrat and Bauer-Grosse [105] that $\langle 111 \rangle$ growth of diamond on Si substrate is considered to be unstable and easily taken over by $\langle 110 \rangle$ growth, because it is difficult to maintain the $\alpha \sim 1$ condition on Si. It was thus concluded that (i) a considerable number of epitaxially oriented diamond particles are created on Pt(111), although non-epitaxial particles coexist in the early stage of CVD and (ii) under proper CVD conditions, the growth conditions represented by $\alpha \sim 1$ are stably realized. Although the real cause for the experimental observation of the stable (111) growth is yet to be made clear, it was inferred that the difference between Pt and Si arises from both the catalytic

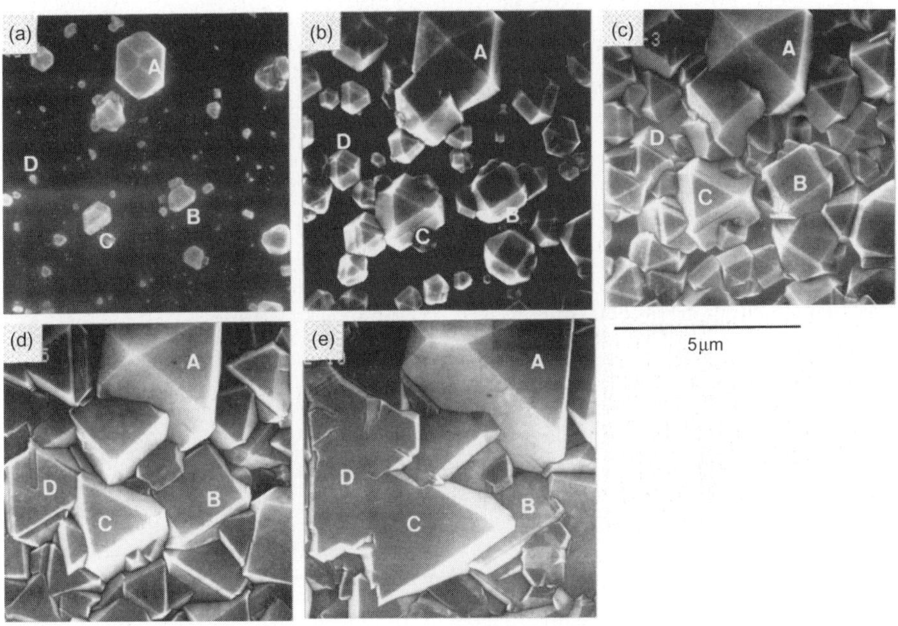

Figure 12.4. SEM images showing diamond growth after (a) 1 h, (b) 2 h, (c) 3 h, (d) 5 h, and (e) 10 h [383].

action of Pt to hydrogen and hydrocarbons at elevated temperatures [166] and the Pt–C–H complex formation.

In certain specimens, there were diamond crystals that were partially embedded in Pt near the peripheral region of the substrate, as seen in Figure 12.5. The diamond CVD was done for 5 h using the standard growth conditions. It appears that the diamond nucleation took place beneath the Pt surface. It is likely that the nucleation site should be the diamond powders that were buried in Pt during the ultrasonic treatment. It was also inferred that a Pt–C–H complex formation plays an important role for the diamond nucleation, as depicted in Figure 12.6.

Figures 12.7 (a) and (b) are SEM images of diamond films grown for 30 and 40 h, respectively, on bulk single crystal Pt(111). In Figure 12.7 (a), the film surface consists of azimuthally oriented (111) faces of diamond, and most faces have coalesced to form a single "network" of (111) faces, while the rest of the region contains randomly oriented diamond crystals. This morphology was uniform over the entire substrate surface of 12 mm in diameter except for the edge. In Figure 12.7 (b), the coalescence developed further and randomly oriented areas decreased, clearly indicating the possibility of forming a totally coalesced (111)-oriented diamond film on bulk single crystal Pt(111). To achieve an *optimized*

Figure 12.5. Diamond crystals embedded in a Pt foil substrate [383].

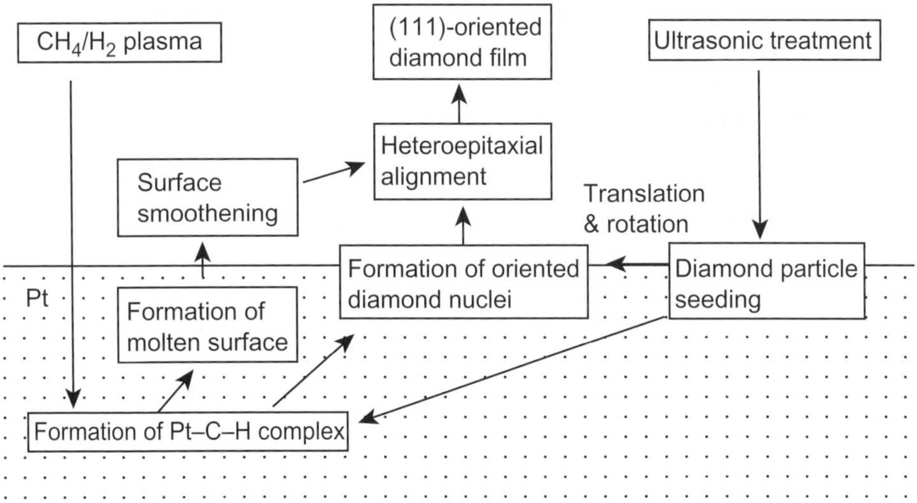

Figure 12.6. Schematic diagram on heteroepitaxial diamond growth on Pt(111) [383].

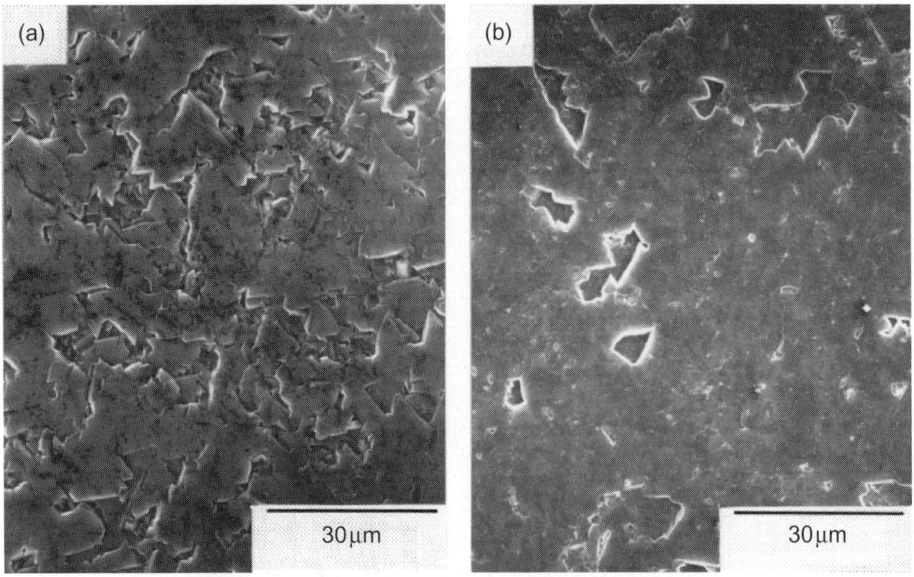

Figure 12.7. SEM images of diamond film grown on bulk Pt(111) substrates for (a) 30 h and (b) 40 h [383].

coalescence of diamond grains at the film surface, the nucleation density should be $0.96–3.1 \times 10^8/cm^2$ [393].

In the latest achievement, a fully coalesced diamond film was deposited on a 10-mmϕ Pt(111) bulk surface, as shown in Figure 12.8. The surface consisted of a grain boundary-less diamond film with numerous bunched steps. Although the detailed analysis of the film has not yet been done, this demonstrated that (111)-oriented single crystal diamond films can be deposited on Pt(111).

12.1.2. XPF

Figures 12.9 (a) and (b) are the XPF spectra from a (111)-oriented diamond film grown for 30 h on single crystal Pt(111) (see Figure 12.7 (a)) and a Pt(111) substrate, respectively [383, 394]. The FWHM of diamond {111} diffraction poles was approximately 4°. This value was similar to the FWHM of highly (100)-oriented diamond films on silicon, i.e. 5°. However, since the FWHM value includes diffraction signals from misoriented diamond particles near the Pt interface, the 4° FWHM should be the upper limit of the misorientation distribution. Note that

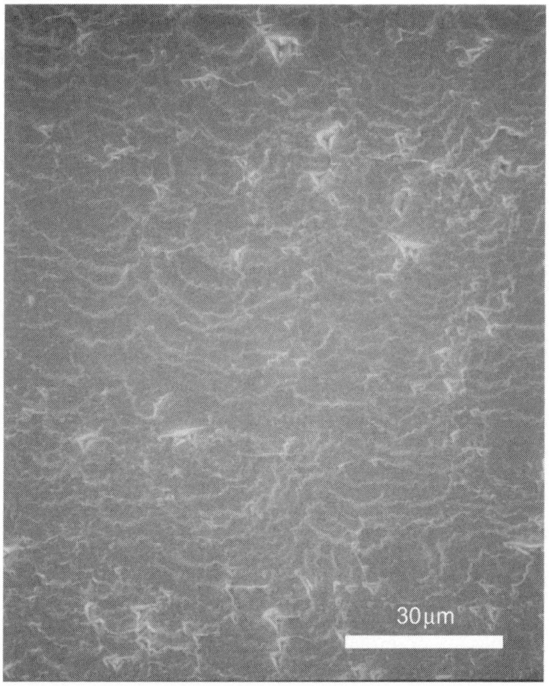

Figure 12.8. Fully coalesced (111)-oriented diamond film on Pt(111) of 10 mm in diameter [14].

the coincidence of the diffraction patterns in Figures 12.9 (a) and (b) is evidence that the cubic crystal axes of the diamond film and the Pt substrate are parallel to each other.

In Figure 12.9 (a), (i) there are six {111} diffraction poles, rather than three, around the central (111) pole, and (ii) three diffraction poles, $(11\bar{1})$, $(1\bar{1}1)$, and $(\bar{1}11)$ are more intense that the other poles, $(1\bar{1}\bar{1})$, $(\bar{1}1\bar{1})$, and $(\bar{1}\bar{1}1)$. The result (i) indicates that there are two possible orientations of diamond crystals that are rotated by 60° in plane with respect to each other. A similar type of orientational structure has been observed for β-SiC(111) epitaxially grown on 6H–SiC(0001) [395]. In this case, incoherent boundaries, called double positioning boundaries, are formed in the film by 60°-rotated domains. On the other hand, the result (ii) indicates that the three orientations, $(11\bar{1})$, $(1\bar{1}1)$, and $(\bar{1}11)$, are preferable to other three orientations. The existence of non-equivalent diamond orientations on the Pt(111) surface was attributed to interactions with the second nearest neighbor Pt atoms. However, the "incoherent boundaries" due to the existence of 60°-rotated diamond crystals have not been identified yet. It is of interest that an azimuthally unidirectional surface was formed as the diamond CVD was continued for a longer time, as seen in Figure 12.7 (b).

Finally, it is noted that the X-ray rocking curve for (111) diffraction line was 2.5°. This value is almost comparable to the corresponding value of HOD films, but much greater than the value of single crystal diamond, 0.03°. Part of the reason for this is that for the diamond–Pt case, the X-ray diffraction picked up the signals from diamonds near the interface, which are more randomly oriented.

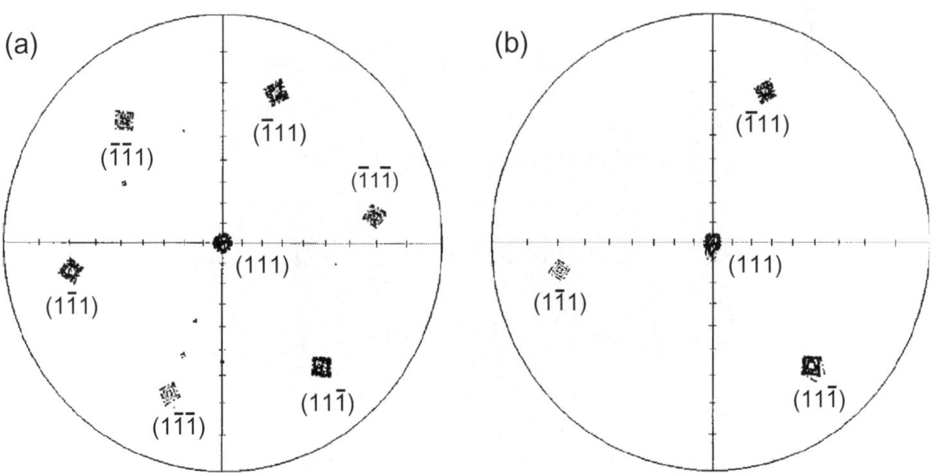

Figure 12.9. XPF spectra of (a) (111)-oriented diamond film with a Pt(111) substrate and (b) Pt(111) substrate [383, 394].

12.1.3. TEM analysis

An XTEM study of diamond film on Pt(111) revealed unexpected results. Figure 12.10 shows a TEM image at the bottom of the diamond film. The Pt substrate was detached during the specimen preparation. In the figure, the transparent areas in the upper part of the figure are diamond, where the linear and spotty dark contrasts are stacking faults and separated dislocation loops, respectively. The dark areas are Pt, and small semitransparent areas are graphite. Despite the presence of graphite, most diamond grains are directly in contact with Pt without any intermediate layer, as shown by A and B. It is amazing that even though the diamond/Pt interface is so rough and even graphite is present, an azimuthally (111)-oriented and spontaneously coalesced diamond film resulted in the end. In certain cases, it was seen that a small Pt particle was trapped at a grain boundary between diamond crystals. This and Figure 12.10 underwrites the previously stated inference that diamond nucleates in Pt bulk and a Pt–C–H complex plays a certain role for diamond nucleation, as shown in Figure 12.6.

A TEM image of the surface region is shown in Figure 12.11. Apart from the grain boundary regions, the defect density is small in the diamond grain. Also, bright areas, containing fewer defects, existed at the grain boundaries, indicating that the misorientations between the adjacent crystals are very small. The arrows in Figure 12.11 indicate low angle grain boundaries (the misalignment is less than 0.01°), and sharp lines within the diamond grain are screw dislocations. This image indicates that the crystal quality of diamond is fairly high at the surface.

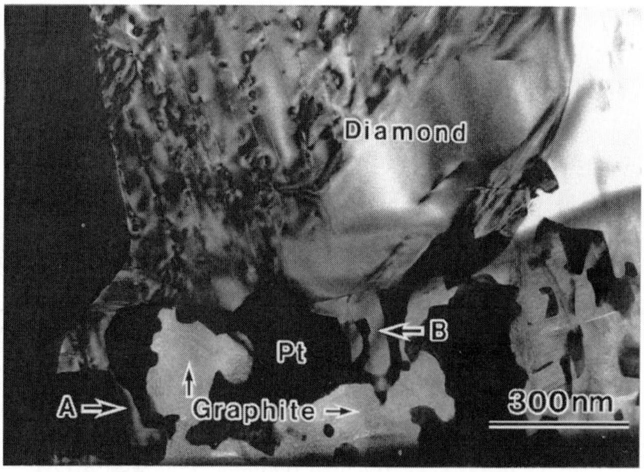

Figure 12.10. XTEM image of diamond/Pt interface [383].

12.1.4. Interface structure

The interface structure between Pt and diamond was investigated in great detail by TEM and ED in Ref. [396]. The results are summarized as follows: at the film surface, (i) the single grain with a (111) face actually consisted of several subgrains, (ii) some diamond grains were rotated within $\sim 10^\circ$, (iii) misfit dislocations existed at the grain boundaries, and furthermore, (iv) many defects such as stacking faults, microtwins, and dislocations were unevenly distributed within the grain. In the middle region of the film, (i) misoriented $\langle 110 \rangle$ grains existed among (111)-oriented grains and (ii) defect densities were higher than in the surface region. In the interface region, (i) it was rough, and diamond, Pt, and graphite were intermixed, (ii) the thickness of the intermingled layer between diamond and Pt was about 700 nm, which included graphite of about 300-nm thickness. Fragments of Pt were trapped within diamond, implying an occurrence of complex reactions between diamond and Pt during the nucleation stage. Even so, diamond seemed to directly nucleate on Pt, and diamond particles with less defects were observed, which were assigned to the residual fragments embedded in Pt during the ultrasonic treatment.

In Ref. [397], it was confirmed that diamond was in direct contact with Pt, as seen in Figure 12.12. Graphite was discretely existing at the interface, and G(0001) planes tended to be parallel to the Pt surface. The orientational relationship was G{0001} // D{111}.

The interface structure of Pt and diamond particle was closely investigated by Takai and Shimizu [398]. As seen in Figure 12.13, the diamond particle nucleated

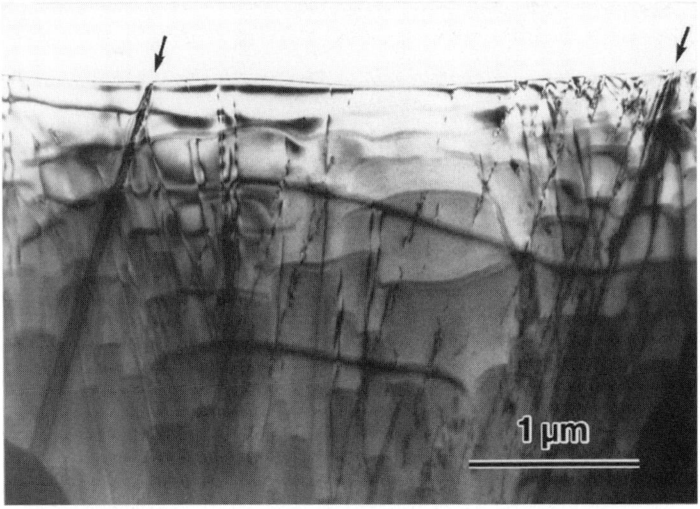

Figure 12.11. XTEM image of a (111)-oriented grain at the surface. The arrows indicate low angle grain boundaries [383].

directly on Pt. Although graphite particles existed discontinuously, they were only a secondary product and did not influence diamond nucleation. In contrast to the proposed mechanism described in Ref. [383], this TEM result indicated that the nuclei of heteroepitaxial diamonds on Pt(111) are not attributed to the diamond powder embedded during the ultrasonic treatment, but the direct nucleation can also occur on Pt(111).

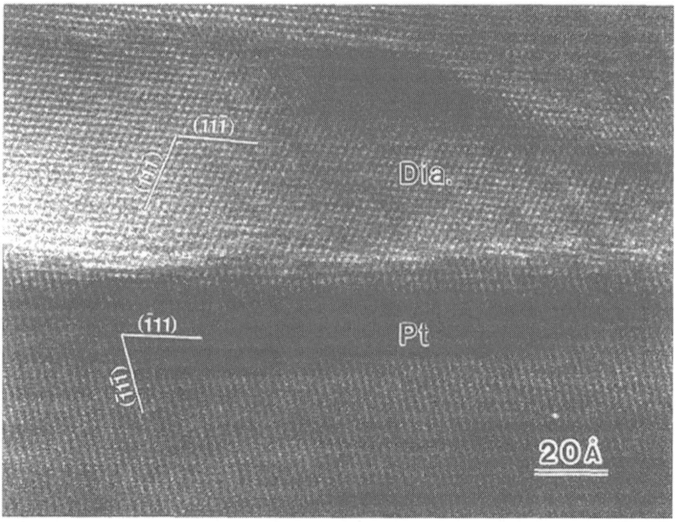

Figure 12.12. HRTEM image of diamond/Pt interface. No intermediate layer is observed [397].

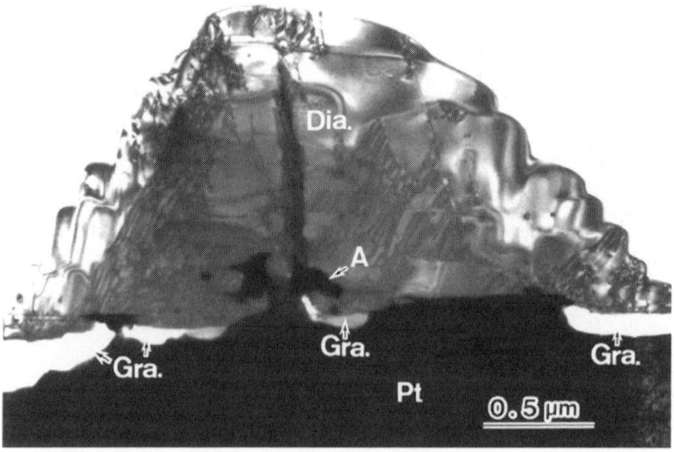

Figure 12.13. XTEM image of an isolated diamond particle heteroepitaxially grown on Pt(111) substrate. "A" indicates a protrusion of Pt into diamond [398].

The interface between diamond and Pt was also studied by HRTEM in Ref. [190]. In this work, the CVD diamond was directly deposited on a Pt TEM grid, which had been ultrasonically scratched with diamond powder. The diamond CVD was done using a NIRIM-type reactor under conditions of $c = 0.5\% CH_4/H_2$, and $P = 35$ Torr. The {0002} planes of the intermediate graphite layer were nearly parallel to the Pt surface. The HRTEM images clearly revealed the presence of a 4-nm thick, continuous graphite intermediate layer between the diamond and the Pt substrate. The appearance of the (111)-textured diamond crystals on Pt was supposed to be closely related to the presence of the graphite intermediate layers. This result is in contradiction with that of Ref. [398]. It is inferred that the differences arise from the different growth conditions of diamond on Pt.

In Ref. [399], the initial stage of (111)-oriented diamonds grown on Pt foils was studied by SEM and TEM. The TEM images showed that the scratching pretreatment with diamond powder severely damaged the Pt(111) substrate surface. Due to the pretreatment, columnar polycrystalline damaged layers were formed in the Pt surface region, whose average depth was 400–500 nm. During the subsequent deposition process, the damaged layers gradually recrystallized to be a single crystal. The nucleation and the initial growth of diamond particles were in fact accompanied by the recrystallization in the Pt surface region.

In Ref. [400], overgrowth characteristics of B-doped diamonds layers on thin Pt films with thicknesses of 100–200 nm were studied by SEM with X-ray emission analysis, Raman scattering spectroscopy, and XRD. The thin Pt film had been sputter-deposited on either polycrystalline or single-crystal diamond substrates. As a result, a continuous (B-doped and undoped) diamond layer was formed on the thin Pt film. The secondary electron emission from the undoped diamond layer deposited on the thin Pt film was weaker than that from an undoped diamond film deposited on an Si substrate because of the higher work function of Pt. By contrast, the specimens with the overgrowth of B-doped diamond layer exhibited no distinction of the secondary electron emission between the two kinds of specimens. For both polycrystalline and single-crystal diamond substrates, a number of islands and grains of Pt with 1–3 μm in size were embedded in the overgrown diamond layer on the thin Pt film. This suggests that significant agglomeration and recrystallization of the thin Pt film occurred at ~900 °C, well below the Pt melting point of 1772 °C, during the diamond overgrowth process. Finally, a considerable amount of B segregation was observed after the overgrowth.

12.1.5. Confocal Raman spectroscopy

The internal stress within the diamond film grown on Pt(111) substrate was characterized by confocal Raman spectroscopy [401, 402]. The film thickness was

8 μm, and the film surface was considerably coalesced, as seen in Figure 12.14. Figure 12.15 shows the Raman peak shift and the FWHM of the 1332 cm^{-1} line of the diamond film as a function of the depth from the film surface. The internal stress was significant up to 6 μm from the Pt/diamond interface, but released near the surface. The observed diamond peak shift and its FWHM for each depth are plotted in Figure 12.16. It is more clearly seen that the diamond film is subject to a compressive stress up to the thickness of 6 μm from the Pt surface, but the stress is gradually released above 6 μm. On the other hand, the FWHM increases rapidly above this point. The broadening of the Raman band width indicates the presence of internal random strain within the oriented grain. By contrast, for the Raman shift and the FWHM in the region B, where diamond grains are randomly oriented, the diamond peak was 1332 cm^{-1} + 1.7 cm^{-1} and the FWHM was 11.6 cm^{-1}, independent of the position.

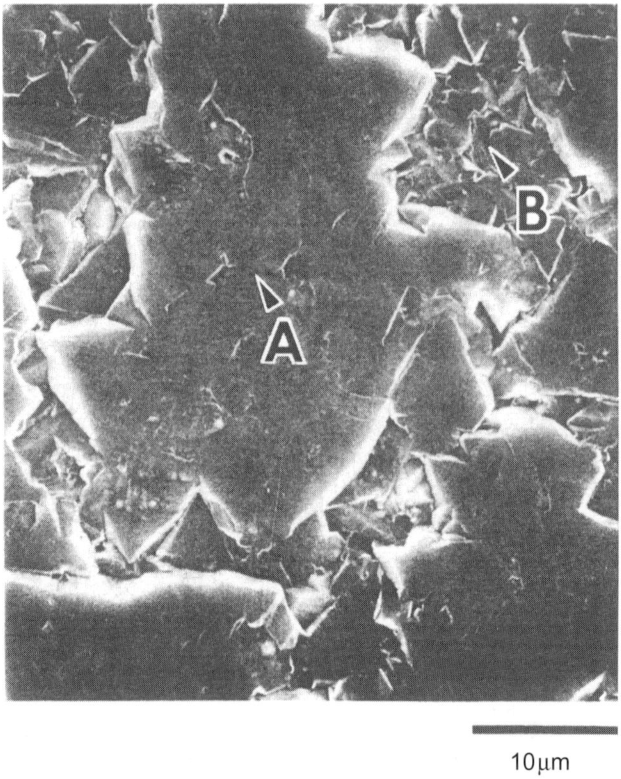

Figure 12.14. A SEM image of the film grown on Pt(111) for 40 h. (A) oriented region and (B) random region [402].

12.1.6. *Various techniques of diamond deposition on Pt*

The diameter of commercially available Pt single crystal is limited to about 1 cm so that to deposit (111)-oriented diamond film on larger areas, one must use single crystal Pt films. To this end, it would be very convenient if Si wafer can be used as the base material. Just depositing Pt on Si did not work, however, because randomly oriented platinum silicide can be very easily formed at high temperatures, and hence this is not an appropriate substrate for heteroepitaxial growth of diamond.

In Ref. [403], a multi-layered structure of Pt(111)/TiO_2/SiOx/Si was used as the substrate, and Pt (25–50 nm thickness) was (111)-oriented, even though SiOx was amorphous. The presence of the TiO_2 layer (20–100 nm thickness) improved

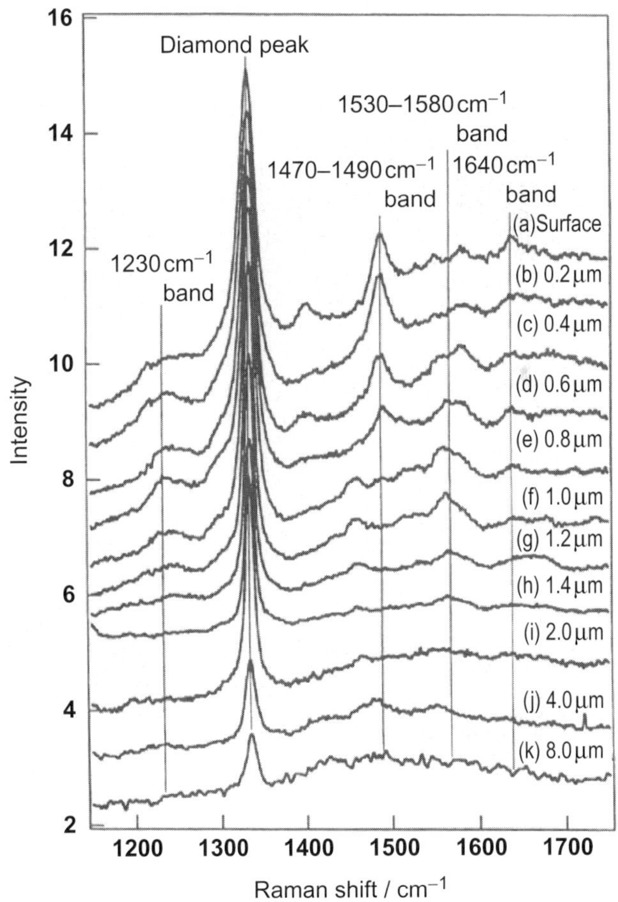

Figure 12.15. Raman depth profile in the region A of Figure 12.14 [402].

the adhesion of the Pt film to SiOx/Si. Diamond was deposited by MPCVD using an ASTeX reactor under conditions of $c = 1\%CH_4/H_2$, $P = 50$ Torr, $P_m = 750$ W, and $T_s = 700$–$850\,°C$. Consequently, uniaxially (111)-oriented diamond film was grown on the substrate. Unfortunately, the azimuthal orientation was random so that no coalescence between diamond grains occurred. Even so, this method seems to be one of the most promising techniques to achieve (111)-oriented, well-coalesced diamond films on larger areas, because Si is used as the base material of the substrate.

Apart from Si, sapphire(0001) is also a promising base material as Pt(111) can be heteroepitaxially grown on it. In Ref. [404], a Pt(111) film was deposited on sapphire of 0.5-mm thickness, and then diamond CVD was undertaken on Pt to 1.5-μm thickness. According to the X-ray rocking curve measurements, the FWHM of Pt on sapphire was 0.16°, significantly smaller than the corresponding value 0.5° for *as received* bulk single crystal Pt. As a consequence of better crystallinity of the Pt film, the FWHM of the X-ray rocking curve of (111) diffraction line was 2.0°, even though the diamond film was only 1.5-μm thick. This technique was further developed to make a substrate consisting of Pt(111)/Ir(111)/Pt(111)/sapphire(0001), where Pt and Ir was roughly 1–2 and 1-μm thick, respectively [386]. The FWHM of the X-ray rocking curve of (111) diffraction line for a 1.5-μm thick diamond film was 1.1°, significantly smaller than

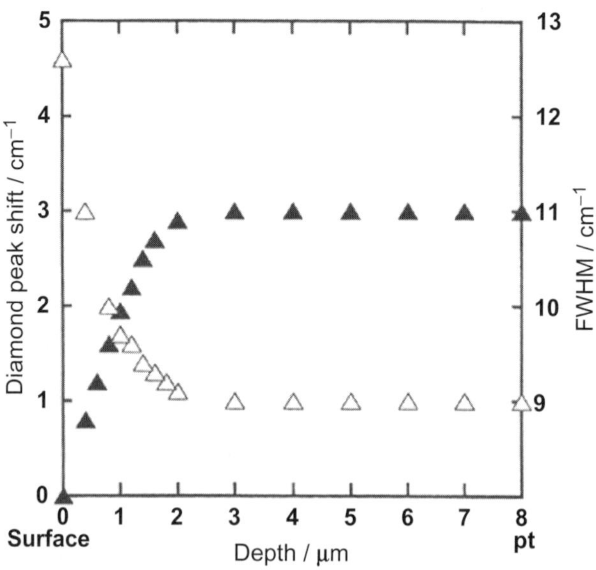

Figure 12.16. The diamond peak shift from 1332 cm^{-1} (solid triangles) and the FWHM at each depth (open triangles) in the region A of Figure 12.14 [402].

when Pt/sapphire or bulk Pt was used as the substrate. Other methods, for instance, using BEN, was not successful so far [387]. This is in strong contrast to the fact that heteroepitaxial growth of diamond on Ir(100) is achieved by using BEN, as will be seen in the next section.

12.2. Ir(100)

It was discovered by Sawabe and co-workers [10, 405] by DC plasma CVD that diamond grows heteroepitaxially on Ir(100). A single crystal Ir(100) film for a substrate was deposited on MgO(100) of 10-mm square by electron beam deposition at the substrate temperature of 750 °C, and then annealed in vacuum for 60 min at 950 °C. The Ir film thickness was 500 nm. The setup of the DC plasma CVD reactor that has the third electrode between the anode and the cathode for bias application has been shown in Figure 4.2.

The DC plasma CVD reactor consisted of a negatively biased Mo electrode 1, and a Mo electrode 2 which was 2.5 cm apart from electrode 1. In addition, a third ring electrode 3 was placed 1 mm above the electrode 1. The experiments were undertaken using the conditions listed in Table 12.1. In the process, 2%CH_4/H_2 gas was introduced in the chamber, and $V_b = -260$ V was applied to the substrate for diamond nucleation for 10 min, which was followed by diamond growth. Unlike standard MPCVD, the gas pressures used for DC plasma CVD were very high. Interestingly, oriented diamond nucleation was possible by BEN (ion irradiation), unlike the Pt case. As a result, azimuthally oriented diamond particles with a pyramidal shape were formed on the substrate with a density of ~ $10^8/cm^2$, as seen in Figure 12.17. The orientational relationship between diamond and Ir{100} was D{100} // Ir(100) and D[1001] // Ir[1001], i.e. a cube-on-cube structure.

Table 12.1. Growth conditions of diamond on Ir(100) by DC plasma CVD [10].

Parameters	Ion irradiation	Growth
Discharging voltage	600 V	1000 V
Discharging current	670 mA	160 mA
Substrate voltage	−260 V	0 V
Current density (at substrate)	70 mA/cm^2	560 mA/cm^2
CH_4/H_2	2%	2%
Pressure	100 Torr	200 Torr
Substrate temperature	900 °C	840 °C
Time	10 min	30 min

Successively, a thin diamond film of 1.5-μm thickness was synthesized, and crystal structure was analyzed in Refs. [13, 72]. In the heteroepitaxially grown diamond film area of 1 mm in diameter, where diamond grains were fully coalesced and no grain boundary was seen by SEM, as shown in Figure 12.18, the film surface was extremely smooth ($R_a = 1$ nm), and the RHEED pattern showed streak lines, indicating that the diamond film was a single crystal and epitaxial to the Ir(100) substrate. Confocal Raman spectra showed no sign on the existence of non-diamond carbon. The FWHM of the 1332 cm^{-1} line was 18 cm^{-1}, as compared with 5 cm^{-1} for a bulk single crystal diamond. This was attributed to the anisotropic internal stress in the diamond film as well as the small film thickness. An XTEM was also observed, as seen in Figures 12.19 (a) and (b) [375]. For the case of the Ir(100) substrate, the diamond nuclei seemed to be so perfectly oriented that they coalesce with each other in the very early stage of the diamond growth, resulting in a fully coalesced diamond film without grain boundary even when the film thickness is only 1.5 μm. This is in strong contrast to the Pt(111) case, where the orientational order of diamond nuclei are less than that of the Ir(100) case, and a significant coalescence of diamond grains is noticeable only after 40 h of diamond CVD.

Figure 12.19 indicated that unlike the Pt case, the Ir/diamond interface was only slightly rough, i.e. there were protuberances of Ir that were approximately

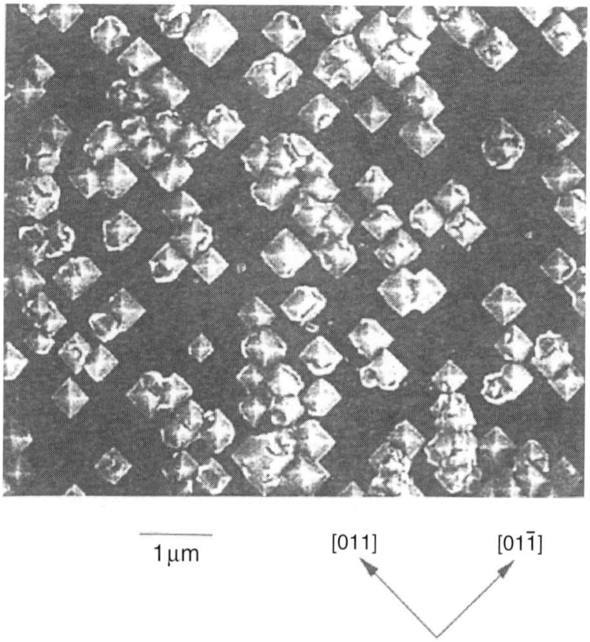

Figure 12.17. Azimuthally oriented diamond crystals on Ir(100) [10].

5 nm in width. In Figure 12.19 (b), it is seen that diamond is clearly in direct contact with Ir. The small arrow shown in Figure 12.19 (b) indicates a misfit dislocation. The presence of the misfit dislocation means that the adhesion, or chemical bonding, between carbon and Ir is fairly strong. Although the nucleation mechanism of diamond on Ir has not been made clear, it was inferred that diamonds nucleate preferentially on the protrusions of Ir, as schematically shown in Figure 12.20.

Based on the DC plasma CVD technique, Sawabe and co-workers [376] were successful in making a free-standing, transparent, 2 mm × 2 mm, 8 μm thick, (100)-oriented diamond platelet. A defect study of the film was done using ESR [406]. In the latest result, a free-standing, transparent diamond plate of 1-inch in diameter has been made by Sawabe's group.

Heteroepitaxial growth of diamond on Ir(100) was also carried out by a NIRIM-type MPCVD reactor using the BEN process [11, 374]. The experimental conditions are listed in Table 12.2. The nucleation density was higher for 740 °C than 650 °C. It had the maximum of approximately $10^8/cm^2$, when $V_b = -150$ V. The deposited diamond were clusters of nuclei, each of which had a pyramidal shape with 100–250 nm in size. A longer deposition time, however, did not lead to a continuous film at that time. Subsequently, the same group was successful in making a continuous heteroepitaxial diamond film on Ir(100) [407], as shown in Figure 12.21. In this case, similar to the previous work, a single crystal Ir(100) layer was sputter-deposited on Mg(100) at 850 °C, and the BEN treatment

Figure 12.18. SEM image of the fractured edge of the diamond film grown on Ir(100) [13].

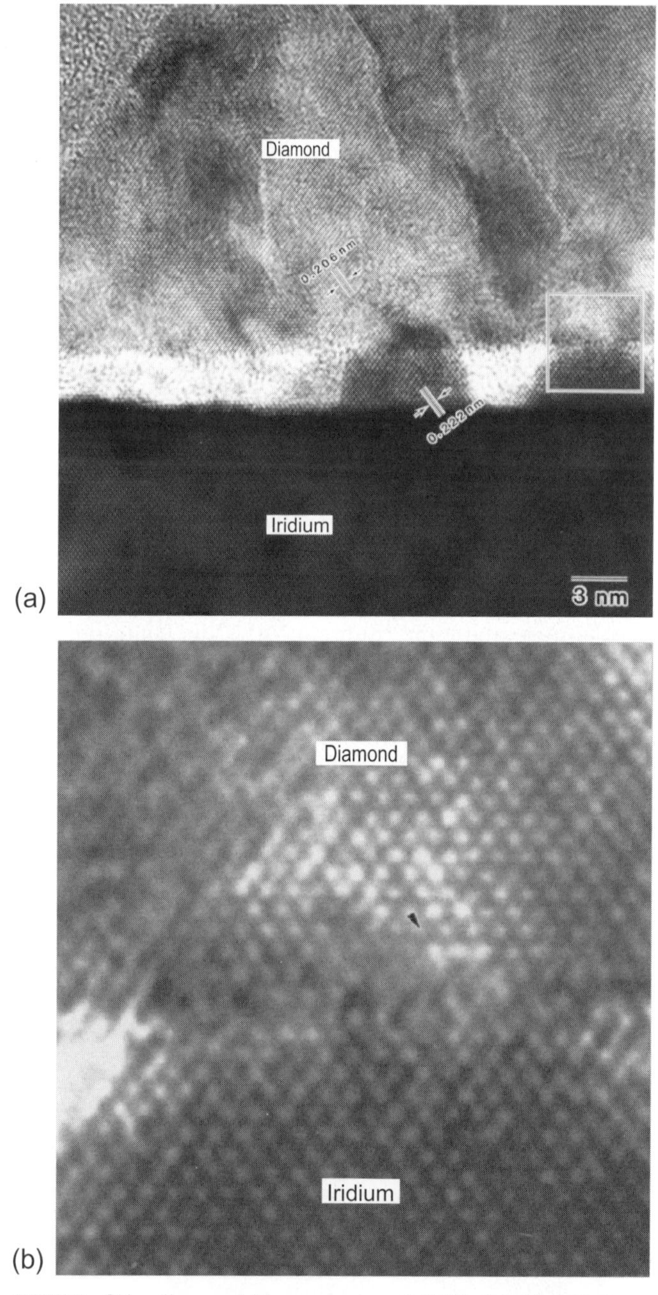

Figure 12.19. XTEM of the diamond film grown on Ir(100). (a) Global view at the interface and (b) a close-up view of the area indicated by square frame in (a) [375].

was done using a needle-shape anode, which was placed 2 mm away from the substrate. The optimized conditions for BEN and diamond growth are listed in Table 12.3 [407].

As seen in Figure 12.21, after the BEN treatment for 30 min, there were round islands of approximately 1.2 μm in diameter on the Ir surface. The diameter became larger as the biasing time was increased. Those islands served as the precursors for oriented diamond particles. Indeed, after the growth for 50 min using the ⟨100⟩ growth conditions in Table 12.3, the Ir surface was almost covered by flat diamond films. After 60 min, it was almost entirely covered with a smooth film. According to X-ray rocking curve measurements of the (400) diffraction line, the FWHM of the diamond film was only 0.16°, comparable with that of single crystal diamond, 0.11°, and much smaller than the value for an HOD film, 2.88°, that was synthesized under the conditions listed in Table 12.4 (the three-step process).

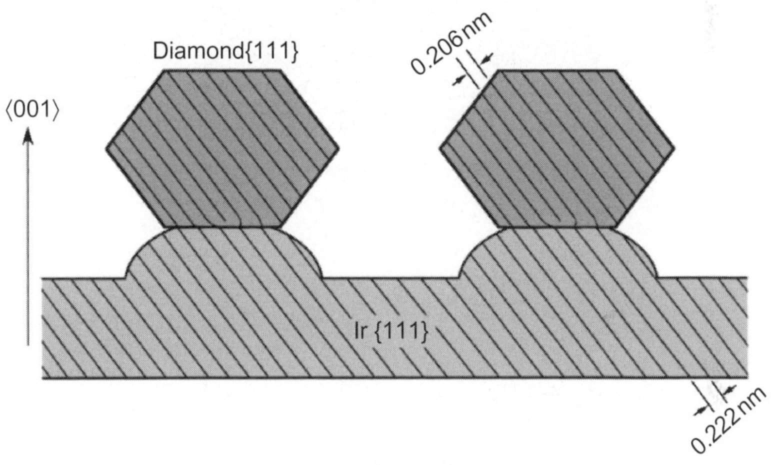

Figure 12.20. Schematic interface structure between Ir and diamond [375].

Table 12.2. Growth conditions of diamond films on Ir(100) [11].

	Diamond nucleation with bias treatment	Diamond growth
Total pressure (kPa)	4.0	5.3
Substrate temperature (°C)	650 ~ 740	700
H_2 flow rate (ccm)	100	100
CH_4 flow rate (ccm)	3.0	2.0
Bias voltage (V)	−100 ~ −200	–
Reaction period (min)	15	30
Microwave power (W)	260 ~ 380	300

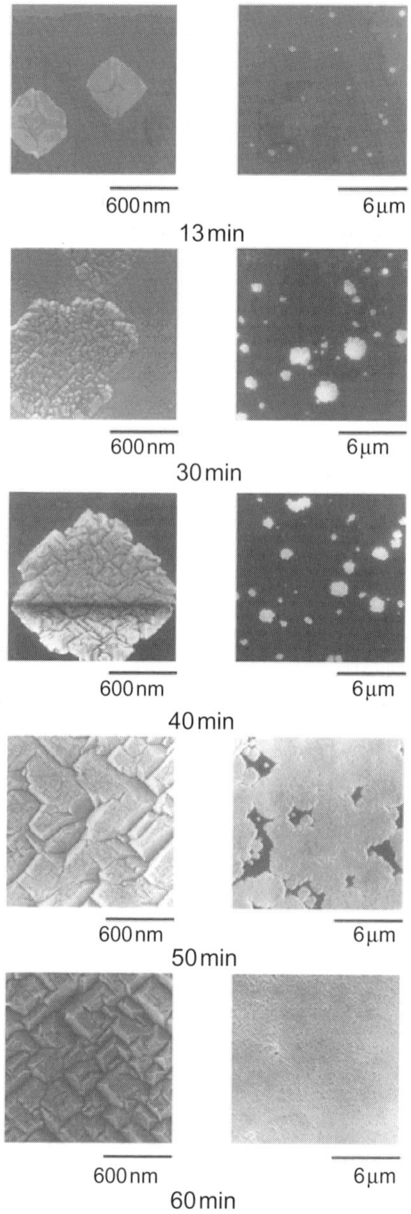

Figure 12.21. Surface morphologies of the Ir substrates, which were BEN-treated and then subjected to diamond formation step. The times are the periods of the BEN treatments [407].

The SEM image of the HOD film is shown in Figure 12.22. It was thus demonstrated that heteroepitaxial growth of diamond film could be achieved on Ir substrate by MPCVD.

In Ref. [377], Ir films were epitaxially deposited on $SrTiO_3(001)$ of 10-mm square by electron-beam evaporation at 950°C. The diamond growth was then carried out using an ASTeX reactor by first biasing at $V_b = -250$ V for 25–60 min under conditions of $c = 5\%CH_4/H_2$, $T_s = 700$ °C, $P = 30$ mbar (22.8 Torr), and $P_m = 1100$ W, and then undertaking diamond growth using $c = 1\%CH_4/H_2$. During the whole process, N_2 of 30–50 ppm was added to the source gas to achieve appropriately textured film growth conditions. After the whole processes of 2 and 20 h, diamond films of 0.6 and 8 μm thicknesses were synthesized. The (111)XPF measurements indicated that the orientational relationships of diamond, Ir, and $SrTiO_3$ were such that D{001} // Ir{001} // $SrTiO_3${001}, and D[100] // Ir[100] // $SrTiO_3$[100]. For the specimen of 2-h growth, azimuthally oriented diamond crystals with a maximum density of $10^9/cm^2$ were grown, as seen in Figure 12.23. For the 8-μm thick diamond film, the FWHM for polar and azimuthal rotations were as small as 0.34° and 0.65°, respectively.

Table 12.3. Optimized conditions of diamond growth on Ir(100) by MPCVD [407].

	BEN	⟨100⟩-growth	⟨111⟩-growth (lateral growth)
P (Torr)	30	30	30
T_s (°C)	800	700	800
c (%)	3	2	1
V_b (V)	−150	0	0
Time (min)	60	30	≥ 120

Table 12.4. Growth conditions of HOD films using the same reactor as that used for diamond growth on Ir(100) [407].

	Carburization	β-SiC formation	Nucleation	⟨100⟩ growth	First ⟨111⟩ growth	Second ⟨111⟩ growth
Total pressure (kPa)	2.7	2.7	2.7	5.4	5.4	5.4
Substrate temp. (K)	1063	1063	1063	973	1123	1043
Bias voltage (V)	0	−130	−40	0	0	0
H_2 flow rate (sccm)	98	98	98	100	100	100
CH_4 flow rate (sccm)	2	2	2	2	3	0.4
CO_2 flow rate (sccm)	0	0	0	0	0.75	0
Reaction time (min)	30	3	7	6	5	8

In Ref. [12], epitaxial Ir layers were deposited on $SrTiO_3(001)$ of 10-mm square by electron-beam evaporation at various deposition temperatures between ~280 and ~950 °C to a thickness of 150 nm for the first group of specimens, and at ~950 °C to thicknesses of ~3–200 nm for the second group of specimens. Since the lattice constants of Ir, $SrTiO_3$, and MgO are 3.8394, 3.9050, and 4.213 Å, respectively, $SrTiO_3$ is more preferable than MgO as the substrate for depositing single crystal Ir(100) films. The BEN was used for diamond nucleation. In experiment I, the BEN conditions were $c = 2\%CH_4/H_2$, $P = 22.5$ Torr, $T_s \sim 920$ °C, $V_b = -240$ V, and $P_m = 1100$ W, and the growth conditions were $c = 1\%CH_4/H_2$, $P = 22.5$ Torr, and

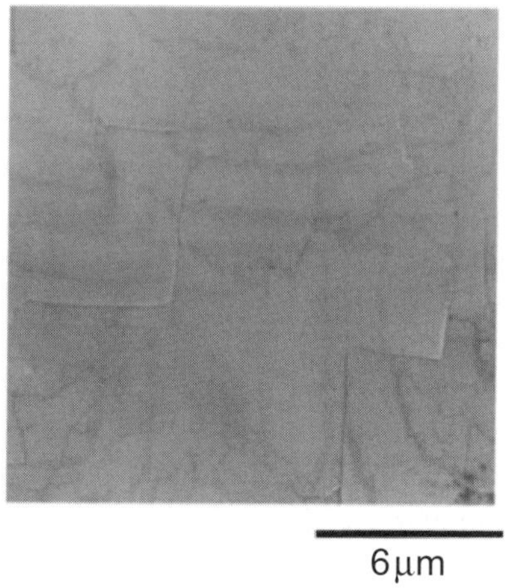

Figure 12.22. HOD film grown using the conditions listed in Table 12.4 [407].

Figure 12.23. SEM image of diamonds on $Ir/SrTiO_3$ [12].

$T_s = 770\,°C$. This process, however, resulted in a very low nucleation density. In experiment II, the BEN conditions were changed to $c = 5\%CH_4/H_2$ and $T_s = 700\,°C$, and consequently, well-aligned heteroepitaxial diamond crystallites were formed after the diamond growth. When the electron-beam evaporation was done for Ir deposition at 950 °C, the deposited Ir became like islands, networks, and a flat film for the Ir thicknesses of 3, 10, and 200 nm, respectively. In all cases, Ir was preferentially oriented along the [110] and [$\bar{1}$10] directions. For the Ir film with 150-nm thickness deposited at 950 °C, the surface had terraces with a typical width of 100 nm and a height of 1–3 nm. Also, some holes were present. At 860 °C, the surface roughness increased, and rounding structures were present but the number of holes decreased. For Ir deposition at lower temperatures, the films were flat but had polycrystalline nature. According to XPF measurements of the Ir films deposited at 950 and 370 °C as well as $SrTiO_3$, the Ir film deposited at 950 °C had sharp peaks in the (100)XPF diagram, while the Ir film deposited at 370 °C had small peaks in addition to the main peaks. The small peaks were attributed to the presence of twins, and their volume fraction was estimated to be approximately 10%. The roughness of the Ir film surface due to holes and terraces did not virtually affect the epitaxy.

The Ir film, after the BEN treatment at 920 °C for 30 min, followed by a 4-h diamond growth at 770 °C, was not covered with diamond, and there were domains of both Ir(001) and Ir(111) adjacent to each other [12]. The reason for this was not clear, but both the higher stability of Ir(111) surface over Ir(001), and the low activation energy of Ir atom transport seemed to contribute to this result. The formation of Ir(111) domains after BEN and diamond growth seems to indicate that a deposition of Ir(111) film on $SrTiO_3$(001) will be possible not simply by electron-beam evaporation but by a post-treatment. If this is actually the case, a growth of (111)-oriented diamond will occur on Ir(111).

In order to more closely study the nucleation step, the Ir surface was observed by SEM and TEM after the BEN treatment [377, 378, 405, 408, 409]. The Ir film thickness was 100 to 200 nm. The surface roughening, consisting of furrows and ridges that were preferentially aligned along [100] and [110], as shown in Figure 12.24, occurred after the BEN treatment under conditions of $c = 5\%CH_4/H_2 + 40$ ppm N_2, $T_s = 700\,°C$, $P = 22.5$ Torr, and $V_b = -250$ V, followed by the diamond growth using $c = 1\%CH_4/H_2$. In addition, dark spots were observed by SEM. The AFM and TEM studies showed that the typical height of the Ir surface structure was 3 nm, and they were {111} facetted. The dark spots were attributed to open cavities created at the Ir surface by the BEN treatment. Although the degree of roughness at the Ir(100) surface is much smaller than that of Pt(111), the creation of surface roughness and cavities are common between them. Diamonds were able to grow after the BEN treatment for 30 min, while no diamond was formed after the BEN treatment for

75 min. Thus, for the case of Ir substrates, the process conditions used for nucleation may not be used for diamond growth.

In Ref. [379], the isotopic carbon (^{13}C) atom distribution in the Ir (150-nm thick) layer, deposited on a $SrTiO_3$ substrate after a BEN step using ^{13}C-methane, was investigated by elastic recoil detection (ERD) of a 170-MeV iodine atomic beam. The ^{13}C atoms existed both at the substrate surface and near the Ir/$SrTiO_3$ interface. The fact that carbon diffuses into Ir to reach $SrTiO_3$ is similar to that of Pt described in Section 12.1, although the diffusion rate was smaller for Ir.

In Refs. [380, 381, 409], a heteroepitaxial diamond growth on Ir was done by MPCVD using an antenna for microwave emission. The distance between the antenna top and the substrate was 7–12 mm, and the area of a smooth coating was roughly 3–5 mm in diameter. The process conditions, including the BEN treatment, a ⟨100⟩-growth, and a lateral growth, used are listed in Table 12.5. It seems that the smooth heteroepitaxial film was made fairly easily based on the experience of HOD film growth. Even though the heteroepitaxial area was small, it is sufficient to fabricate diamond electronic devices.

Most recently, heteroepitaxial diamond films were deposited on Ir(100)/α-sapphire$\{11\bar{2}0\}$. The orientational relationships between Ir and α-sapphire, and diamond and Ir were as follows: Ir{100} // $Al_2O_3\{11\bar{2}0\}$, Ir[011] // $Al_2O_3[1\bar{1}00]$, D{001} // Ir{001} and D[100] // Ir[100]. The heteroepitaxial area was 50 mm^2 [410, 411].

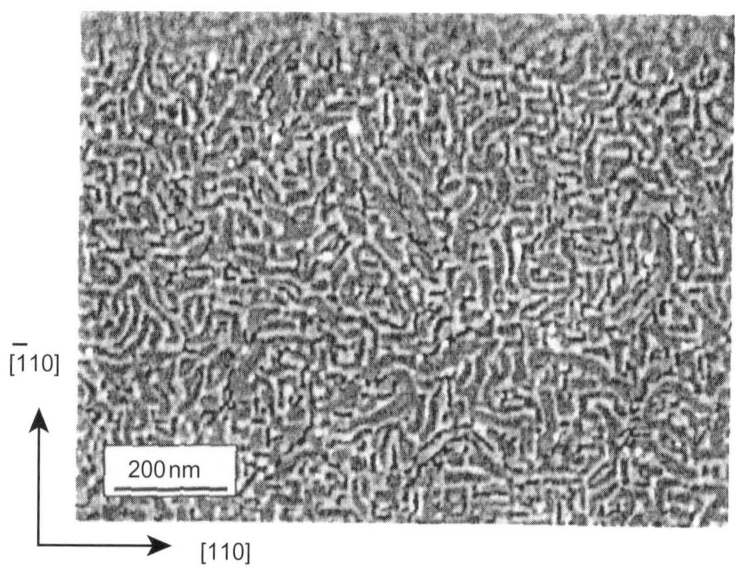

Figure 12.24. Surface of Ir layer after the 45-min BEN treatment [378].

12.3. PALLADIUM

An attempt of growing diamond on Pd(111) was done in Ref. [382], and (111)-oriented diamond films were deposited. The (111) surface of Pd was prepared from a commercially available Pd foil by repeating cold-rolling followed by annealing in H_2. The Pd substrate was scratched ultrasonically using diamond powder suspended in ethanol. The conventional MPCVD method was employed for diamond CVD using CH_4/H_2 as the source gas. X-ray diffraction and Raman spectroscopy measurements revealed that a small amount of (0001)-oriented graphite was formed together with (111)-oriented diamond particles. It seems that like the case of Pt, diamond fragments were embedded in the Pd substrate, and the growth mechanism is similar to the Pt case.

12.4. LOW PRESSURE SOLID STATE SOURCE (LPSSS)

Although the LPSSS method, invented by Roy's group in 1992, has not generated single crystal diamond films, it provides with useful information for diamond creation in metals. Indeed, the concept was already described in Section 9.2.1 for diamond heteroepitaxy on Ni. It seems that the LPSSS method will find practical applications in the future. It would thus be worthwhile to review the LPSSS method here. The LPSSS method using various carbon sources and metals (Mn, Fe, Co, Ni, Pt, Ag, Au, Cu) were published in Refs. [175, 391, 392, 412, 413], but only typical two techniques will be described below.

The first is the method to use carbon, diamond powder, and hydrogen plasma [392] but without metals. The initial material was a gel that consisted of a mixture

Table 12.5. Process conditions for the antenna-type MPCVD reactor [380, 381, 409].

	Bias nucleation (antenna-edge CVD)	⟨001⟩ Selective growth (ASTeX AX52000)	Smoothing growth (ASTeX AX52000)
Pressure (Torr)	50–150	25	35
Microwave power (W)	150	1100	750
Substrate temperature (°C)	880–920	800	830–870
Bias voltage (V)	−100 to −150	–	–
Current density (mA/cm^2)	150–200	–	–
H_2 (sccm)	95	180	400
CH_4 (sccm)	5 (5%)	2 (1%)	0.4 (0.1%)
N_2 (500 ppm)/H_2 (sccm)	0	20	–
Process time (h)	0.25–0.5	10–40	10–30

of carbon (1.8 g, Aquadag) and diamond powder (0.2 g) with a particle size of $<1\,\mu m$. To this mixture, 50 ml of 1,1,1 trichloroethane and 2-propanol mixture as well as a binder (0.1 g) were added, and an ultrasonic treatment was undertaken for 2 min. A couple of drops of the specimen was put on a Mo or Si substrate, and dried at 100 °C. The specimen was then set in an MPCVD system using pure H_2 as a reaction gas under conditions of $P=90$ Torr, $T_s=985$ °C for 5 h. In some cases, $c=1\%CH_4/H_2$ was used instead of pure H_2. Consequently, diamond grains with a diameter of approximately 10 μm were formed. Its Raman spectrum consisted of an intense 1333 cm^{-1} line due to diamond, and a weak broad band at 1528.0 cm^{-1} due to sp^2 carbon components. Namely, carbon was converted to diamond in hydrogen plasma. It seems that carbon reacted preferentially with hydrogen to generate chemically active hydrocarbon species, which were then redeposited on diamond powder. Thus, this method is perhaps considered to be a standard diamond CVD using a solid carbon source.

The second is an LPSSS method using Ni [412]. In this case, carbon (Aquadag), 10 wt% Ni powder, and 1 wt% diamond of $<1\,\mu m$ in size were mixed, and a less than 10-μm thick layer of this material was applied on an Si or Mo substrate. By treating the specimen in a plasma of $1\%CH_4/H_2$, well-facetted diamond particles were grown. By contrast, no diamond was formed, if Ni had not been added. This experiment is different from the first case in that metallic catalyst is used. It is thus inferred that Ni plays an important role in the second case.

In Ref. [414], a slurry of graphite powder was sprayed on a polycrystalline Pt substrate to form a 1-mm thick carbon layer. The specimen was first treated in hydrogen plasma at $P=40$ Torr, and $T_s=900$ °C for 30 min, which was followed by a high temperature treatment up to 1400 °C for 3 min to make a Pt–H–C phase. The specimen was then treated in the plasma of $0.5\%CH_4/H_2$ for 10 h to grow out diamonds. Diamond particles were grown on the Pt substrate, and notably, azimuthally (111)-oriented diamond particles were formed in local areas.

The mechanism of diamond formation by the LPSSS method using metal catalyst is considered to be similar to that of heteroepitaxial growth of diamond on Pt(111) and Ir(100) described in Sections 12.1 and 12.2, because a metal–C–H complex seems to be involved in the nucleation process. An advantage of the LPSSS method in view of engineering is that substrates with complex shapes could be coated with diamond.

Chapter 13

Properties and Applications of Heteroepitaxial Diamond Films

Chapter 13

Properties and Applications of Heteroepitaxial Diamond Films

So far, growth processes of oriented diamonds and diamond films have been reviewed by focusing on the process conditions. In this section, properties and applications of heteroepitaxial diamond films thus produced will be reviewed.

13.1. B-DOPING

In order to investigate the spatial distribution of B atoms in a B-doped HOD film, Graham et al. [415] deposited a 4-μm thick B-doped diamond layer on an undoped HOD film of 30-μm thickness. The B-doped layer was deposited using CH_4, H_2, and B_2H_2, where $B/C = 44$ ppm in the source gas. The specimen was thinned from the HOD film side so that TEM and CL measurements could be done for the same position of the specimen. In the CL spectrum related with dislocations, there were two bands at 2.87 eV (431 nm) and 2.32 eV (535 nm) due to bound excitons. A comparison between the TEM image and the monochromatic CL images for 2.87 and 2.32 eV indicated that the B dopants were distributed uniformly within the film on the submicron level. Furthermore, the incorporation of B dopants created dislocations in the film.

The electrical properties of single crystal diamond will be useful to study those of heteroepitaxial diamond films. As a reference, Figures 13.1 (a)–(c) show the resistivity, mobility, and carrier density of single crystal diamond as a function of temperature [107]. Figures 13.2 (a) and (b) are the resistivity and mobility as a function of the carrier density. A more thorough study on B-doped homoepitaxial diamond is presented in Ref. [416], where AFM observation of the layer surface, Hall measurements at different temperatures, and other data are presented.

The carrier concentrations and hole mobilities in HOD films were measured in Ref. [294]. In this case, B-doped diamond layers of 1.7-μm thickness were simultaneously deposited on a polycrystalline diamond film, an undoped HOD film of roughly 30-μm thickness, and a type IIa single crystal (100) using a NIRIM-type reactor. The measured carrier concentrations and the mobilities of the polycrystalline diamond film, the HOD film, and the homoepitaxial layer are shown in Figures 13.3 (a) and (b), respectively. The B concentration determined by SIMS was $4 \times 10^{18}\,cm^{-3}$ [294]. Interestingly, the surface of the B-doped layer (deposited on the HOD film) was much smoother than that of the basal HOD film, presumably

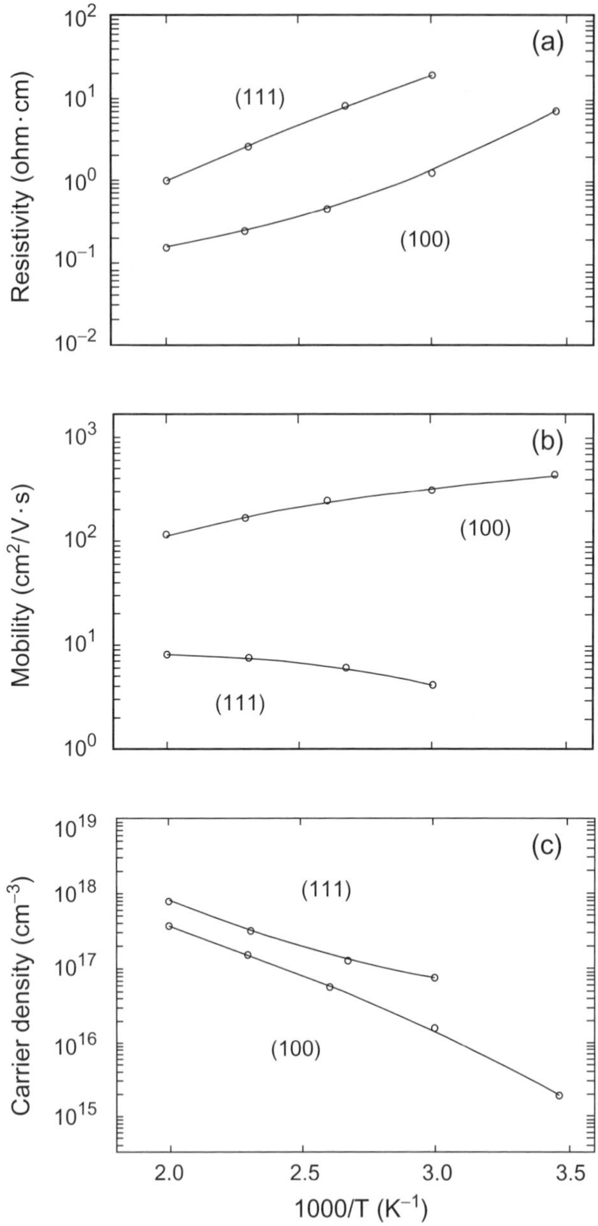

Figure 13.1. (a) Resistivity, (b) Hall mobility, (c) carrier density as a function of temperature for B-doped (100)- and (111)-oriented homoepitaxial diamond layers. The B/C ratio in the source gas of diamond CVD was 100 ppm [107].

because of the preferential (111) growth (lateral growth) of B-doped diamond layer. The upward bending of the carrier concentration at low temperatures is indicative of the occurrence of hopping conduction rather than impurity conduction that governs the conduction at higher temperatures.

In Refs. [4, 244], a B-doped layer of 1.5-μm thickness was deposited on an HOD film. For the source gas, $CH_4 + B_2H_6/H_2$ was used, where the B–C ratio was 44 ppm. The carrier concentration and the hole mobility were measured as a function of temperature, and the results are shown in Figure 13.4. In this case, there was no bending in the carrier concentration at low temperatures, meaning that the concentrations of defects, impurity atoms, and hence the carrier compensation were minimal.

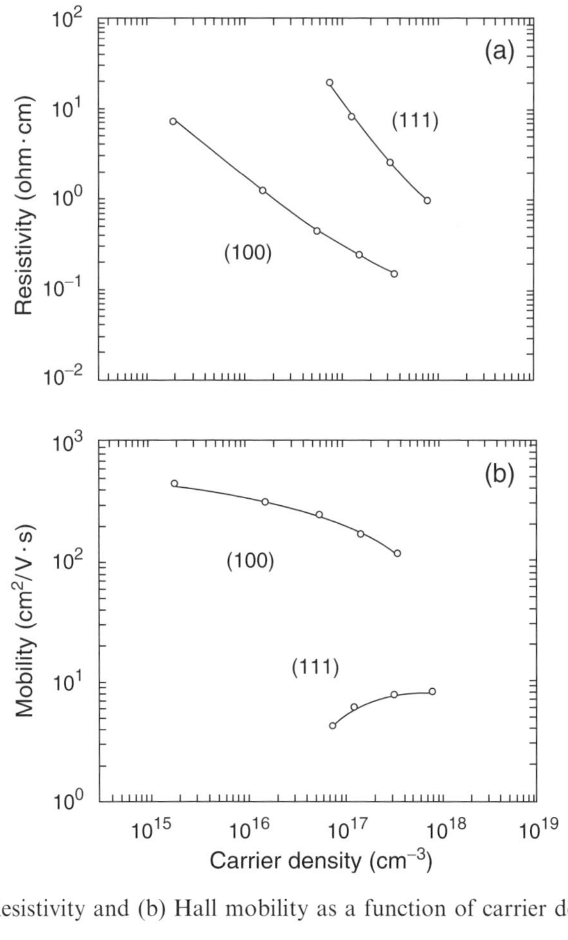

Figure 13.2. (a) Resistivity and (b) Hall mobility as a function of carrier density [107].

Table 13.1 summarizes the experimental results of carrier concentrations and mobilities of various diamond films at room temperature. The electrical conduction of B-doped diamond films (polycrystalline diamond films, HOD films, and homoepitaxial diamond layers) was also investigated in Ref. [417] in the temperature

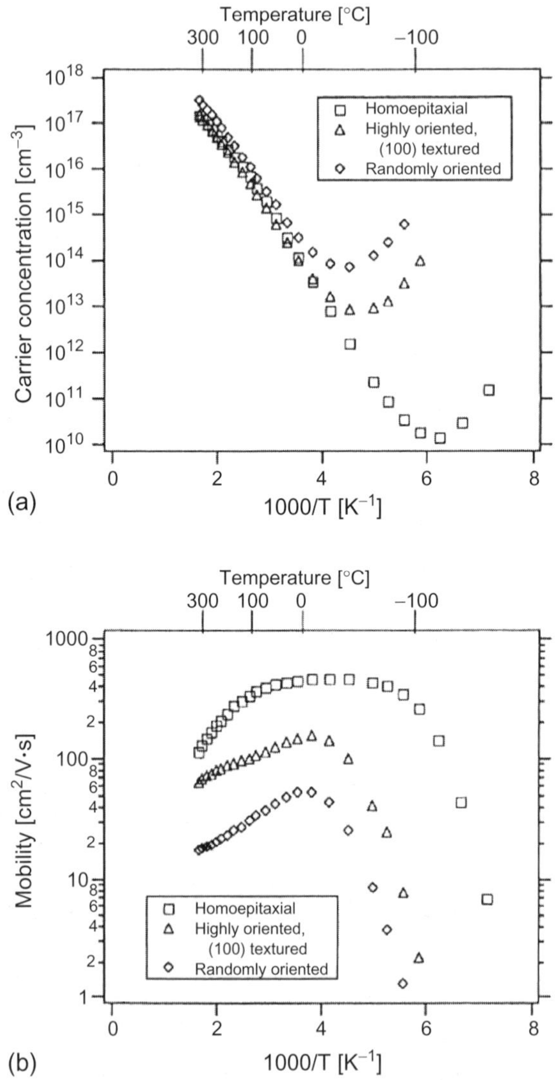

Figure 13.3. (a) The carrier concentrations and (b) the mobilities of simultaneously deposited randomly oriented polycrystalline, highly oriented (100)-textured, and homoepitaxial diamond films [294].

range of 80–600 K and the B concentrations of 1.5×10^{17} to $1.5 \times 10^{20}/cm^3$. Among nine specimens, an HOD film exhibited a mobility of 135 $cm^2/V \cdot s$ at room temperature, where the B concentration was $6 \times 10^{18}/cm^3$, and the hole concentration was $2.4 \times 10^{14}/cm^3$. This paper contains a comprehensive analysis of the electrical conduction in diamond films, and the readers are recommended to read it to know the details. In Ref. [245], an HOD film growth was done, and the Hall measurements resulted in a hole mobility of 309 $cm^2/V \cdot s$.

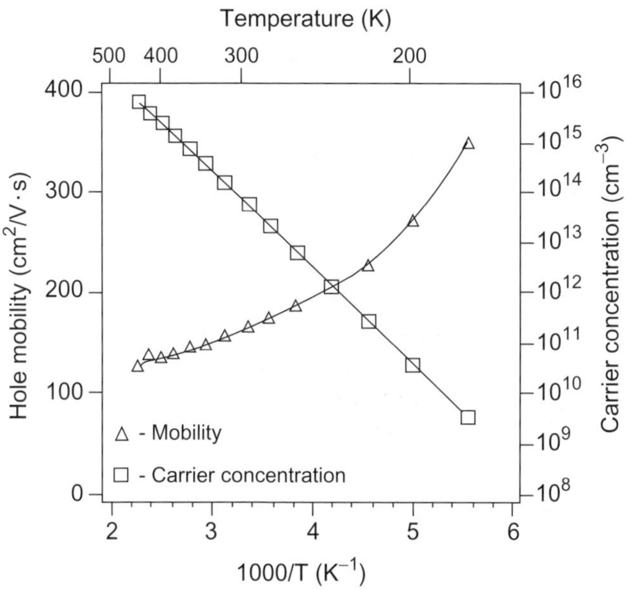

Figure 13.4. Carrier concentration and hole mobility as a function of temperature for an HOD film with a B-doped layer at the surface [4].

Table 13.1. Carrier concentrations and hole mobilities of diamond films at room temperature.

Material	Carrier concentration ($10^{14}/cm^3$)	Hole mobility ($cm^2/V{\cdot}s$)	Activation energy (eV)	Reference
Polycrystalline diamond film	6.8	47.8	Typical value	[294]
		50	is 0.356 eV	[244]
HOD film	2.4	135	0.34 eV	[294]
		165	0.356 eV	[4]
	0.6	278		[244]
		309		[245]
Homoepitaxial diamond layer	3.1	427	0.356 eV	[294]
		1340		[244]

13.2. PIEZORESISTIVE DEVICES

B-doped p-type diamond films are known to exhibit piezoresistivity. Namely, the resistance increases under compressive stress [418]. The piezoresistivity is expressed by the gauge factor K, which is defined by:

$$K = \frac{(\Delta R/R)}{\varepsilon} \tag{13.1}$$

where R is the electrical resistance under strain ε and ΔR is defined by $\Delta R = R - R_0$, where R_0 is the resistance without strain. The strain ε can be evaluated by:

$$\varepsilon = \frac{3hs}{2l^3}\left(l - \frac{x_1}{2}\right) \tag{13.2}$$

The notations in Eq. (13.2) are shown in Figure 13.5 [419].

The gauge factor K can also be calculated as follows: assuming that the biaxial stress σ is present in the film, it is evaluated from the shift of the Raman peak position $\Delta\nu$ (in units of cm^{-1}) from the corresponding peak position of natural diamond at $1333\,cm^{-1}$ by the following relationship [420]:

$$\sigma = -2.63 \times 10^{10} \Delta\nu \text{ (dyn/cm}^2\text{)} \tag{13.3}$$

The $-$ and $+$ signs correspond to the compressive and tensile stresses, respectively. The strain ε can be given by:

$$\varepsilon = (S_{11} + S_{12})\,\sigma, \tag{13.4}$$

where S_{11} and S_{12} are the elastic compliances of diamond, i.e. $S_{11} = 9.524 \times 10^{-14}\,cm^2/dyn$ and $S_{12} = -0.9913 \times 10^{-14}\,cm^2/dyn$. Thus, the gauge factor K is calculated using Eq. (13.1).

In Ref. [421], an HOD film of about 6-μm thickness were deposited on p-type Si by the two-step process using a custom-made CVD reactor under the conditions given in Table H.4. Then, B ions were implanted at $-76\,°C$, which was followed by a rapid thermal annealing. The B concentration was $10^{18}/cm^3$, the hole mobility was $80\,cm^2/V \cdot s$, and the hole concentration was $10^{16}/cm^3$ at $20\,°C$. The gauge factor K was as high as 1200 at 100 μ strain at room temperature, which is far superior to 6 for polycrystalline diamond films and 550 for homoepitaxial film.

It was demonstrated by Kohn et al. that HOD films are useful for sensor and MEMS applications because of the high mechanical strength, the high thermal conductivity, the low thermal expansion coefficient, and the chemical inertness of diamond. In particular, HOD films are advantageous over polycrystalline films, because their surfaces are flat enough to fabricate sensors and MEMS structures on them. Piezoresistive sensor applications of HOD film are presented in Refs. [16, 419, 422]. The HOD film was grown on a 3-inch Si(100) wafer using AC-BEN [344]. Figure 13.6 shows free-standing diamond cantilevers, where a B-doped piezoresistors were fabricated at the root of the beam, just like the one shown in Figure 13.5. The gauge factor was $K=5.1$, significantly smaller than the value of 1200 observed previously. A most recent comprehensive work of piezoresistivity [419] of free-standing B-doped diamond films, including HOD films, unidirectionally oriented films, and polycrystalline films indicated that the values of K for HOD films at room temperature were only between 4 and 28. The highest K-value was obtained from a micrometer-size piezoresistor that included only a single (100) grain. It therefore seems that previous values around $K=1200$ arose from problems

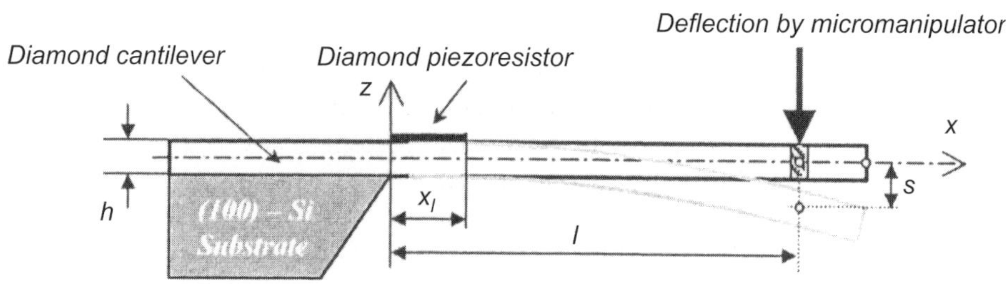

Figure 13.5. Test structure for gauge factor measurements [419].

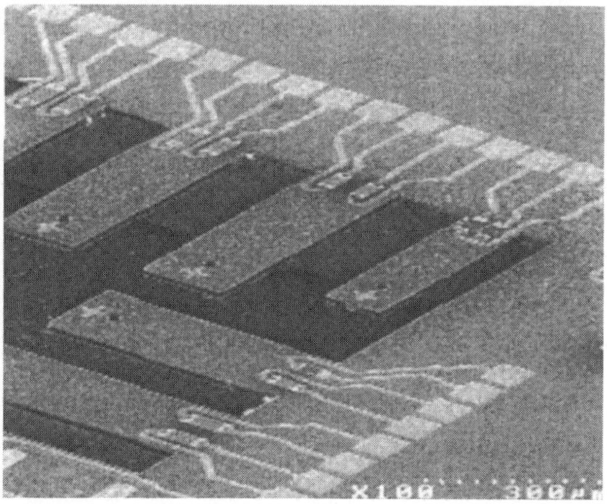

Figure 13.6. Diamond cantilever structures of different geometries made by dry etching an HOD film. It is seen that piezoresistors are integrated at the suspension point [419].

in data analysis, and the *K*-values obtained by Kohn et al. may be more reliable. The gauge factor was dependent on temperature, and its activation energy was 177 meV for HOD and ⟨100⟩-textured films.

Figures 13.7 (a) and (b) show an acceleration sensor also made using an HOD film of 15-μm thickness, where the seismic mass was Si [422]. Its resonance frequency was 4.7 kHz, the fracture strength was 4.5 GPa, and the estimated shock tolerance was 7000 G. The decay time of the seismic mass oscillation was within seconds.

13.3. INK-JET MODULE

Diamond has a low heat capacity, a low thermal expansion coefficient, and a high mechanical and thermal stability. These properties are very useful for devices using high dynamic thermal stress such as ink-jet heads. Indeed, ink-jet heads were fabricated using diamond films [17, 422]. Figure 13.8 shows a thermal actuator

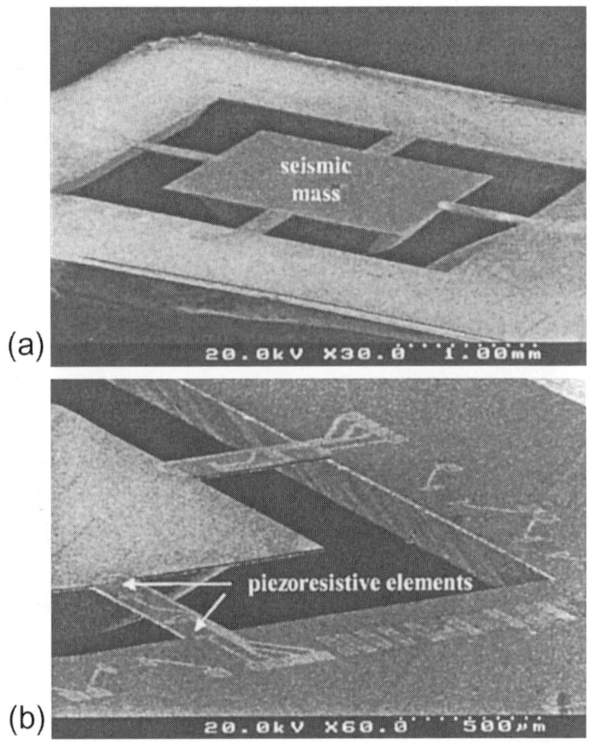

Figure 13.7. (a) Acceleration sensor made of HOD film and (b) piezoresistive sensors attached at the suspension points [422].

made by B-doped polycrystalline diamond films that are used both for a heater (a large film in the center) and thermistors (temperature sensors) on both sides of the heater. The B-doped diamond film is conducting and a test device showed that 10 W could be applied for 10 μs to a 60 × 60 μm diamond heater, yielding an electrothermal conversion of 0.3 MW/cm^{-2}. Moderately B-doped diamond films can be used for thermistor as studied in Refs. [418, 423–426]. In the ink-jet device, free-standing HOD film was used as a nozzle plate because the HOD film was flat [17]. Figure 13.9 shows a cross-sectional diagram of the ink-jet device and Figure 13.10 shows a fabricated module and its components. It should be noted that this device could also be used as a reactor for DNA chain synthesis, because a precise control of reagent dosage is possible [427].

Figure 13.8. Diamond thermal actuator with two thermistors at each side to monitor the thermal transient [17, 422].

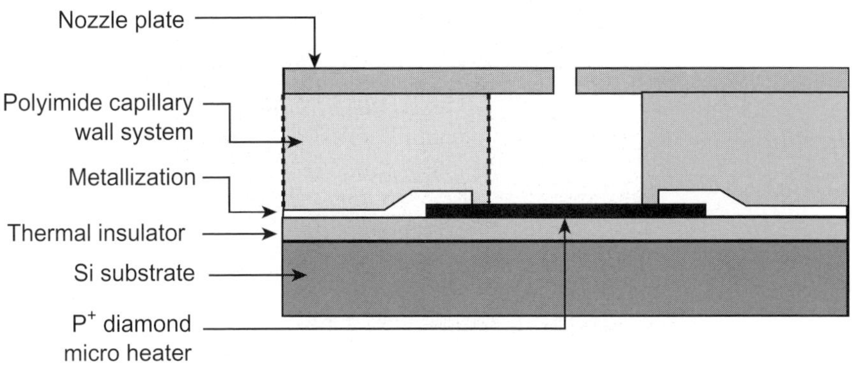

Figure 13.9. Cross-sectional diagram of ink-jet device [17].

Since the surface of HOD film is fairly flat, micro electro-mechanical system (MEMS) structures can be fabricated on the film even without polishing the surface. This advantage was actually used for diamond switching devices. Figures 13.11 (a) and (b) show SEM images of a micro-relay and a magnified view, and Figure 13.12 shows a fabrication step of the cantilever beam microswitch [428]. The switching time of a few millisecond was achieved, and the device worked even at 650 °C [17, 428, 429].

13.4. pH SENSORS

As will be stated in Section 13.7, the H-terminated diamond surface possesses a 3–10 nm thin layer that is a p-type semiconductor with a hole density of

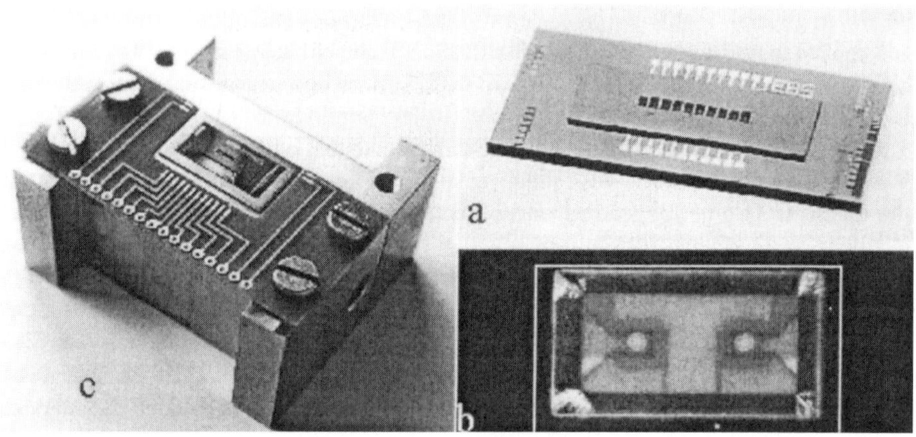

Figure 13.10. (a) Ink-jet array with 10 devices, (b) dual nozzle part of the device, and (c) ink-jet module [17].

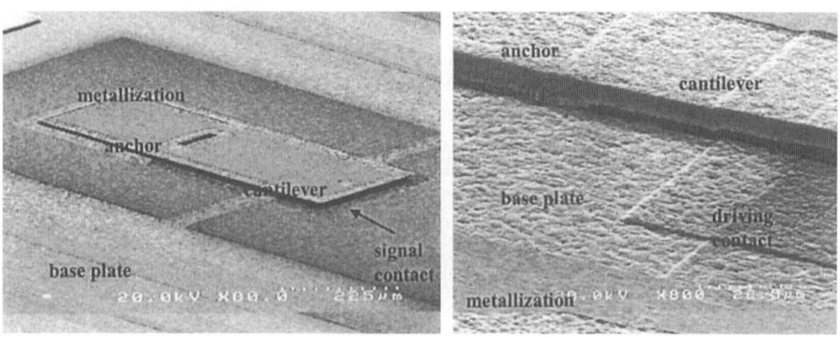

Figure 13.11. (a) Diamond relay and (b) its magnified view around the anchor, cantilever, and capacitive driving contact area. The base layer is an HOD film [17].

approximately $10^{13}/cm^2$. Using the surface conducting layer, a pH sensor was fabricated in Ref. [17]. The sensor material was a H-terminated, lightly B-doped HOD film, and the sensor had a structure of field effect transistor (FET) modified for a pH sensor, as shown in Figure 13.13. It has source and drain electrodes fabricated on the HOD film. The sensor channel was open and in contact with the

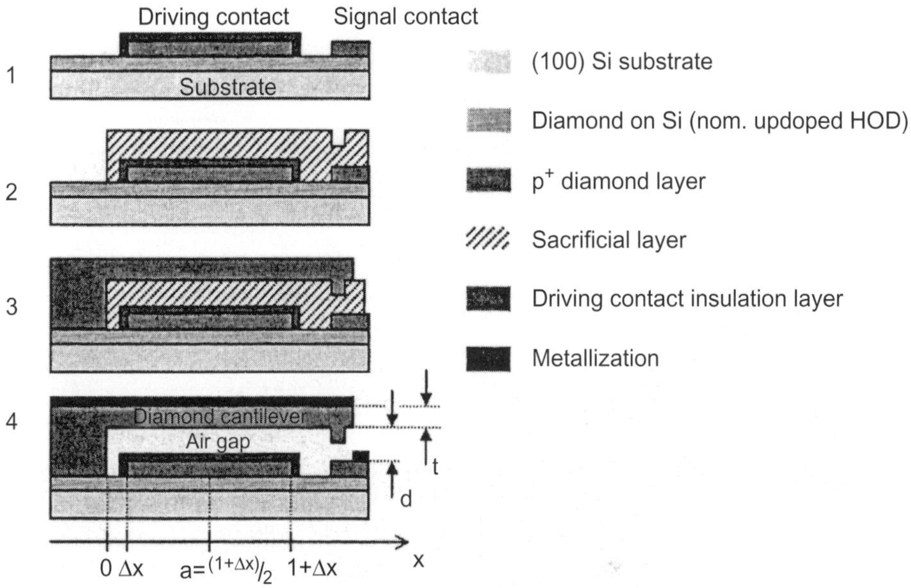

Figure 13.12. Fabrication step of the cantilever beam microswitch [428].

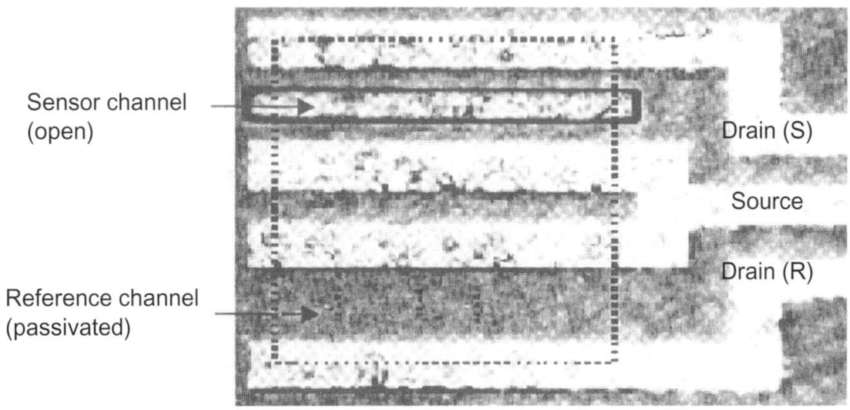

Figure 13.13. pH sensor made of HOD film [17].

solution to be measured, and otherwise it was passivated to protect the sensor from strong acid and alkaline solutions. The channel corresponds to the gate of FET. The pH of the solution influences the electric field in the surface conducting layer, thus changing the electrical resistance between the source and drain electrodes. Figure 13.14 shows the drain current against the bias applied to a reference electrode. The drain current changes as a function of the gate voltage, indicating that this device works as an ion sensitive FET.

13.5. ULTRAVIOLET SENSOR

Since the band gap of diamond is 5.47 eV, it absorbs lights with wavelengths λ shorter than 227 nm, generating electron-hole pairs. In an electric field, the electrons and the holes are separated to move and reach the electrodes with opposite voltages. In other words, diamond is photoconducting for ultraviolet (UV) lights with $\lambda \leq 227$ nm. The photoconducting properties of diamond can be used for solar light-insensitive UV sensor, and indeed UV sensors were fabricated in Refs. [430–434].

In Ref. [430], a pair of interdigited electrodes (Al or Pt) with a gap of 5 to 15 μm were fabricated at the surfaces of 10-μm thick HOD films. Figure 13.15 (a) is a cross-sectional diagram of the UV sensor, and Figure 13.15 (b) is an optical microscopic image of the electrode structure, where the dark lines are the open surface of the HOD film that receive the light, while the bright areas are the Pt thin film electrodes. Figure 13.16 shows a SEM micrograph of the sensor surface [430]. It is seen that Pt electrodes are formed on an HOD film consisting of (100) diamond faces.

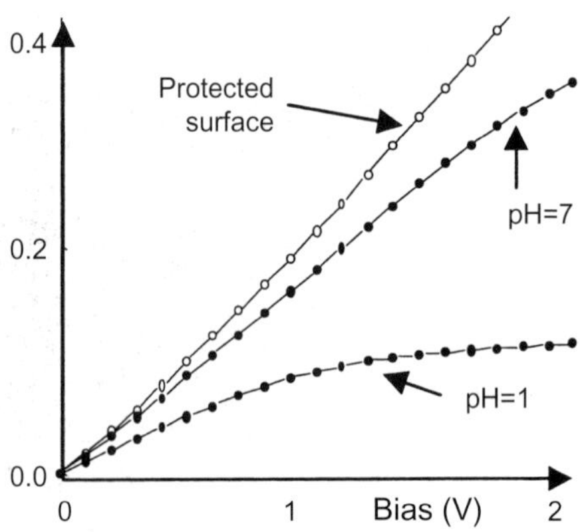

Figure 13.14. Drain current vs gate bias voltage characteristics of the pH sensor [17].

The temporal response of the sensor (in units of output voltage V) made of an HOD film is shown by solid squares in Figure 13.17 [431]. The sensor was irradiated by an ArF laser (193 nm), where a DC bias of 10 V was applied between the interdigited electrodes. The FWHM of the laser pulse was roughly 5 ns, and it is seen that the sensor followed almost exactly the laser pulse in nanosecond time scale. A direct irradiation of the ArF laser on the UV sensor caused no damages at all. The open circles in Figure 13.17 correspond to the temporal response of an UV sensor made of polycrystalline diamond film. The sensitivity is at least five times inferior to that of the HOD UV sensor. The poor sensitivity of the UV sensor made of polycrystalline diamond film can

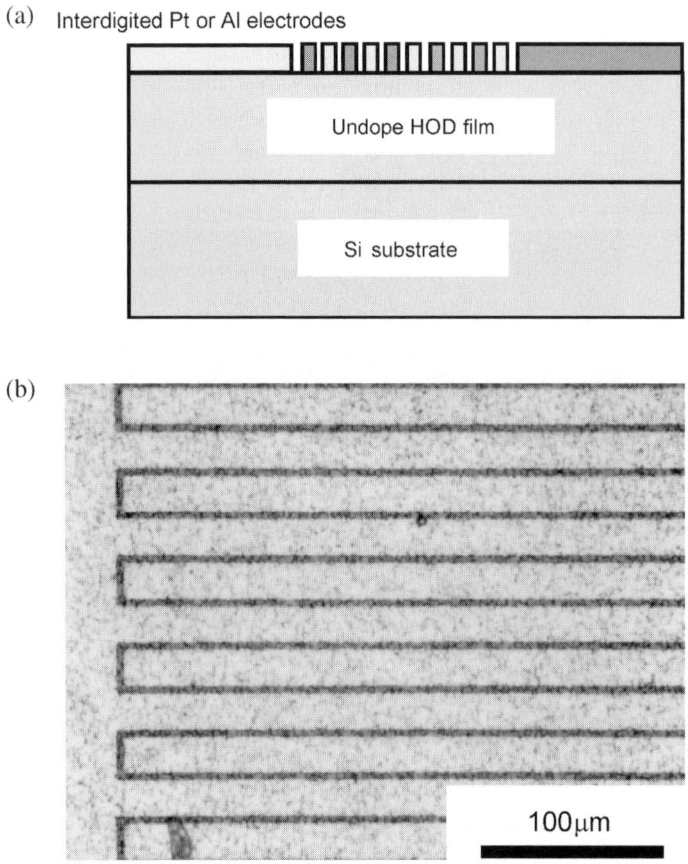

Figure 13.15. (a) Schematic diagram of the cross section of the UV sensor and (b) the interdigited electrodes fabricated on the HOD film surface. The dark lines are open gaps that are 5 to 15 μm width [430].

be attributed to the high density of grain boundaries that prevent the transport of photogenerated carriers.

The absorption coefficient as a function of light wavelength was different between HOD and polycrystalline diamond films, as seen in Figure 13.18. Note that the two data for HOD and polycrystalline diamond films are separately normalized at the band gap position. The photocurrent of the HOD film sharply increases as the wavelength approaches the band gap, while it is opposite for the polycrystalline diamond film. This makes a large difference in the sensitivity against the laser light with different wavelength. Indeed, the UV sensor made of HOD film exhibited a marked selectivity to the photoresponse intensity against laser lights with different wavelength. Figure 13.19 shows a temporal response of an HOD UV sensor. Just like Figure 13.17, it exhibited a high sensitivity to the ArF laser light (193 nm). By contrast, the sensitivity to the dye laser (313 nm), shown in the inset of Figure 13.19, was roughly 1% of that to the ArF laser. It was thus demonstrated that the UV sensor made of HOD film has a fast temporal response to the ArF laser pulse, a high selectivity to the light for a wavelength of $\lambda \leq 227$ nm, and a robustness against a direct laser irradiation. All these features arise from the

Figure 13.16. SEM micrograph of Pt electrodes on an HOD film [431].

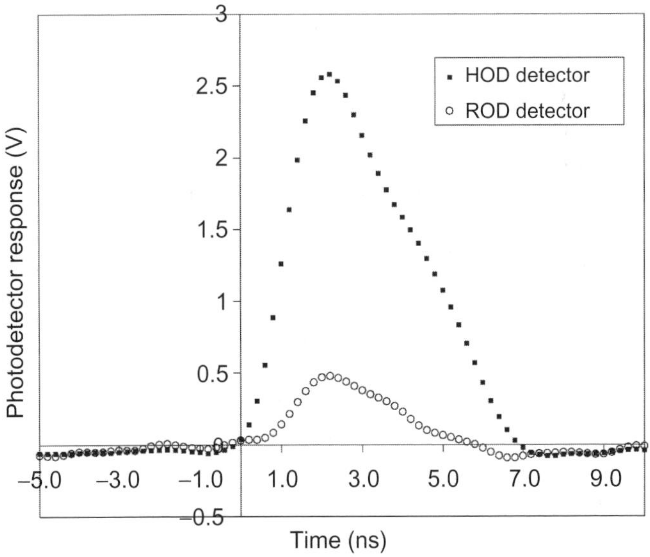

Figure 13.17. Temporal responses of the UV sensors made of an HOD film (squares) and a randomly-oriented polycrystalline diamond film (open circles) [431].

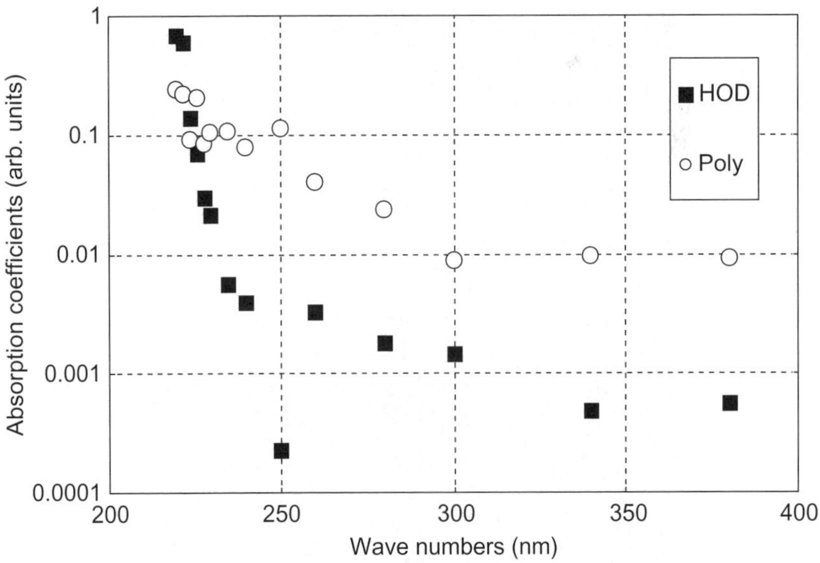

Figure 13.18. Photoresponses (output voltages) of HOD film (solid squares) and polycrystalline diamond film (open circles) as a function of wavelength [431].

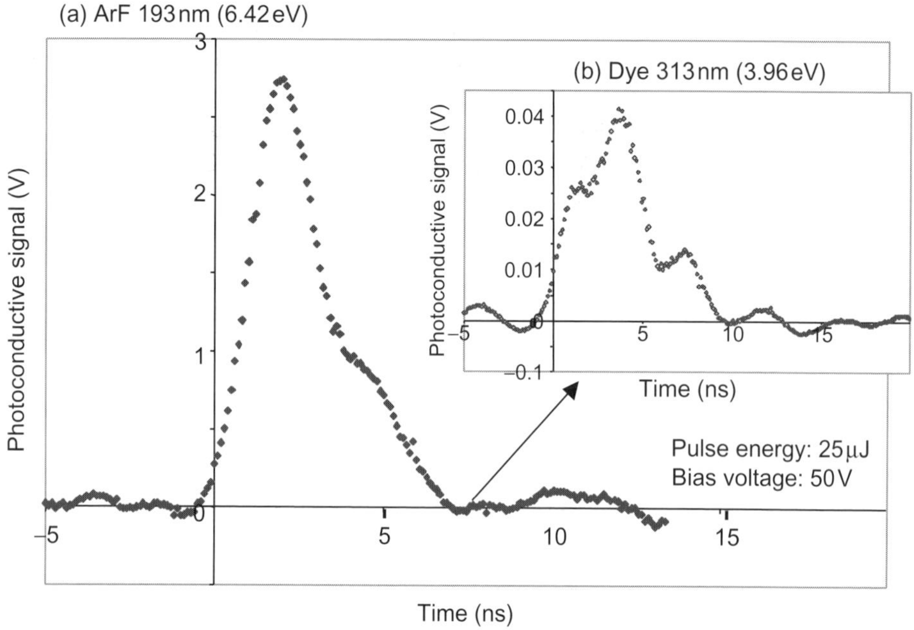

Figure 13.19. Photoresponse (output voltage) of diamond UV sensor against irradiations of ArF and dye lasers. The sensitivity to the ArF laser is 100 times better than to the dye laser [430].

fact that the density of grain boundaries in HOD films are significantly smaller than that of polycrystalline diamond films, and its crystal quality is better. In this sense, HOD films are advantageous over polycrystalline diamond films for sensor applications.

13.6. CATHODOLUMINESCENCE

The cathodoluminescence spectra of heavily B-doped (111) coalesced films on Pt(111) had an edge emission band at 5.0 eV at room temperature [385, 435]. In Figure 13.20 [435], B-doped films were synthesized by MPCVD with B/C = 3300 ppm in the source gas. When the film thickness was 1.8 μm, there was an intense band A in the wavelength region of 400 to 600 nm in addition to a weak band at 248 nm (5.0 eV). When the film thickness was increased to 16 μm, band A entirely disappeared, and the CL spectrum consisted of the 248-nm band only. The major difference between the two films was that the 1.8-μm thick film consisted of (111)-oriented diamond grains, while the (111) faces were significantly coalesced in the 16-μm thick film. Thus, it seems that the significant reduction in grain boundaries suppressed the band A emission process, and enhanced the 248-nm emission process.

The temperature dependence of the 248-nm band exhibited a curious behavior, as seen in Figure 13.21 [435]. The band intensity at 300 K was stronger than that at 89 K. The cause of this behavior has not been made clear, but this study suggested that an extremely heavy doping was attained for the (111)-oriented diamond film on Pt(111) as (111) faces of diamond can incorporate more B atoms than other faces.

The origin of the intense 248-nm emission without band A has not been elucidated yet. A speculation is that the B impurity band is extended deep inside the Brillouin zone, and a direct transition from the conduction band minimum to the impurity band is involved in addition to the ordinal indirect transition

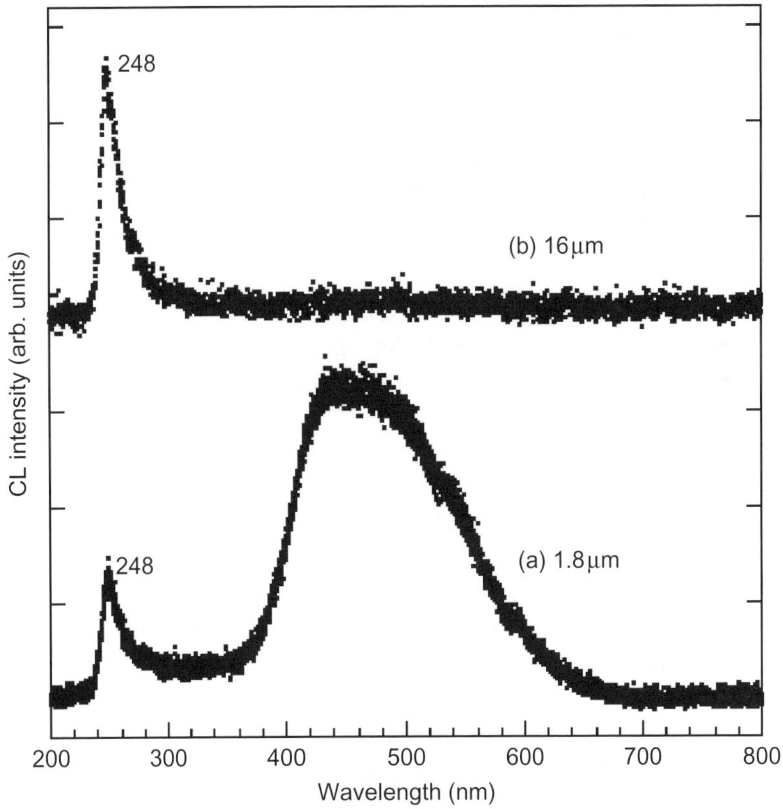

Figure 13.20. CL spectra of heavily B-doped diamond films of (a) 1.8 μm and (b) 16 μm thicknesses. The B/C ratio in the source gas was 3300 ppm [435].

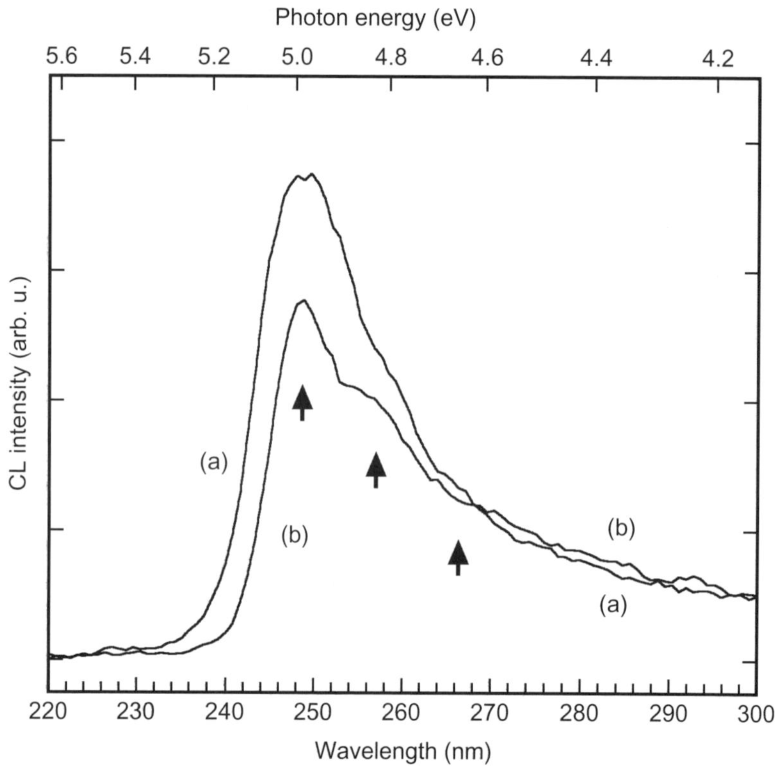

Figure 13.21. CL spectra of heavily B-doped diamond films at (a) 89 K and (b) 300 K. Note that the intensity at room temperature is higher than the intensity at 89 K. The B/C ratio in the source gas was 3300 ppm [435, 437].

assisted by phonons. This phenomenon suggests that UV emission diodes can be made using a heavily B-doped diamond film. Indeed, Yokota et al. [436] has made a diode with metal/undoped diamond/heavily B-doped diamond on Pt(111), and observed an UV emission at 246.8 nm, as shown in Figure 13.22. The UV emission was made possible because the heteroepitaxial diamond film grown on Pt(111) consisted of diamond (111) faces that could be more heavily doped with B than (100) faces.

13.7. HOLE CONDUCTION AT H-TERMINATED SURFACE

Diamond is electrically a good insulator because of its large band gap (5.47 eV), but the diamond surface grown by CVD was found to be p-type and conducting.

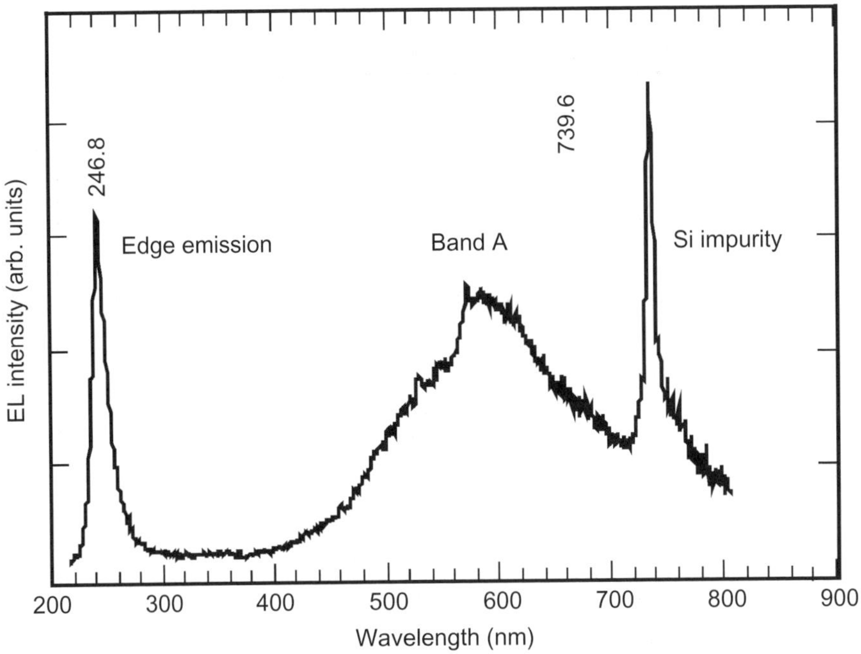

Figure 13.22. Emission spectrum from a UV emission diode [436].

The hydrogen-terminated diamond surface also exhibits a p-type conduction. The surface structure, electronic properties, etc. have been extensively studied. This subject has a long history of research, and the readers can refer to Ref. [137] on this topic. Simply summarizing the results obtained so far, the thickness of the Surface conducting layer is 30–100 Å, and the hole density at room temperature is approximately $10^{13}/cm^2$.

13.8. FIELD EFFECT TRANSISTORS

In Ref. [438], a metal/semiconductor field effect transistor (MESFET) was fabricated on an HOD film using a 4°-off, 20-μm thick β-SiC film/Si as a substrate. Despite the large off-angle, the tilt angle of the HOD film measured by X-ray rocking curve was only 1°. The fabricated MESFET with the gate length of 5 μm exhibited a transconductance of 7 mS/mm. So far, this is the only work of diamond FET using HOD and other heteroepitaxial films. However, a FET work is ongoing in Kobe Steel using an HOD film as a substrate, in which a p-i-p structure is used as the FET structure. For FET works, it is recommended to read Refs. [15, 137, 439, 440].

13.9. FIELD EMISSION

It is empirically known that *as grown* diamond films do not exhibit a charge-up under SEM observation, while it is not so for single crystal diamonds. This is most likely because of the effect of surface conduction by H-termination. It is also known that H-terminated undoped and B-doped diamonds exhibit negative electron affinity (NEA), where the vacuum level is energetically lower than the conduction band. It is thus expected that once electrons are excited to the conduction band of diamond, they are spontaneously emitted to the vacuum presumably across a small barrier at the surface.

Field emission from a heteroepitaxial undoped (100)-oriented diamond film of 20-μm thickness deposited on Ir(100) substrate was investigated in Ref. [441] using a W needle as an anode. As compared with a polycrystalline diamond film also deposited on Ir(100), the heteroepitaxial film emitted electrons at lower voltage, roughly 1/3 of the voltage needed for the polycrystalline film. The emission was uniform over the entire area of 3 mm in diameter. The threshold voltage when the emission current was 10^{-11} A was estimated to be 40 V/μm, which was fairly high, because the film surface was flat and the film was thick.

Field emission measurements were also done using a B-doped, heteroepitaxial, (100)-oriented diamond film grown on Ir(100) [442]. The observed results are

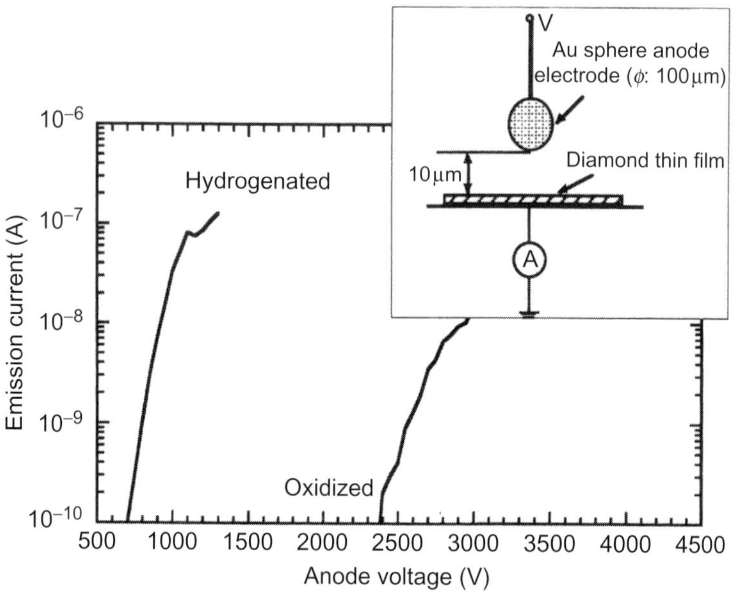

Figure 13.23. Field emission properties from a heteroepitaxial diamond film grown on Ir(111). The inset shows the experimental setup [442].

shown in Figure 13.23, where the inset shows the experimental setup. In this case, the anode was an Au sphere of 100 μm in diameter and the separation from the diamond film surface was 100 μm. The field emission with a current of 10^{-10} A from a hydrogen-terminated film occurred at 600 V (6 V/μm). By contrast, the emission voltage was 2400 V (24 V/μm) for an oxygen-terminated heteroepitaxial film. Notice that in both Refs. [441, 442], the diamond surfaces were very flat. Assuming that the specimens have the NEA property, it follows that electrons excited to the conduction band are emitted without high voltage. These experiments indicated that the electron excitation to the conduction band is not an easy task.

Chapter 14
Conclusions

Chapter 14

Conclusions

One of the ultimate goals of the research of heteroepitaxial growth of diamond is to make single crystal diamond wafers by CVD. Once such wafers become possible to make at a reasonable cost, high value-added products such as electronic and optical devices will be quickly developed and put on the market to use for various applications that have never been achieved by conventional materials. Table 14.1 shows applications of diamond films currently under study or already on production. Since each item consists of a number of segmented areas, and there are more items not shown in the table, it is expected that diamond films will create a market of billions of dollars worldwide in the coming years. The source material of diamond is carbon that inexhaustibly exists on Earth. Most importantly, diamond is biocompatible and chemically inert, thus it does not pollute our environments unlike other semiconductor elements. In this sense, diamond is a gift from the heaven for human beings in the twenty-first century [443].

Among diamond films with various morphologies, HOD and heteroepitaxial diamond films will provide us with the best material for highest value-added, high performance applications. In this sense, the works reviewed in this monograph have given a great contribution.

However, we still have to establish several technologies on heteroepitaxial growth of diamond, as listed below:

1) Increase the grain size to reduce the grain boundaries,
2) Increase the growth rate,
3) Increase the uniform quality area,
4) Reduce the atomic-level defects in each grain of the film,
5) Synthesize oriented and heteroepitaxial films on large area without bow,
6) Synthesize transparent films,
7) Synthesize thick free-standing films, and
8) Create production-type CVD reactors.

In terms of science, it is expected that diamond still has some new properties that have never been exploited yet. Since the establishment of diamond CVD technology, the controlled doping for p- and n-type semiconductors, the presence of the ultrathin conducting layers of hydrogen-terminated surface, the negative electron affinity, and the heteroepitaxial growth are discovered. Most recently, the presence of high-density exciton was found, and also B-doped diamond was found to be superconducting. More new findings are expected to follow as diamond is continued to be studied.

Under those circumstances, oriented and heteroepitaxial growth of diamond will certainly play an important role for both science and technology.

Table 14.1. Materials properties of diamond and its future applications.

Materials properties	Applications
Hardness and anti-abrasiveness	AFM probe, MEMS (micro-relays, acceleration sensors, micro-gripper), coating on surgery scalpels, coating on cutting tools, tab tools
High thermal conductivity	Heat sinks for semiconductor lasers, power ICs, and multi-yip modules, heater for inkjet head, flow sensors
Transparency	Windows for infrared lasers and gyrotrons, optical lenses
Absorption of UV light	UV sensor
Radiation hardness	Detectors for neutrons and a particles, X-ray lithography masks
Large band gap	UV radiation devices for sterilization, light sources for UV microscopes and displays
High electronic properties	High-power, high-frequency transistors, high-voltage diodes, ion sensors, active heat sinks, gas sensors, and thermistors
Field emission of electrons	Flat panel displays, electron beam lithography, transmission wave tubes, electron guns, micro X-ray sources
Chemical electrodes	Materials sensors, biosensors, electrochemical decomposition of organic materials, electrodes for organic synthesis
Biocompatibility	DNA tips, bioreactors, micro-reactors

Appendices

Appendix A

Notations and Units

A.1. ORIENTATION

It is sometimes very confusing to express the orientation of diamond faces and growth directions. In the present article, a crystal face will be expressed by such a notation as (100), and an expression of "(100)-oriented films" will be used to describe diamond films with azimuthally oriented faces at the film surfaces. To express a unidirectional growth orientation, such a notation as ⟨100⟩ will be used. This expression will be used particularly to express the direction of uniaxial growth. Some authors use this notation to denote a direction normal to a crystal face. Since the figures in the present book were copied from the literature, such notations are also used in some figures. To show a direction in a certain portion of a crystal, such notation as [100] will be used. A notation, {100}, will be used to express (100) lattice planes in general. On the other hand, to show that the {100} lattice planes of diamond are parallel to those of SiC, such a notation as D{100} // SiC{100} will be used, where D stands for diamond. Similarly, G{0001} will be used to express (0001) lattice planes in graphite.

A.2. PROCESS CONDITIONS FOR CHEMICAL VAPOR DEPOSITION (CVD)

For diamond CVD, various growth parameters are used. They are listed below along with their symbols:

c *Composition of the source gas*: When 1 vol%CH_4 diluted with H_2 is used for diamond growth, it will be written as $c = 1\%CH_4/H_2$. In most cases, the unit for the gas concentration is vol%, unless noted, and thus "vol" will be omitted. The unit, sccm, stands for the gas flow rate, i.e. standard cubic cm per minute.

P *Gas pressure of the source gas*: Its unit used in the present book is Torr, which is not recommended by IUPAC. This unit however has been most frequently used by diamond researchers. Note that 1 Torr = 133.3 Pa, 1 hPa = 100 Pa, and 1 mbar = 0.76 Torr = 101.3 Pa.

T_s *Substrate temperature*: The substrate temperature is most frequently "measured" by a single-wavelength optical pyrometer without correction of emissivity. The emissivity from specimen is usually unknown so that it is assumed to be unity. Thus, T_s is only a read of the pyrometer display. Even so, the numbers are useful to reproduce the substrate temperature in the process conditions.

P_m Input microwave power (W)

V_b Bias voltage (V)

$I_b(t)$ Bias current (mA) as a function of time t

A.3. STRUCTURAL PARAMETERS

a, *b* and *c* Lattice constants of crystals in units of nm or Å (1 nm = 10 Å)

A.4. ANALYTICAL TECHNIQUES

AES	Auger electron spectroscopy
AFM	Atomic force microscopy
CL	Cathodoluminescence
ED	Electron diffraction
EELS	Electron energy loss spectroscopy
FT-IR	Fourier transform technique of infrared spectroscopy
HRTEM	High resolution TEM
IR	Infrared spectroscopy
LEED	Low energy electron diffraction
RHEED	Reflection high-energy electron diffraction
SEM	Scanning electron microscopy
STM	Scanning tunneling microscopy
TEM	Transmission electron microscopy
XPF	X-ray pole figure
XPS	X-ray photoelectron spectroscopy
XPS-EELS	Electron energy loss spectroscopy based on XPS spectra
XRD	X-ray diffraction
XPD	X-ray photoelectron diffraction
XTEM	Cross-sectional transmission electron microscopy

A.5. RAMAN SPECTROSCOPY

Raman spectroscopy is a standard method for diamond film research [444]. For the laser source, usually an argon ion laser with the wavelength of either 514.5 nm (red) or 488.0 nm (blue). The Raman frequency shift of the TO phonon for single crystal diamond is located at $1333\,cm^{-1}$. CVD diamond films contain non-diamond components such as graphite and *a*-C, particularly along the grain boundaries, so that their Raman spectra has a sharp band due to diamond around $1333\,cm^{-1}$, and a broad band around $1550\,cm^{-1}$ due to non-diamond components. A typical Raman spectrum is shown in Figure A.1.

It should be noted that the diamond band is usually shifted from $1333\,cm^{-1}$ because of internal stress or strain in the CVD diamond films. On the other hand, carbon has Raman bands at $1340\,cm^{-1}$ (D-band) and $1520\,cm^{-1}$ (G-band), as shown in Figure A.2 (Band positions are shifted).

A.6. CATHODOLUMINESCENCE (CL) SPECTROSCOPY

In the CL spectrum of diamond, there is usually a broad band in the visible region, 400–700 nm, which is called as "band A." The peak positions of band A for undoped and B-doped diamonds are 435 and 510–512 nm, respectively. The origin

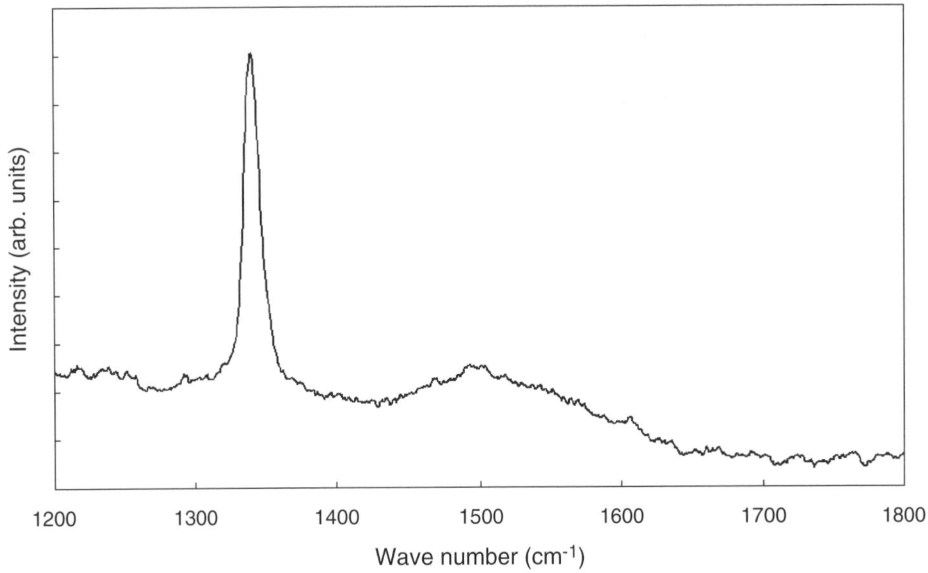

Figure A.1. Typical Raman spectrum of polycrystalline diamond [325].

of band A has not been known for a long time but considered to arise from defects and internal stress, but recent studies show that it arises from penetration twins [126, 445]. Presumably, there are other defects that are the origins of band A [446].

For a detailed database of optical properties of diamond, the reader can refer to an excellent monograph by Zaitsev [444, 447].

A.7. CVD REACTORS

A.7.1. Microwave plasma CVD reactors

There are several types of microwave plasma CVD (MPCVD) reactors designed by several organizations. The widely known reactors are presented in Section 3:

NIRIM-type A microwave plasma CVD reactor developed by National Institute for Research in Inorganic Materials (NIRIM), Japan. This laboratory is now called National Institute for Materials Science (NIMS).

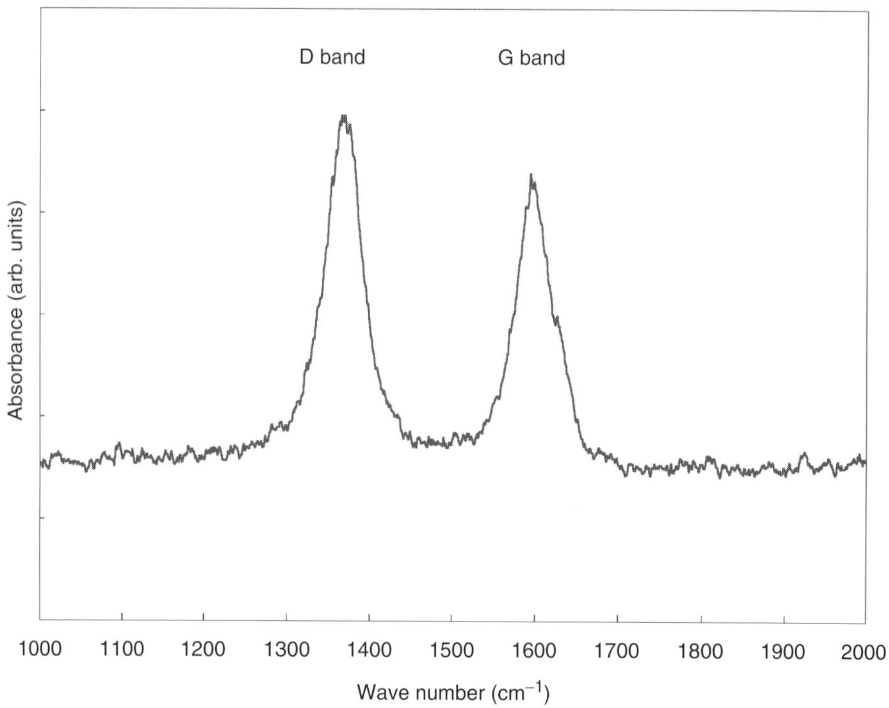

Figure A.2. Typical Raman spectrum of graphite [325].

The chamber is a quartz tube that perpendicularly penetrates the waveguide, and the plasma is generated at the cross section in the quartz chamber. The microwave frequency is 2.45 GHz, and usually $P_m = 300$–500 W is used.

ASTeX-type ASTeX, Inc., USA, has developed many types of MPCVD reactors. The most frequently used is a 1.5-kW system with a stainless steel chamber with an inner diameter of 5 inches. A 5-kW system also has the similar structure as the 1.5-kW system. The 8-kW and 100-kW systems have a different basic design from the previous reactors. The MPCVD reactor division of ASTeX, Inc. has been recently transferred to Seki-Technotron, Inc., Japan.

A.7.2. Hot filament CVD reactor

This reactor was also invented at NIRIM (NIMS). The structure is very simple. It has a hot filament of W or Ta that is heated to 1900–2200 °C, and placed 5–10 mm above the substrate. The gas pressure P is usually 20–30 Torr. In the electron-assisted hot filament CVD (EACVD) reactor, a negative bias is applied to the filament [219, 448].

A.7.3. DC plasma CVD reactor

Diamond films were synthesized by DC plasma for the first time by Sawabe et al. [449]. The negative electrode with about 1-kV voltage is positioned at 20–25 mm above the grounded substrate. The gas pressure is about 200 Torr, so that the growth rate of diamond film is very high, typically 20 μm/h.

A.8. CRYSTAL GROWTH MODES

It is known that there are three modes of film growth:

1) Frank van der Merwe mode: The atoms are deposited on the substrate layer by layer.
2) Volmer-Weber mode: Islands of atoms are discretely nucleated on the substrate.
3) Stranski–Krastanov mode: Islands of atoms are discretely nucleated only after at least an atomic monolayer has been deposited on the substrate.

For the case of diamond CVD, the growth mode is Volmer-Weber mode. This is because the surface energy of diamond is so high that the wetting of diamond on any materials is very difficult. The Frank van der Merwe mode occurs only when diamond is used as the substrate, i.e. homoepitaxial growth.

A.9. CARBON MATERIALS

Typically, there are following carbon materials to be considered in the present book:

a-C amorphous carbon: Carbon atoms are randomly distributed in the film. Although each carbon atom has four covalent bonds, there are several-member rings as well as dangling bonds to make random atomic arrangements.

a-C:H hydrogenated amorphous carbon, or usually called diamond-like carbon (DLC): The carbon arrangement in the film is random just like *a*-C, and there are carbon atoms with bonds terminated by hydrogen atoms.

ta-C tetrahedrally coordinated amorphous carbon: The *ta*-C film is amorphous but it has a band gap and hence transparent. It appears that even though it is amorphous, there are a high density of local structures that have the diamond structure.

a-SiC amorphous silicon carbide

A.10. MISCELLANEOUS NOTATIONS

BEN Bias-enhanced nucleation invented by Yugo et al. [3].

HOD Highly oriented diamond film invented by Stoner et al. [1].

LRI Laser reflection interferometry. See Refs. [330, 331].

Appendix B
Plasma

The plasma under DC bias is categorized into three regions, depending on the voltage as shown in Figure B.1. The first is the random burst pulse region for $V_b < 30\,V$. The second is the saturation current region for $50\,V < V_b < 250\,V$, where a luminous glow appears along the edge of the substrate and substrate holder. For $V_b > 250\,V$, the bias current increased rapidly with the bias voltage. This is associated with the Townsend discharge region.

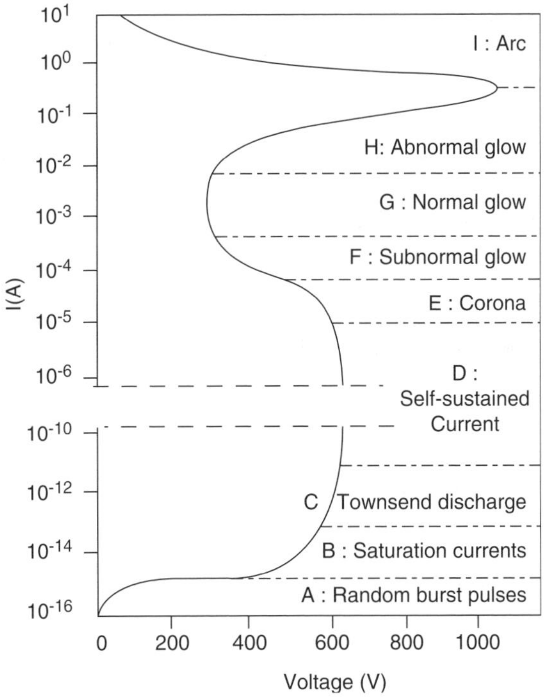

Figure B.1. Current–voltage characteristic for a DC discharge [326].

Appendix C

Properties of Diamond and Other Semiconductors

Table C.1. Properties of diamond and other semiconductors.

		Si	GaAs	SiC (3C)	SiC (4H)	SiC (6H)	GaN	Diamond
Band gap (eV)		1.11	1.43	2.23	3.26	2.86	3.39	5.47
Transition type		Indirect	Direct	Indirect	Indirect	Indirect	Direct	Indirect
Lattice constant (Å)		5.431	5.653	4.359	$a = 3.073$	$a = 3.08$ $c = 15.1$	$a = 3.19$ $c = 5.19$	3.567
Density (g/cm^3)		2.42	5.32	3.16	3.16	3.16	6.1	3.52
Thermal conductivity ($W/cm \cdot K$)		1.51	0.54	4.9	4.9	4.9	1.3	20.9
Dielectric constant		11.8	12.8	9.72	9.6	a: 9.66 c: 10.03	a: 9.5 c: 10.4	5.68
Mobility ($cm^2/V \cdot s$)	Electron Hole	1350 480	8600 400	1000 50	900 100	450 50	900 400	1800 1600
Saturation velocity (10^7 cm/s)		1.0	2.0	2.7	2.7	2.0	2.7	2.5 (electron) 1 (hole)
Breakdown field (10^5 V/cm)		3	4	30	30	30	20	100
Refractive index		3.45	3.4	2.48		2.65–2.69	2.4	2.42

Appendix D

Reconstruction of Diamond Surfaces

Surface reconstruction of diamond is described in Section 8. This section will just give additional information. Possible reconstructed structures of diamond (100) surface are shown in Figure D.1 [139], where (a) and (b) are hydrogen-terminated, while (c) and (d) contain oxygen atoms.

In Ref. [139], S_A and S_B are observed and discussed. Here, S_A expresses a single-atom step whose upper terrace contains dimer rows directed parallel to the step edge, while S_B is a single-atom step with dimer rows on the upper terrace directed perpendicularly to the step edge. S_A steps can extend onto the layer underneath, as seen in Figure 8.2, while S_B steps are directed parallel to the upper rows [450].

Theoretical studies on the structure and growth of S_A and S_B steps were done in Refs. [451–460]. Also, electronic calculations on step structures were presented in Ref. [450].

The possibility of diamond surface reconstruction without hydrogen termination has been discussed in Ref. [461] based on an analogy with Si and Ge. The surface reconstruction models are: Hanemann's (111) 2×1 buckling model [462], Pandy's (111) 2×1 model, Pandy's π-bonded model [463], and Chadi's π-bonded (111) 2×1 model [464]. It was suggested that the buckling and π-bonded structures are in equilibrium at temperatures above 1000 °C. In Ref. [465], a reconstruction from an H-terminated (111) 1×1 to an H-free (111) 2×1 structures after heating at 1000 °C is described.

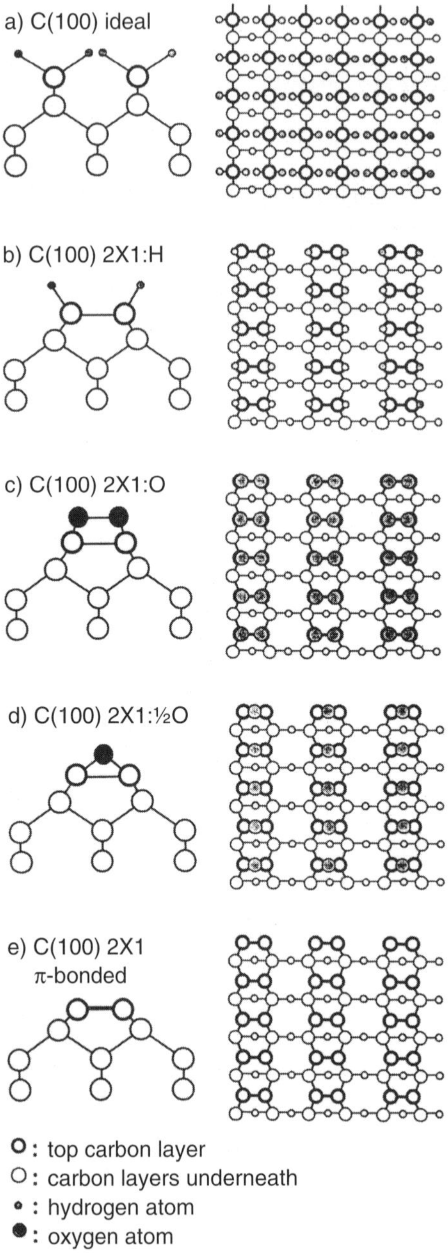

Figure D.1. Possible reconstructed structures of diamond (100) surface. (a) unrelaxed surface with the dangling bonds being terminated by H atoms, (b) hydrogen-terminated 2×1 structure, (c) oxygen-terminated 2×1 structure, (d) single oxygen-terminated 2×1 structure, and (e) π-bonded 2×1 structure [139].

Appendix E
Materials Constants

Table E.1. Crystal structures of diamond and substrate materials for heteroepitaxial diamond growth (data taken from Ref. [205]).

Material	Crystal structure	Lattice constant (Å)	Difference from diamond (%)	Melting point (°C)	Thermal expansion coefficient (10^{-6} /K)
Diamond	Diamond	3.567	–	3057	0.8
Hexagonal diamond	Hexagonal diamond P63/mmc	$a=2.52$ $c=4.12$	–	–	–
cBN	Zinc blende	3.616	1.4	2727	0.59
Ni	Face-centered-cubic (fcc)	3.517	1.2	1455	13.3
Co	Hexagonal < 417°C fcc > 417°C	 3.554	 0.6	1494	–
BeO	Wurzite	$a=2.696$ $c=4.379$	–		–
β-SiC	Zinc blende	4.359	22	2697	4.63
Si	Diamond	5.482	52	1412	7.6
Graphite	Graphite	$a=2.46$ $c=6.708$	–	3797	–
Cu	fcc	3.608	1.1	1084	17
Pt	fcc	3.92	9.8	1770	9.0
Ir	fcc	3.84	7.6	2410	6
Au	fcc	4.07	14	1064	14.1
SiO_2	Hexagonal	3.464 4.382	–	1710	0.55
Si_3N_4	Hexagonal	7.603 2.909	–	2442	2.11
TiC	fcc	4.32	21	3160	6.52
Al_2O_3	Hexagonal	4.785 12.991	–	2049	7.5
Mo	Body-centered-cubic (bcc)	3.14	−12	2615	5.1
Nb	bcc	3.294	−8.2	2467	7.2
AlN	Hexagonal	3.10 4.97	–	2202	4.84
α-W	bcc	3.16	−12	3387	4.5
WC	Hexagonal closed packed	2.90 2.83	–	2627	4.42

Appendix F

Phase Diagrams of Carbon and Metals

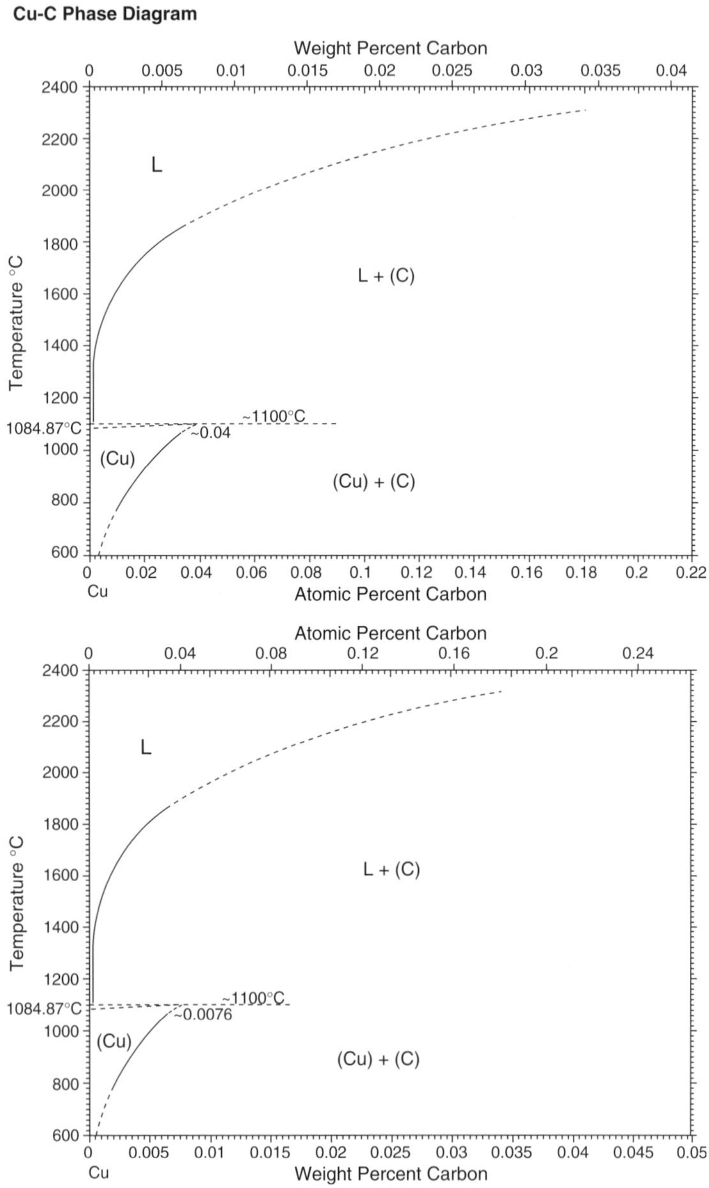

Figure F.1. Cu–C phase diagram [390].

Pt-C Phase Diagram

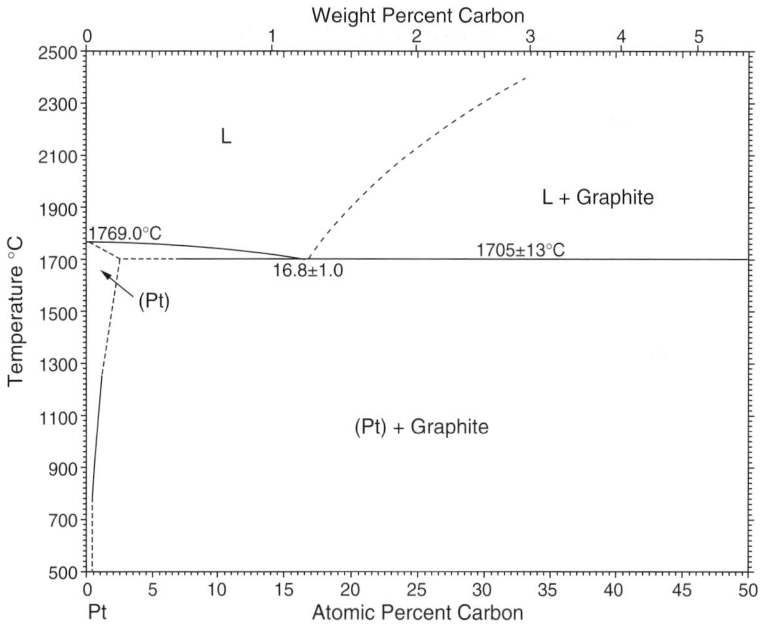

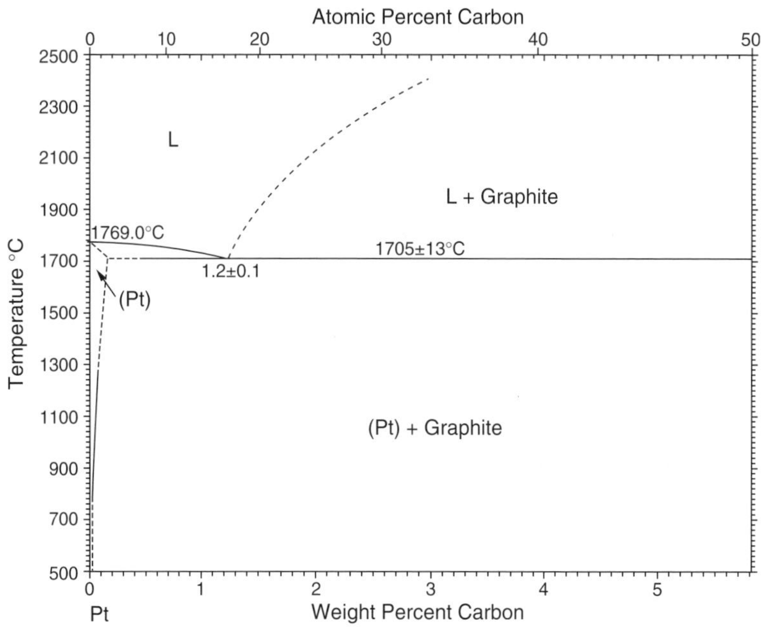

Figure F.2. Pt–C phase diagram [390].

Ir-C Phase Diagram

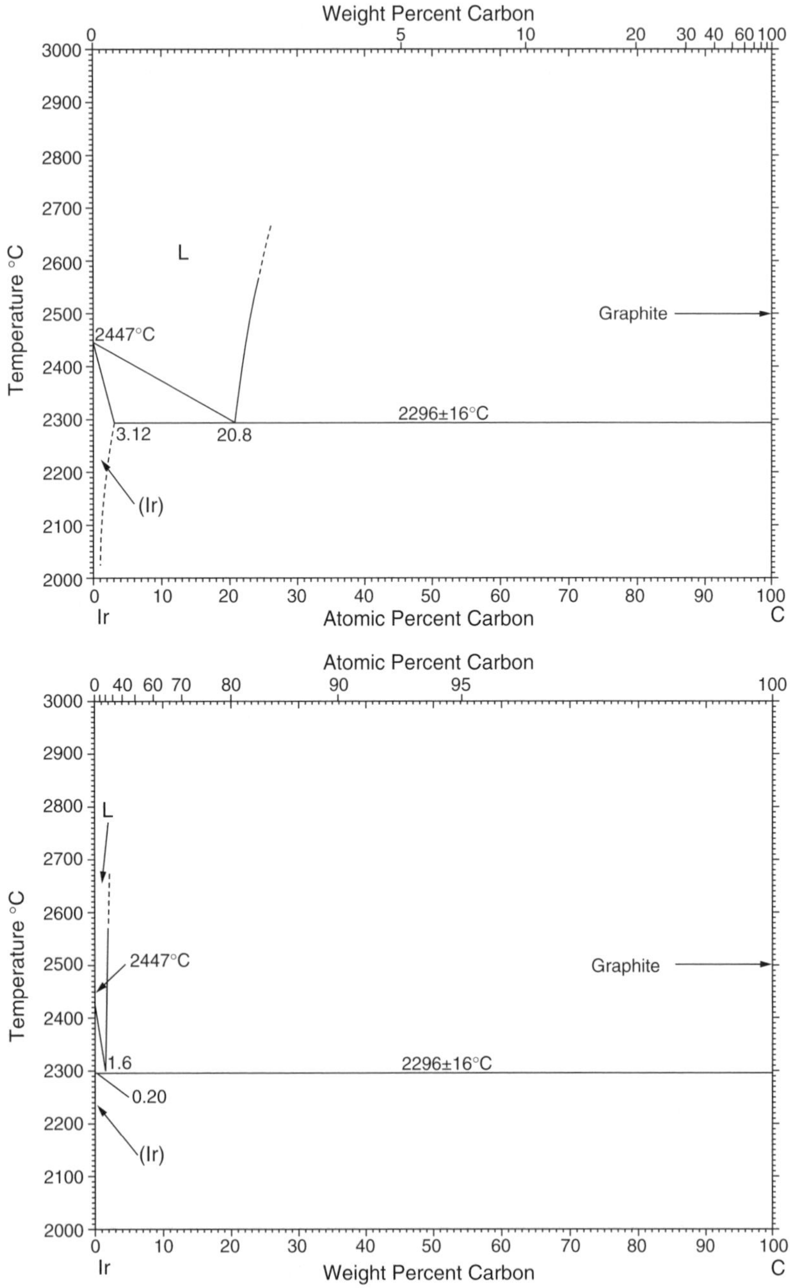

Figure F.3. Ir–C phase diagram [390].

Pd-C Phase Diagram

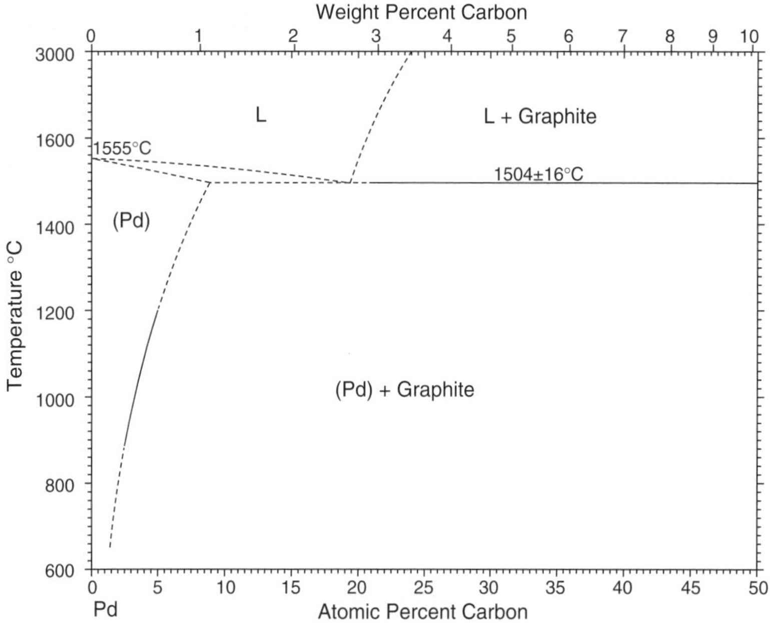

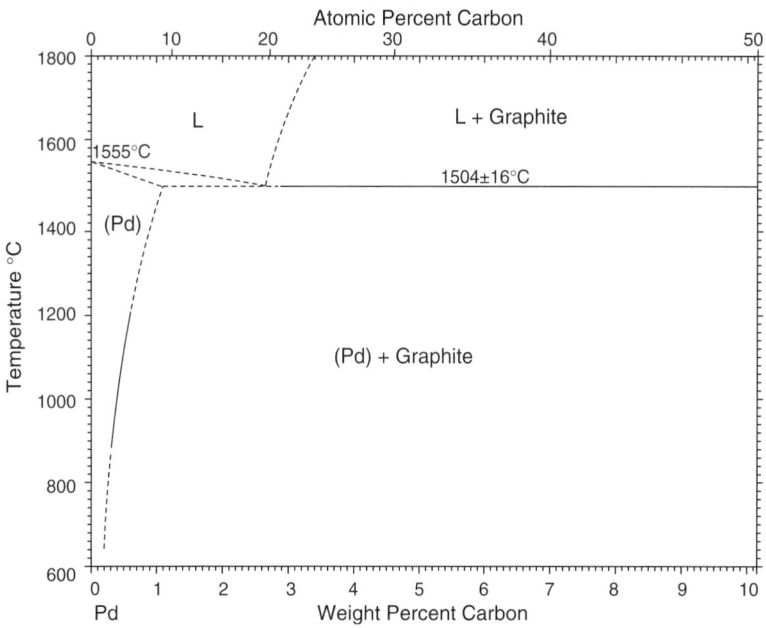

Figure F.4. Pd–C phase diagram [390].

Ni-C (Graphite) Phase Diagram

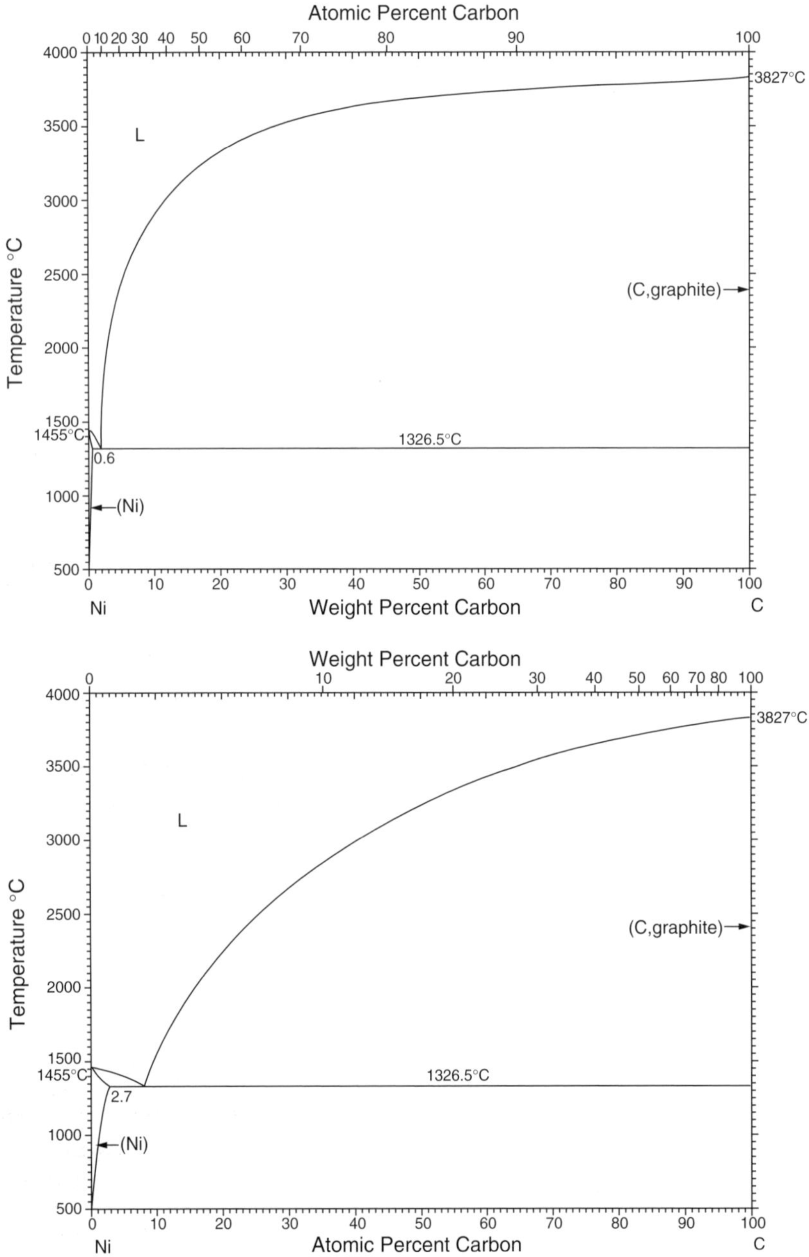

Figure F.5. Ni–C phase diagram [390].

Appendix G
Carbon Solubility in Metals

The carbon solubilities in metals at high temperatures are listed below [371].

Table G.1. Carbon solubilities in metals.

Metals	C solubility at 900 °C (wt%)
Fe	1.3
Co	0.1
Ni	0.2
Pt	0.047
Pd	0.379
Au	0.0021

Appendix H

Process Conditions for Biasing and HOD Film Growth

The numbers in the parentheses are based on author's assumption.
N: NIRIM-type quartz tube reactor for MPCVD
A: ASTeX-type 1.5 kW reactor for MPCVD
ASTeX-type: Custom made reactor whose structure is similar to ASTeX
B: Bell jar-type
ECR: ECR plasma CVD reactor for MPCVD
ND: Nucleation density

Table H.1. Biasing and growth conditions using Si substrates.

		Bias pretreatment						Diamond growth				
Reference	CVD reactor	Gas composition c	Gas pressure P (Torr)	MW Power P_m (W)	Substrate temp. T_s (°C)	Bias voltage V_b (V)	Time (min)	Gas composition c	Gas pressure P (Torr)	MW Power P_m (W)	Substrate temp. T_s (°C)	Results and comments
[3]	N	40% CH_4/H_2	15	350	900	− 70 to − 100	2–5	0.5% CH_4/H_2	30	(350)	(900)	ND = $10^{10}/cm^2$
[206] [208]	N	40% CH_4/H_2	15	400	850	− 100	5	0.5% CH_4/H_2	(30)	(400)	(900)	Tree-like structure
[2]	A	2% CH_4/H_2	15	600	(650)	− 250	≤120	1% CH_4/H_2	25	(600)	(650)	Polycrystalline films
[213]	A	≤10% CH_4/H_2	14.4	1000	600–950	− 250 to + 70	20	0.5% CH_4/H_2	14.4	1000	850	Polycrystalline films
[229]	A (Model HPM/M)	4 sccm CH_4 496 sccm H_2 30 sccm Ar	25–65	(1063)	865	0 to − 180	(<60)	(4 sccm CH_4 496 sccm H_2 30 sccm Ar)	(25–65)	(1063)	(865)	Particles
[216]	A	1.6% CH_4/H_2	37.5	1063	865 (thermo-couple)	− 200	≤180	1.6% CH_4/H_2	37.5	(1063)	865 (thermo-couple)	Particles
[241]	A	5% CH_4/H_2	15	900	725	− 250	–	1% CH_4/H_2	25	900	700	Refractory metal substrates are used.

(*Continued*)

Table H.1. Continued

		Bias pretreatment						Diamond growth				
Reference	CVD reactor	Gas composition c	Gas pressure P (Torr)	MW Power P_m (W)	Substrate temp. T_s (°C)	Bias voltage V_b (V)	Time (min)	Gas composition c	Gas pressure P (Torr)	MW Power P_m (W)	Substrate temp. T_s (°C)	Results and comments
[232]	N	1.7% CH_4/H_2	30	360–560	690–820	−250	15	0.5% CH_4/H_2	30	360–560	690–820	Ball-like surface morphology
[222]	A	2% CH_4/H_2	20*1	1000	800	−200	<30	0.5% CH_4/H_2	20*1	1000	800	
[231]	N	5.5% CH_4/H_2	20	650	750	≤ 120	15	1% CH_4/H_2	40	650	800	Polycrystalline films
[311]	B	5% CH_4/H_2	15	900	810	−230	17, 38	0.9% CH_4/H_2	15	900	810	Particles and films
[233]	N	1.4% CH_4/H_2	30	340–570	600–760	−250	15	0.5% CH_4/H_2	30.4	580	820	Particles and films
[362]	N	3 sccm CH_4 100 sccm H_2 100 ppm N_2	38	500	900	250	<60	1 sccm CH_4 100 sccm H_2 80 ppm N_2	37.5	500	800	Locally HOD region

*1 Ref. [222] describes that the gas pressure was 2.6 hPa, which seems to be 26 hPa, or 20 Torr, in reference to [223] and [224].

Table H.2. Biasing and growth conditions for HOD film growth on β-SiC(100) layer.

		Bias pretreatment						Diamond growth				
Reference	CVD reactor	Gas composition c	Gas pressure P (Torr)	MW Power P_m (W)	Substrate temp. T_s (°C)	Bias voltage V_b (V)	Time (min)	Gas composition c	Gas pressure P (Torr)	MW Power P_m (W)	Substrate temp. T_s (°C)	Results and comments
[1]	A	2% CH_4/H_2	15	600	650	− 250	30	0.5% CH_4/H_2	25	600	650–700	Diamond growth rate was 0.05 µm/h
[225]	A	2–4% CH_4/H_2	15	600–900	600–700	− 250	10–100	0.5–2% CH_4/H_2	25	900–1200	600–800	HOD films
[306]	A	4% CH_4/H_2	15	600	650	− 250	30	1% CH_4/H_2	25	(600)	650–700	Diamond growth rate was 0.05 µm/h
[257]	N	4% CH_4/H_2	38	–	700–800	− 180	5	1–2% CH_4/H_2	38	–	(700)	20-µm thick HOD films
[254]	ECR	40 sccm CH_4 440 sccm H_2	0.3–2	1400	750	− 100 ~0	0 ~120	10 sccm CH_4 5 sccmO_2 485 sccm H_2	0.3	1400 (+30 V DC bias)	750	30% oriented, ND = 10^7–10^8/cm^2
[249]	N	2% CH_4/H_2	50	–	900	− 50	5	5% CO + 5% CH_4 /H_2	50	–	840	A 6-µm thick HOD film, lateral growth was also used
[438]	N	2% CH_4/H_2	38		800	0 to −180	(5)	1.2% CH_4/H2 50 ppm N_2	(37.5)	–	(1) 675 (2) 700–800	Transistors were fabricated
[6]	N	1.2–2% CH_4/H_2	37.5	–	800	0 to −180	(5)	(1) 1.2% CH_4/H_2 50 ppm N_2 (2) 0.8–1.2% CH_4/H_2 50 ppm N_2	(37.5)	–	(1) 675–700 (2) 725–800	ND = 5×10^{10} to 1×10^{11}/cm^2, the FWHM of X-ray rocking curve was 0.6°

Table H.3. Biasing and growth conditions for HOD film growth on Si(100) substrate using the three-step process.

Reference	CVD reactor	Carburization: Gas composition c	Carburization: Gas pressure P(Torr)	Carburization: MW Power P_m(W)	Carburization: Substrate temp. T_s(°C)	Carburization: Time (min)	Bias pretreatment: Gas composition c	Bias pretreatment: Gas pressure P(Torr)	Bias pretreatment: MW Power P_m(W)	Bias pretreatment: Substrate temp. T_s(°C)	Bias pretreatment: Bias voltage V_b(V)	Bias pretreatment: Time (min)	Diamond growth: Gas composition c	Diamond growth: Gas pressure P(Torr)	Diamond growth: MW Power P_m(W)	Diamond growth: Substrate temp. T_s(°C)	Results and comments
[261]	A	2%CH_4/H_2	20	1000	800	180	2%CH_4/H_2	20	1000	800 (heater temp.)	−170 to −200	15	7.5%CH_4 + 2.5%CO/H_2	**35**	1300	650 (heater temp.)	HOD films
[260]	A (horizontal)	2%CH_4/H_2	25	1200	~900	60	5%CH_4/H_2	15	900	~750	−250	3	0.5%CH_4/H_2	25	900	600–700	HOD films (Table 11.3.1)
	A (vertical)	2%CH_4/H_2	15	800	~800	60	2%CH_4/H_2	15	800	~700	−170	9	0.5%CH_4/H_2	25	900	600–700	HOD films (Table 11.3.1)
[351]	Custom-made	1.8%CH_4/H_2	20	175	~835	360	1.8%CH_4/H_2	20	175	835	−340	25	1.3%CH_4 + 44.6% CO/H_2	45	800	–	HOD films
[228]	Custom-made	1.8%CH_4/H_2	20	175	840–850	360	1.7%CH_4/H_2	20	175	840–850	−250 to −350	0–35	1.3%CH_4 44.6% CO/H_2	45	800	670	HOD films
[329]	Custom-made	1.8%CH_4/H_2	21.2	175	839	360	1.8%CH_4/H_2	21.2	175	840	−320	15	1.3%CH_4 44.6% CO/H_2	46.1	800	656	HOD films
[352]	Custom-made	1.8%CH_4/H_2	20	175	~850	360	1.8%CH_4/H_2	20	175	850	−320	15	1.1%CH_4 + 44.6% CO/H_2	45	800	670	HOD films
[267]	N	2%CH_4/H_2	20	570	830	15–120	2%CH_4/H_2	20	570	830	−150	1–15	0.5–2% CH_4/H_2	30	460 –670	800 –900	HOD films
[159] [86]	N	2%CH_4/H_2	40	–	830	30	2%CH_4/H_2	40	–	770	−120	6	2%CH_4/H_2	40	–	700	HOD films, Lateral growth was also used
[268]	A	2%CH_4/H_2	(20)	(1000)	780	60	0.5–2%CH_4/H_2	20	1000	780	−225 to −275	5–15	0.5%CO + 1.5% CH_4/H_2	32	1300	850	HOD films
[22]	N	1%CH_4/H_2	60	600	850	–	1%CH_4/H_2	60	600	850	−250	15	1%CH_4/H_2	60	600	850	HOD films

[226]	B	2%CH_4/H_2	15	700	~800	30–60	2%CH_4/H_2	15	700	~800	−250	10	0.8%CH_4/H_2	25	700	~850	HOD particles
[269]	(A)	2%CH_4/H_2	20	1000	850	120	2%CH_4/H_2	20	1040	810	−200	10	1%CH_4/H_2	20	500	750	HOD films Hexagonal diamond
[303]	A	2%CH_4/H_2	20	1000	850	120	2%CH_4/H_2	20	1040	810	−150 to −300	10	1%CH_4/H_2	20	500	750	HOD films
[269] [305]	A	2%CH_4/H_2	20				2%CH_4/H_2	20			−200		1%CH_4/H_2	15			HOD films
[335] [336]	(A)	(2%CH_4/H_2)	20	900	860	120	2%CH_4/H_2 for 3 min, and 100% H_2 for 0.5 min (Two cyclic processes)	20	900	860	−200	6–7 (Two cyclic processes)	2%CH_4/H_2	25	1000	700	HOD films
[227]	B	2%CH_4/H_2	15	700	~800	30–60	2–3%CH_4/H_2	10–20	700	800	−100 to −250	10	0.8–2.0% CH_4/H_2	25	800	850	HOD films
[361]	Custom-made	2%CH_4/H_2	19	800	850	120	3%CH_4/H_2	19	1000	750	−250	10	1% CH_4/H_2	25	1000	800	HOD films
[337]	A	2%CH_4/H_2	19	800	850	120	2%CH_4/H_2 for 30–90 sec, and 100% H_2 for 0–180 sec (periodic)	19	800	750	−250	6	1% CH_4/H_2	25	1000	800	HOD films
[312]	A	1%CH_4/H_2	30	850	797	120–180	5%CH_4/H_2	15	650	722	−200 to −250	6–7	0.6%CH_4 + 25% Ar/H_2	30–34	300–400	677	HOD films
[355]	N	2%CH_4/H_2	20	580–610	(880)	0–60	2%CH_4/H_2	20	(600)	(880)	−200 to +200	5–30	0.5%CH_4/H_2	40	300	880	HOD films
[466]	A	4%CH_4/H_2	–	–	–	120	1.5%CH_4/H_2	25	–	900	−200	25	1%CO + 4%CH_4/H_2	35	–	800	HOD films
[332]	Reactor made by RFA, Ltd.	2%CH_4/H_2	20	–	800	20	5%CH_4/H_2	20	800	800	−200	30	4%CH_4 + 3% Ar/H_2	30	800	600	HOD films
[333]	A	0.5%CH_4/H_2	20	1000	850	180	0.5%CH_4/H_2	20	1000	850	−200	12	0.5%CH_4/H_2	20	600	700	HOD films

Table H.4. Biasing and growth conditions for HOD film growth on Si(100) substrate using the two-step process.

Reference	CVD reactor	Bias pretreatment: Gas composition c	Gas pressure P (Torr)	MW Power P_m (W)	Substrate temp. T_s (°C)	Bias voltage V_b (V)	Time (min)	Diamond growth: Gas composition c	Gas pressure P (Torr)	MW Power P_m (W)	Substrate temp. T_s (°C)	Results and comments
[261]	A	2%CH_4/H_2	15	800	700 (heater temp.)	200	25	2.5–7.5% CH_4 + 7.5–2.5% CO/H_2	50, 35	1300	650–750 (heater temp.)	HOD films
[296]	A	(0.5% CH_4/H_2)	–	900	700–900	– 100 to – 200	–	(0.5% CH_4/H_2)	–	(600–1000)	–	AFM study
[263]	A	(0.5% CH_4/H_2)	–	(900)	800	– 100 to – 300	–	0.5% CH_4/H_2		600–1000		
[276]	A	3%CH_4/H_2	15	1200	850	– 150	15–45	5%CH_4 + 1% CO_2/H_2	15	1200	750	HOD films
[321]	A	3%CH_4/H_2	80	2700	900	– 160	30	4–6% CH_4/H_2	80	2700 (with a biasing voltage of 0 to –100 V)	900	HOD films
[285]	N	2%CH_4/H_2	15	–	780	– 225	8	1%CH_4/H_2	30	–	845	HOD films
[285]	N	2%CH_4/H_2	30	–	810	– 225	8	1%CH_4/H_2	30	–	870	HOD films
[274]											845	
[287]	N	2%CH_4/H_2	15	480	780	– 225	8	1%CH_4/H_2	30	480	845	HOD films

[289] [290]	A	0.3–1% CH_4/H_2	15	600–900	780–950	– 200	5–30					Nucleation
[315]								0.5% CH_4/H_2	15	900	660–720	Grow out
								1.25% CO_2 + 8.75% CH_4/H_2	34	1300	780	HOD films
[275] [297] [284]	(A)	0.88% CH_4/H_2	34	1300	790	– 200	5–10	0.88% CH_4 + 0.13% CO_2/H_2	34	1300	790	HOD films
[467]	(A)	8.8% CH_4/H_2	34	1300	780	– 200	1–5	8.8% CH_4 1.25% CO_2/H_2	34	1300	780	Thin films
[292]	A	2.2% CH_4/H_2 8.9% CH_4/H_2	15, 34, 45	900 1300	675–975	– 180 to – 300	15 s to 30 min	8.8% CH_4 + 1.25% CO_2/H_2	34	1300	790	HOD films
[293]	A	8.9% CH_4/H_2	34	–	790	– 180 to – 300	5	7.6% CH_4 + 1.25% CO_2/H_2	34	–	790	HOD films
[364] [363]	N	3% CH_4 + 100 ppm N_2 / H_2	38	500	900	– 250	< 60	1% CH_4 + 80 ppm N_2 / H_2	38	500	800	HOD films
[313]	–	0.5–2.5% CH_4/H_2	11–30	700–2000	650–850	– 70 to – 300	–	0.5–2.5% CH_4/H_2	11–30	700–2000	650–850	HOD films
[342]	A	4% CH_4/H_2	20	1000	850	200 (AC 60 Hz)	25	2% CH_4/H_2	30	1500	820	HOD films
[316]	A	4% CH_4/H_2	20	700	600	200	25	2% CH_4/H_2	90	700	970	HOD films

(Continued)

Table H.4. Continued

Reference	CVD reactor	Bias pretreatment: Gas composition c	Gas pressure P (Torr)	MW Power P_m (W)	Substrate temp. T_s (°C)	Bias voltage V_b (V)	Time (min)	Diamond growth: Gas composition c	Gas pressure P (Torr)	MW Power P_m (W)	Substrate temp. T_s (°C)	Results and comments
[343]	A	4% CH_4/H_2 8% CH_4/H_2	20	1000	~850	200 (60 Hz)	15–40	2% CH_4/H_2	30	1500	820	HOD films
[344]	Custom-made	1% CH_4/H_2 + 50 ppm N_2	(24.3)		860–880	−220 (pulse bias)	17	See Table 11.8 for growth conditions				HOD films, see also [345]
[353]	N	1% CH_4/H_2	60	800	850	−250	15	1% CH_4/H_2	800	800	850	HOD films
[354]												
[421]	Custom-made	2% CH_4/H_2	14	800	850	−250	20	0.5% CH_4/H_2	14	800	750	HOD films
[327]	A	4% CH_4/H_2	7.5	300	730	−120	60 s	0.8% CH_4 + 0.2% O_2/H_2	7.5	300	730	HOD by short biasing time
[328]	A	4% CH_4/H_2	9.8	300	840	−150	30 s	1% CH_4/H_2	9.8	300	800	HOD films

References

1. Stoner B.R. & Glass, J.T. (1992) *Appl. Phys. Lett.*, **60**, 698.
2. Stoner, B.R., Ma, G.-H.M., Wolter, S.D. & Glass, J.T. (1992-I) *Phys. Rev. B*, **45**, 11067.
3. Yugo, S., Kanai, T., Kimura, T. & Muto, T. (1991) *Appl. Phys. Lett.*, **58**, 1036.
4. Stoner, B.R., Kao, C., Malta, D.M. & Glass, R.C. (1993) *Appl. Phys. Lett.*, **62**, 2347.
5. Tachibana, T., Hayashi, K. & Kobashi, K. (1994) in *Advances in New Diamond Science and Technology*, Eds. Saito, S., Fujimori, N., Fukunaga, O., Kamo, M., Kobashi, K. & Yoshikawa, M. (MYU, Tokyo), p. 279.
6. Kawarada, H., Wild, C., Herres, N., Locher, R., Koidl, P. & Nagasawa, H. (1997) *J. Appl. Phys.*, **81**, 3490.
7. Kawarada, H., Ref. [31] p. 139.
8. Shintani, Y. (1996) *J. Mater. Res.*, **11**, 2955.
9. Tachibana, T., Yokota, Y., Nishimura, K., Miyata, K., Kobashi, K. & Shintani, Y. (1996) *Diamond and Related Mater.*, **5**, 197.
10. Ohtsuka, K., Suzuki, K., Sawabe, A. & Inuzuka, T. (1996) *Jpn. J. Appl. Phys.*, **35**, L1072.
11. Saito, T., Tsuruga, Sh., Ohya, N., Kusakabe, K., Morooka, Sh., Maeda, H., Sawabe, A. & Suzuki, K. (1998) *Diamond and Related Mater.*, **7**, 1381.
12. Hörmann, F., Roll, H., Schreck, M. & Stritzker, B. (2000) *Diamond and Related Mater.*, **9**, 256.
13. Ohtsuka, K., Fukuda, H., Suzuki, K. & Sawabe, A. (1997) *Jpn. J. Appl. Phys.*, **36**, L1214.
14. Tachibana, T. (2003) *Annual Report of the FCT Project*, NEDO, in Japanese.
15. *Diamond: Electronic Properties and Applications* (1995) Eds. Pan, L.S. & Kania, D.R. (Kluwer, Boston).
16. Gluche, P., Adamschik, M., Vescan, A., Ebert, W., Szücs, F., Fecht, H.J., Flöter, A., Zachai, R. & Kohn, E. (1998) *Diamond and Related Mater.*, **7**, 779.
17. Kohn, E., Ebert, W., Adamschik, M., Schmid, P. & Denisenko, A. (2001) *New Diamond Science and Technology* (MYU, Tokyo), **11**, 81.
18. Suzuki, T. & Argoitia, A. (1996) *phys. stat. sol (a)*, **154**, 239.
19. Jiang, X. & Klages, C.-P. (1996) *Ceramics International*, **22**, 443.
20. Arnault, J.C. (2003) *Surf. Rev. Lett.*, **10**, 127.
21. Lee, S.-T., Lin, Z. & Jiang, X. (1999) *Mater. Sci. Eng., Reports: A Review Journal*, **25**, 123.
22. von Kaenel, Y., Stiegler, J., Blank, E., Chauvet, O., Hellwig, Ch. & Plamann, K. (1996) *phys. stat. sol. (a)*, **154**, 219.
23. Kobashi, K., Nishimura, K., Kawate, Y. & Horiuchi, T. (1988-II) *Phys. Rev. B*, **38**, 4067.
24. *Handbook of Industrial Diamond and Diamond Films* (1998) Eds. Prelas, M.A., Popovici, G. & Bigelow L.K. (Marcel Dekker, New York).
25. *The Properties of Natural and Synthetic Diamond* (1992) Eds. Field, J.E. Academic Press, London.
26. *Synthetic Diamond: Emerging CVD Science and Technology* (1984) Eds. Spear, K.E. & Dismukes, J.P. (J. Wiley & Sons).
27. Davies, G. (1984) *Diamond* (Adam Hilger, Bristol).
28. *Thin Film Diamond* (1994) Eds. Lettington, A. & Steeds, J.W. (Chapman & Hill, London).

29. *Diamond Films and Coatings* (1993) Eds. Davis, R.F. (Noyes Publication, Park Ridge, NJ).
30. *The Properties of Diamond* (1979) Eds. Field, J.E. (Academic, London).
31. *Low Pressure Synthetic Diamond Manufacturing and Applications* (1998) Eds. Dischler, B. & Wild, C. (Springer-Verlag, Berlin).
32. Nishimori, T., Sakamoto, H., Takakuwa, Y. & Kono, S. (1995) *Jpn. J. Appl. Phys.*, **34**, Part 2, L1297.
33. Chu, C.J., Hauge, R.H., Margrave, J.L. & D'Evelyn, M.P. (1992) *Appl. Phys. Lett.*, **61**, 1393.
34. Clausing, R.E., Heatherly, L., Horton, L.L., Specht, E.D., Begun, G.M. & Wang, Z.L. (1992) *Diamond and Related Mater.*, **1**, 411.
35. Fu, G.S., Wang, X.H., Yu, W., Han, L., Dong, L.F. & Li, X.W. (1997) *Appl. Phys. Lett.*, **70**, 1965.
36. Li, Z., Wang, L., Suzuki, T., Argoitia, A., Pirouz, P. & Angus, J.C. (1993) *J. Appl. Phys.*, **73**, 711.
37. Yang, J., Lin, Z., Wang, L.-X., Jin, S. & Zhang, Z. (1994) *Appl. Phys. Lett.*, **65**, 3203.
38. Stubhan, F., Ferguson, M., Füsser, H.-J. & Behm, R.J. (1995) *Appl. Phys. Lett.*, **66**, 1900.
39. Yang, J., Lin, Z., Wang, L.-X., Jin, S. & Zhang, Z. (1995) *Mat. Res. Soc. Symp., Proc.*, **357**, 231.
40. Chen, Q., Yang, J. & Lin, Z. (1995) *Appl. Phys. Lett.*, **67**, 1853.
41. Yang, J., Lin, Z., Wang, L., Jin, S. & Zhang, Z. (1995) *J. Phys. D: Appl. Phys.* **28**, 1153.
42. Feng, K., Yang, J. & Lin, Z. (1995-II) *Phys. Rev. B*, **51**, 2264.
43. Chen, Q., Wang, L.-X., Zhang, Z., Yang, J. & Lin, Z. (1996) *Appl. Phys. Lett.*, **68**, 176.
44. Li, X., Shirafuji, T., Hayashi, Y., Lilov, S. & Nishino, S. (1997) *Jpn. J. Appl. Phys.*, **36**, 6295.
45. Li, X., Hayashi, Y., Lilov, S. & Nishino, S. (1997) *Jpn. J. Appl. Phys.*, **36**, Part 2, L1370.
46. Gao, Q.J., You, L.P., Pan, X.H., Zhang, F., Peng, X.F. & Lin, Z.D. (1997) *Diamond and Related Mater.*, **6**, 1836.
47. Li, X., Hayashi, Y. & Nishino, S. (1997) *Jpn. J. Appl. Phys.*, **36**, Part 1, 5197.
48. Li, X., Hayashi, Y. & Nishino, S. (1997) *Diamond and Related Mater.*, **6**, 1117.
49. Hayashi, Y., Li, X. & Nishino, S. (1997) *Appl. Phys. Lett.*, **71**, 2913.
50. Liao, K.J., Wang, W.L. & Feng, B. (1998) *phys. stat. sol. (a)*, **167**, 117.
51. Zhou, X.T., Lai, H.L., Peng, H.Y., Sun, C., Zhang, W.J., Wang, N., Bello, I., Lee, C.S. & Lee, S.T. (2000) *Diamond and Related Mater.*, **9**, 134.
52. Yokonaga, N., Katsu, Y., Machida, T., Inuzuka, T., Koizumi, S. & Suzuki, K. (1996) *Diamond and Related Mater.*, **5**, 43.
53. Qimin, W., Qinggui, C., Rihua, Shi., Weibing, Z., Qiwei, D. & Rongkang, D. (1993) *J. Crystal Growth*, **129**, 774.
54. Qimin, W. & Qinggui, C. (1994) *J. Cryst. Growth*, **142**, 349.
55. Watanabe, H., Takeuchi, D., Yamanaka, S., Okushi, H., Kajimura, K. & Sekiguchi, T. (1999) *Diamond and Related Mater.*, **8**, 1272.
56. Takeuchi, D., Watanabe, H., Yamanaka, S., Okushi, H. & Kajimura, K., *Diamond and Related Mater.*, **9**, 231 (2000)
57. Okushi, H. (2001) *Diamond and Related Mater.*, **10**, 281.
58. Ri, S.-G., Yoshida, H., Yamaoka, S., Watanabe, H., Takeuchi, D. & Okushi, H. (2002) *J. Cryst. Growth*, **235**, 300.
59. Sternshulte, H., Albrecht, T., Thonke, K., Saue, R., Grießer, M. & Grasserbauer, M. (1996) *Mater. Res. Soc. Symp. Proc.*, **423**, 693.

60. Remes, Z., Uzan-Saguy, C., Baskin, E., Kalish, R., Avigal, Y., Nesládek, M. & Koizumi, S. (2004) *Diamond and Related Mater.*, **13**, 713.
61. Sauer, R., Teofilov, N., Thonke, K. & Koizumi, S. (2004) *Diamond and Related Mater.*, **13**, 727.
62. Iijima, S., Aikawa, Y. & Baba, K. (1990) *Appl. Phys. Lett.*, **57**, 2646.
63. Silva, S.R.P., Carey, J.D., Khan, R.V.A., Gerstner, E.G. & Anguita, J.V. (2002) *Handbook of Thin Film Materials*, Eds. Nalwa, H.S. (Academic Press, London) **4**, 403.
64. Kamo, M., Sato, Y., Matsumoto, S. & Setaka, N. (1983) *J. Cryst. Growth*, **62**, 642.
65. Kamo, M., Matsumoto, S., Sato, Y. & Setaka, N. (1984) US patent No. 4,434,188.
66. Sevillano, E. (1998) *Low Pressure Synthetic Diamond Manufacturing and Applications*, Eds. Dischler, B. & Wild, C. (Springer-Verlag, Berlin), 11.
67. Ando, Y., Tachibana, T., Kobashi, K. (2001) *Diamond and Related Mater.*, **10**, 312.
68. Tachibana, T., Ando, Y., Watanabe, A., Nishibayashi, Y., Kobashi, K., Hirao, T. & Oura, K. (2001) *Diamond and Related Mater.*, **10**, 1569.
69. Ando, Y., Yokota, Y., Tachibana, T., Watanabe, A., Nishibayashi, Y., Kobashi, K., Hirao, T. & Oura, K. (2002) *Diamond and Related Mater.*, **11**, 596.
70. Yokota, Y., Ando, Y., Kobashi, K., Hirao, T. & Oura, K. (2003) *Diamond and Related Mater.*, **12**, 295.
71. Gicquel, A., Hassouni, K., Farhat, S., Breton, Y., Scott, C.D., Lefebre, M. & Pealat, M. (1994) *Diamond and Related Mater.*, **3**, 581.
72. Suzuki, K., Fukuda, H. & Sawabe, A. (1998) in *Proc. the 5th NIRIM Int. Symp on Advanced Materials (ISAM '98)*, Tsukuba, Japan, 89.
73. Baik, Y.-J., Lee, J.-K., Lee, W.-S. & Eun, K.Y. (1998) *J. Mater. Res.*, **13**, 944.
74. Lee, J.-K., Eun, K. Y., Baik, Y.-J., Cheon, H. J., Rhyu, J. W., Shin, T. J. & Park, J.-W. (2002) *Diamond and Related Mater.*, **11**, 463.
75. Takueuchi, D., Watanabe, H., Sawada, H., Yamanaka, S., Ichinose, H., Sekiguchi, T. & Okushi, H. (2001) *Diamond and Related Mater.*, **10**, 526.
76. Wild, Ch., Herres, N. & Koidl, P. (1990) *J. Appl. Phys.*, **68**, 973.
77. Van der Drift, A. (1967) *Philips Res. Reports*, **22**, 267.
78. Kobashi, K., Nishimura, K., Miyata, K., Kumagai, K. & Kawate, Y. (1989) *Diamond and Diamond-Like Films* (The Electrochemical Society), Proc. Vol. **89–12**, p. 296.
79. Kobashi, K., Nishimura, K., Miyata, K., Kumagai, K. & Nakaue, A. (1990) *J. Mater. Res.*, **5**, 2469.
80. Kumagai, K., Miyata, K., Nishimura, K. & Kobashi, K. (1993) *J. Mater. Res.*, **8**, 314.
81. Wild, C., Koidl, P., Herres, N., Müller-Sebert, W. & Eckermann, T. (1991) *Electrochem. Soc. Proc.*, **91–8**, 224.
82. Wild, Ch., Koidl, P., Müller-Sebert, W., Walcher, H., Kohl, R., Herres, N., Locher, R., Samlenski, R. & Brenn, R. (1993) *Diamond and Related Mater.*, **2**, 158.
83. Wild, C., Herres, N., Wagner, J., Koidl, P. & Anthony, T.R. (1989) *Electrochem. Soc. Proc.*, **89–12**, 283.
84. Wild, Ch., Kohl, R., Herres, N., Müller-Sebert, W. & Koidl, P. (1994) *Diamond and-Related Mater.*, **3**, 373.
85. Zhu, W., Badzian, A.R. & Messier, R. (1990) *Proc. SPIE*, **1325**, 187.
86. Maeda, H., Ohtsubo, K., Irie, M., Ohya, N., Kusakabe, K. & Morooka, S. (1995) *J. Mater. Res.*, **10**, 3115.
87. Rigby, S.D., Cherian, K.A., Lister, J.D., Rudolph, V. & White, E.T. (2000) *Diamond and Related Mater.*, **9**, 328.

88. Miyata, K., Kumagai, K., Nishimura, K. & Kobashi, K. (1993) *J. Mater. Res.*, **8**, 2845.
89. Miyata, K., unpublished data.
90. Locher, R., Wild, Ch., Herres, N., Behr, D. & Koidl, P. (1994) *Appl. Phys. Lett.*, **65**, 34.
91. Jin, S. & Moustakas, T.D. (1994) *Appl. Phys. Lett.*, **65**, 403.
92. Müller-Sebert, W., Wörner, E., Fuchs, F., Wild, C. & Koidl, P. (1996) *Appl. Phys. Lett.*, **68**, 759.
93. Ayres, V.M., Bieler, T.R., Kanatzidis, M.G., Spano, J., Hagopian, S., Balhareth, H., Wright, B.F., Farhan, M., Abdul Majeed, J., Spach, D., Wright, B.L. & Asmussen, J. (2000) *Diamond and Related Mater.*, **9**, 236.
94. Eccles, A. J., Steele, T. A., Afzal, A., Rego, C. A., Ahmed, W., May, P. W. & Lees, S. M. (1999) *Thin Solid Films*, **343–344**, 627.
95. Bohr, S., Haubner, R. & Lux, B. (1996) *Appl. Phys. Lett.*, **68**, 1075.
96. Haubner, R. & Lux, B. (1999) *Diamond and Related Mater.*, **8**, 171.
97. Wild, C., Locher, R. & Koidl, P. (1996) *Diamond for Electronic Applications, MRS-Proc. (MRS, Pittsburgh, PA)*, **416**, 75.
98. Zhang, Y., Zhang, F. & Chen, G. (1994) *Mater. Lett.* **20**, 39.
99. Everson, M.P., Tamor, M.A., Scholl, D., Stoner, B.R., Sahaida, S.R. & Bade, J.P. (1994) *J. Appl. Phys.*, **75**, 169.
100. Tamor, M.A. & Everson, M.P. (1994) *J. Mater. Res.*, **9**, 1839.
101. Tsuno, T., Imai, T. & Fujimori, N. (1994) *Jpn. J. Appl. Phys.*, **33**, 4039.
102. Williams, B.E., Kong, H.S. & Glass, J.T. (1990) *J. Mater. Res.*, **5**, 801.
103. Narayan, J., Srivatsa, A.R., Peters, M., Yokota, S. & Ravi, K.R. (1988) *Appl. Phys. Lett.*, **53**, 1823.
104. Hirabayashi, K. (1992) *J. Appl. Phys.*, **72**, 4083.
105. Barrat, S. & Bauer-Grosse, E. (1995) *Diamond and Related Mater.*, **4**, 419.
106. Yan, C.-S., Vohra, Y.K., Mao, H.-K. & Hemley, R.J. (2002) *Proc. Natl. Acad. Sci. USA*, **99**, 12523.
107. Mori, Y., Yagi, H., Deguchi, M., Sogi, T., Yokota, Y., Eimori, N., Yagyu, H., Ohnishi, H., Kitabatake, M., Nishimura, K., Hatta, A., Ito, T., Hirao, T., Sasaki, T. & Hiraki, A. (1993) *Jpn. J. Appl. Phys.*, **32**, 4661.
108. Tsuno, T., Shiomi, H., Kumazawa, Y., Shikata, S. & Akai, S. (1996) *Jpn. J. Appl. Phys.*, **35**, 4724.
109. Lee, N. & Badzian, A. (1997) *Diamond and Related Mater.*, **6**, 130.
110. Hayashi, K., Yamanaka, S., Watanabe, H., Sekiguchi, T., Ohkushi, H. & Kajiyama, K. (1998) *J. Cryst. Growth*, **183**, 338.
111. Suzuki, C.K., Shinohara, A.H., Godoy, P.H., Watanabe, N. & Kamo, M. (1998) *Diamond and Related Mater.*, **7**, 289.
112. Tarutani, M., Takai, Y., Shimizu, R. & Ando, T., Kamo, M. & Bando, Y. (1996) *Appl. Phys. Lett.*, **68**, 2070.
113. Tarutani, M., Takai, Y. & Shimizu, R. (1996) *Diamond Films and Technol.* **6**, 365.
114. Hayashi, K., Yamanaka, S., Okushi, H. & Kajimura, K. (1996) *Diamond and Related Mater.*, **5**, 1002.
115. Hayashi, K., Yamanaka, S., Watanabe, H., Sekiguchi, T., Okushi, H. & Kajimura, K. (1997) *J. Appl. Phys.*, **81**, 744.
116. Watanabe, H., Takeuchi, D., Hara, S., Sekiguchi, T., Okushi, H. & Kajimura, K. (1997) *Diamond Films and Technol.* (MYU, Tokyo), **7**, 277.

117. Takeuchi, D., Watanabe, H., Sawada, H., Yamanaka, S., Ichinose, H., Sekiguchi, T. & Okushi, H. (2001) *Diamond and Related Mater.*, **10**, 526.
118. Takeuchi, D., Watanabe, H., Yamanaka, S., Okushi, H., Sawada, H., Ichinose, H., Sekiguchi, T. & Kajimura, K. (2001) *Phys. Rev. B*, **63**, 245328.
119. Yamanaka, S., Watanabe, H., Masai, S., Takeuchi, D., Okushi, H. & Kajimura, K. (1998) *Jpn. J. Appl. Phys.*, **37**, L1129.
120. Yamanaka, S., Watanabe, H., Masai, S., Kawata, S., k. Hayashi, Takeuchi, D., Okushi, H. & Kajimura, K. (1988) *J. Appl. Phys.*, **84**, 6095.
121. Takeuchi, D., Yamanaka, S., Watanabe, H., Okushi, H. & Kajimura, K. (2000) *Appl. Surf. Sci.* **159–160**, 572.
122. Yamanaka, S., Takeuchi, D., Watanabe, H., Okushi, H. & Kajimura, K. (1999) *phys. stat. sol. (a)*, **174**, 59.
123. Takeuchi, D., Yamanaka, S., Watanabe, H. & Okushi, H. (2001) *phys. stat. sol. (a)*, **186**, 269.
124. Watanabe, H., Sekiguchi, T. & Okushi, H. (2001) *Solid State Phenomena*, **78–79**, 165.
125. Watanabe, H. & Okushi, H. (2000) *Jpn. J. Appl. Phys.*, **39**, L835.
126. Watanabe, H., Hayashi, K., Takeuchi, D., Yamanaka, S., Okushi, H. & Kajimura, K. (1998) *Appl. Phys. Lett.*, **73**, 981.
127. Takueuchi, D., Yamanaka, S. & Okushi, H. (2002) *Diamond and Related Mater.*, **11**, 355.
128. Yamanaka, S., Takeuchi, D., Watanabe, H., Okushi, H. & Kajimura, K. (2000) *Diamond and Related Mater.*, **9**, 956.
129. Geis, M.W. (1992) *Diamond and Related Mater.*, **1**, 684.
130. Janssen, G. & Giling, L.J. (1995) *Diamond and Related Mater.*, **4**, 1025.
131. Findeling-Dufour, C. & Gicquel, A. (1997) *Thin Solid Films*, **308–309**, 178.
132. Findeling-Dufour, C., Gicquel, A. & Chiron, R. (1998) *Diamond and Related Mater.*, **7**, 986.
133. Geis, M.W., Smith, H.I., Argoitia, A., Angus, J., Ma, G.-H.M., Glass, J.T., Butler, J., Robinson, C.J. & Pryor, R. (1991) *Appl. Phys. Lett.*, **58**, 2485.
134. Geis, M.W., Efremow, N.N., Susalka, R., Twichell, J.C., Snail, K.A., Spiro, C., Sweeting, B. & Holly, S. (1994) *Diamond and Related Mater.*, **4**, 76.
135. Meguro, K. Ando, Y., Nishibayashi, Y., Ishibashi, K., Yamamoto, Y. & Imai, T. (2002) *New Diamond and Frontier Carbon Technology* (MYU, Japan), **12**, 125.
136. Kobashi, K., Nishibayashi, Y., Yokota, Y., Ando, Y., Tachibana, T. et al. (2003) *Diamond and Related Mater.*, **12**, 233.
137. Kawarada, H. (1996) *Surf. Sci. Rep.* **26**, 205.
138. Tsuno, T., Imai, T., Nishibayashi, Y., Hamada, K. & Fujimori, N. (1991) *Jpn. J. Appl. Phys.*, **30**, 1063.
139. Busmann, H.-G., Zimmermann-Edling, W., Sprang, H., Güntherodt, H.-J. & Hertel, I.V. (1992) *Diamond and Related Mater.*, **1**, 979.
140. Frauenheim, Th., Stephan, U., Blaudeck, P., Porezag, D., Busmann, H.-G., Zimmermann-Edling, W. & Lauer, S. (1993-II) *Phys. Rev. B*, **48**, 18189.
141. Kawarada, H., Sasaki, H. & Sato, A. (1995) *Phys. Rev. B*, **52**, 11351.
142. Tsuno, T., Tomikawa, T., Shikata, S. & Fujimori, N. (1994) *J. Appl. Phys.*, **75**, 1526.
143. Tsuno, T., Tomikawa, T., Shikata, S., Imai, T. & Fujimori, N. (1994) *Appl. Phys. Lett.*, **64**, 572.
144. Everson, M.P., Tamor, M.A. & Wu, C.H. (1994) *New Diamond Science and Technology*, Eds. Messier, R., Glass, J.T., Butler, J.E. & Roy, R. *Mater. Res. Soc.* PA, p. 613.

145. Sutcu, L.F., Chu, C.J., Thompson, M.S., Hauge, R.H., Margrave, J.L. & D'Evelyn, M.P. (1992) *J. Appl. Phys.*, **71**, 5930.
146. Busmann, H.-G., Lauer, S., Hertel, I.V., Zimmermann-Edling, W., Güntherodt, H.J., Frauenheim, Th., Blaudeck, P. & Porezag, D. (1993) *Surf. Sci.*, **295**, 340.
147. Sasaki, H. & Kawarada, H. (1993) *Jpn. J. Appl. Phys.*, **32**, L1771.
148. Waclawski, B.J., Pierce, D.T., Swanson, N. & Celotta, R.J. (1982) *J. Vac. Sci. Technol.*, **21**, 368.
149. Aizawa, T., Ando, T., Kamo, M. & Sato, Y. (1993) *Phys. Rev. B*, **48**, 18348.
150. Koizumi, S., Murakami, T., Inuzuka, T. & Suzuki, K. (1990) *Appl. Phys. Lett.*, **57**, 563.
151. Yoshikawa, M., Ishida, H., Ishitanai, A., Murakami, T., Koizumi, S. & Inuzuka, T. (1990) *Appl. Phys. Lett.*, **57**, 428.
152. Yoshikawa, M., Ishida, H., Ishitani, A., Koizumi, S. & Inuzuka, T. (1991) *Appl. Phys. Lett.*, **58**, 1387.
153. Koizumi, S. & Inuzuka, T. (1993) *Jpn. J. Appl. Phys.*, **32**, 3920.
154. Mishima, O. (1991) Applications of Diamond Films and Related Materials (Elsevier, Amsterdam), *Materials Science Monographs*, **73**, 647.
155. Mishima, O. (1990) Diamond, Silicon Carbide and Related Wide Bandgap Semiconductors, *Mater. Res. Soc. Symp. Proc.*, **162**, 543.
156. Haubner, R. (1990) *Int. J. Refractory Metals & Hard Mater.*, **9**, 70.
157. Wang, L., Pirouz, P., Argoitia, A., Ma, J.S. & Angus, J.C. (1993) *Appl. Phys. Lett.*, **63**, 1336.
158. Argoitia, A., Angus, J.C., Ma, J.S., Wang, L., Pirouz, P. & Lambrecht, W.R.L. (1994) *J. Mater. Res.*, **9**, 1849.
159. Maeda, H., Masuda, S., Kusakabe, K. & Morooka, S. (1994) *Diamond and Related Mater.*, **3**, 398.
160. Braun, M.W.H., Kong, H.S., Glass, J.T. & Davis, R.F. (1991) *J. Appl. Phys.*, **69**, 2679.
161. Chai, W.P., Gu, Y.S., Li, M., Mai, Z.H., Li, Q.Z., Yuan, L. & Pang, S.J. (1994) *Appl. Phys. Lett.*, **64**, 1941.
162. Mishima, O., Tanaka, J., Yamaoka, S. & Fukunaga, O. (1987) *Science*, **238**, 181.
163. Komatsu, S. (1997) *J. Mater. Res.*, **12**, 1675.
164. Yanagihara, T. & Yomogita, K. (2000) *Jpn. J. Appl. Phys.*, **39**, Part 1, 5229.
165. Nistor, L., Buschmann, V., Ralchenko, V., Dinca, G., Vlasov, I., Van Landuyt, J. & Fuess, H. (2000) *Diamond and Related Mater.*, **9**, 269.
166. Zhu, W., Yang, P.C., Glass, J.T. & Arezzo, F. (1995) *J. Mater. Res.*, **10**, 1455.
167. Belton, D.N. & Schmieg, S.J. (1989) *J. Appl. Phys.*, **66**, 4223.
168. Sato, Y., Yamashita, I., Fujita, H. Ando, T. & Kamo, M. (1991) in *New Diamond Science and Technology*, Eds. Messier, R., Glass, J.T., Butler, J.E. & Roy, R. *Mater. Res. Symp. Proc.*, NDST2-C3, Pittsburgh, PA, p. 371.
169. Sato, Y., Fujita, H. Ando, T., Tanaka, T. & Kamo, M., *Phil. Trans. R. Soc. London*, **A 342**, 225 (1993)
170. Yang, P.C., Zhu, W. & Glass, J.T. (1993) *J. Mater. Res.*, **8**, 1773.
171. Zhu, W., Yang, P.C. & Glass, J.T. (1993) *Appl. Phys. Lett.*, **63**, 1640.
172. Yang, P.C., Zhu, W. & Glass, J.T. (1994) *J. Mater. Res.*, **9**, 1063.
173. Yang, P.C., Wolden, C.A., Liu, W., Schlesser, R., Davis, R.F., Prater, J.T. & Sitar, Z. (1998) *J. Mater. Res.*, **13**, 1120.
174. Sitar, Z., Liu, W., Yang, P.C., Wolden, C.A., Schlesser, R. & Prater, J.T. (1998) *Diamond and Related Mater.*, **7**, 276.

175. Badzian, A. & Badzian, T. (1996) *Diamond and Related Mater.*, **5**, 93.
176. Yang, P.C., Schlesser, R., Wolden, C.A., Liu, W., Davis, R.F., Sitar, Z. & Prater, J.T. (1997) *Appl. Phys. Lett.*, **70**, 2960.
177. Yang, P.C., Liu, W., Schlesser, R., Wolden, C.A., Davis, R.F., Prater, J.T. & Sitar, Z. (1998) *J. Cryst. Growth*, **187**, 81.
178. Liu, W., Yang, P.C., Wolden, C.A., Davis, R.F., Prater, J.T. & Sitar, Z. (1998) *J. Appl. Phys.*, **83**, 7658.
179. Porter, D.A. & Easterling, K.E. (1992) *Phase Transformation in Metals and Alloys*, Chapman & Hall, London, p. 390.
180. *Handbook of Chemistry and Physics* (1990) 71st edition. Eds. Lide, D. (CRC, Boca Raton), 12.
181. Singleton, M. & Nash, P. (1989) *Bulletin of Alloy Phase Diagrams*, **10**, 121.
182. Liu, W., Tucker, D.A., Yang, P. & Glass, J.T. (1995) *J. Appl. Phys.*, **78**, 1291.
183. Hartsell, M.L. & Plano, L.S. (1994) *J. Mater. Res.*, **9**, 921.
184. Ong, T.P., Xiong, F., Chang, R.P.H. & White, C.W. (1992) *Appl. Phys. Lett.*, **60**, 2083.
185. Narayan, J. & Chen, X. (1992) *J. Appl. Phys.*, **71**, 3795.
186. Ojika, S., Yamashita, S., Kataoka, K., Ishikura, T., Yamaguchi, A. & Kawarada, H. (1993) *Jpn. J. Appl. Phys.*, **32**, Part 2, L200.
187. Ojika, S., Yamashita, S., Kataoka, K. & Ishikura, T. (1993) *Jpn. J. Appl. Phys.*, **32**, Part 2, L1681.
188. Wolter, S.D., Stoner, B.R., Ma, C.-H.M. & Glass, J.T.(1992) *Mat. Res. Symp. Proc.*, Vol. **270**, 342.
189. Wolter, S.D., Stoner, B.R. & Glass, J.T. (1994) *Diamond and Related Mater.*, **3**, 1188.
190. Jiang, N., Wang, C.L., Won, J.H., Joen, M.H., Mori, Y., Hatta, A., Ito, T., Sasaki, T. & Hiraki, A. (1997) *Appl. Surf. Sci.*, **117/118**, 587.
191. Plano, L., unpublished.
192. Wolter, S.D., McClure, M.T., Glass, J. T. & Stoner, B. R. (1995) *Appl. Phys. Lett.*, **66**, 2810.
193. Lambrecht, W.R.L. & Segall, B. (1992) *J. Mater. Res.*, **7**, 696.
194. Argoitia, A., Angus, J.C., Wang, L., Ning, X.I. & Pirouz, P. (1993) *J. Appl. Phys.*, **73**, 4305.
195. Tucker, D.A., Seo, D.-K., Whangbo, M.-H., Sivazlian, F.R., Stoner, B.R., Bozeman, S.P., Sowers, A.T., Nemanich, R.J. & Glass, J.T. (1995) *Surf. Sci.*, **334**, 179.
196. Li, Z., Wang, L., Suzuki, T., Argoitia, A., Pirouz, P. & Angus, J.C. (1993) *J. Appl. Phys.*, **73**, 711.
197. Lambrecht, W.R.L., Lee, C.H., Segall, B., Angus, J.C., Li, Z. & Sunkara, M. (1993) *Nature*, **364**, 607.
198. Suzuki, T., Yagi, M., Shibuki, K. & Hasemi, M. (1994) *Appl. Phys. Lett.*, **65**, 540.
199. Yoshimoto, M., Yoshida, K., Murata, H., Hishitani, Y., Koinuma, H., Nishio, S., Kakihara, M. & Tachibana, T. (1999) *Nature*, **398**, 340.
200. Jeng, D.G., Tuan, H.S., Salat, R.F. & Fricano, G.J. (1990) *Appl. Phys. Lett.*, **56**, 1968.
201. Qimin, W., Qinggui, C., Rihua, S., Weibing, Z., Qiwei, D. & Rongkang, D. (1993) *J. Cryst. Growth*, **129**, 774.
202. Belton, D.N., Harris, S.J., Schmieg, S.J., Weiner, A.M. & Perry, T.A. (1989) *Appl. Phys. Lett.*, **54**, 416.
203. Williams, B. & Glass, J.T. (1989) *J. Mater. Res.*, **4**, 373.
204. Hirose, Y., unpublished.

205. Liu, H. & Dandy, D.S. (1995) *Diamond and Related Mater.*, **4**, 1173.
206. Yugo, S., Kimura, T. & Kanai, T. (1992) *Diamond and Related Mater.*, **2**, 328.
207. Yugo, S., Nakamura, N. & Kimura, T. (1998) *Diamond and Related Mater.*, **7**, 1017.
208. Yugo, S., Semoto, K., Hoshina, K., Kimura, T. & Nakai, H. (1995) *Diamond and Related Mater.*, **4**, 903.
209. Ishigaki, N. & Yugo, S. (2000) *Diamond and Related Mater.*, **9**, 1646.
210. Saito, R., Sano, J., Ishigaki, N., Kimura, T. & Yugo, S. (2001) *J. Appl. Phys.*, **90**, 2559.
211. Kinchin, G.H. & Peace, R.S. (1955) *Rep. Prog. Phys.*, **18**, 1.
212. Friedlich, M., Morley, S., Mainz, B., Hinneberg, H.-J. & Zahn, D.R.T. (1995) *Diamond and Related Mater.*, **4**, 944.
213. Jiang, X., Six, R., Klages, C.-P., Zachai, R., Hartweg, M. & Füßer, H.-J. (1992) *Diamond and Related Mater.*, **2**, 407.
214. Bar-Yam, Y. & Moustakas, D. (1989) *Nature (London)*, **342**, 786.
215. Bar-Yam, Y., Lei, T., Moustakas, T.D., Allen, D.C. & Teter, M.P. (1992) *MRS Symposium Proceedings*, **Vol. 242**, Eds. Moustakas, T.D., Pankov, J.I. & Hamakawa, Y. *Materials Research Society*, Pittsburgh, PA, 335.
216. Sheldon, B.W., Csencsits, R., Rankin, J., Boekenhauer, R.E. & Shigesato, Y. (1994) *J. Appl. Phys.*, **75**, 5001.
217. Boppart, H., van Straaten, J. & Silvera, I.F. (1985) *Phys. Rev. B*, **32**, 1423.
218. Katoh, M., Aoki, M. & Kawarada, H. (1994) *Jpn. J. Appl. Phys.*, **33**, L194.
219. Sawabe, A. & Inuzuka, T. (1985) *Appl. Phys. Lett.*, **46**, 146.
220. Lannon, J.M., Jr., Gold, J.S. & Stinespring, C.D. (1995) *J. Appl. Phys.*, **77**, 3823.
221. Gerber, J., Sattel, S., Jung, K., Ehrhardt, H. & Robertson, J. (1995) *Diamond and Related Mater.*, **4**, 559.
222. Kulisch, W., Ackermann, L. & Sobisch, B. (1996) *phys. stat. sol. (a)*, **154**, 155.
223. Kulisch, W., Sobisch, B., Kuhr, M. & Beckmann, R. (1995) *Diamond and Related Mater.*, **4**, 401.
224. Beckmann, R., Sobisch, B., Kulisch, W. & Rau, C. (1994) *Diamond and Related Mater.*, **3**, 555.
225. Stoner, B.R., Ma, G.H., Wolter, S.D., Zhu, W., Wang, Y.-C., Davis, R.F. & Glass, J.T. (1993) *Diamond and Related Mater.*, **2**, 142.
226. Kim, Y.-K., Lee, K.-Y. & Lee, J.-Y. (1996) *Appl. Phys. Lett.*, **68**, 756,.
227. Kim, Y.-K. & Lee, J.-Y. (1997) *J. Appl. Phys.*, **81**, 3660.
228. Milne, D.K., Roberts, P.G., John, P., Jubber, M.G., Liehr, M. & Wilson, J.I.B. (1995) *Diamond and Related Mater.*, **4**, 394.
229. Shigesato, Y., Boekenhauer, R.E. & Sheldon, B.W. (1993) *Appl. Phys. Lett.*, **63**, 314.
230. Jiang, X. & Lia, C.L. (1995) *Appl. Phys. Lett.*, **67**, 1197.
231. Barshilia, H.C., Mehta, B.R. & Vankar, V.D. (1996) *J. Mater. Res.*, **11**, 2852.
232. Reinke, P., Kania, P. & Oelhafen, P. (1995) *Thin Solid Films*, **270**, 124.
233. Reinke, P., Kania, P., Oelhafen, P. & Guggenheim, R. (1996) *Appl. Phys. Lett.*, **68**, 22.
234. Schreck, M., Th. Bauer, & Stritzker, B. (1995) *Diamond and Related Mater.*, **4**, 553.
235. Whitfield, M.D., Jackman, R.B., Rodway, D., Savage, J.A. & Foord, J.S. (1996) *J. Appl. Phys.*, **80**, 3710.
236. Lang, T., Stiegler, J., von Kaenel, Y. & Blank, E. (1996) *Diamond and Related Mater.*, **5**, 1174.
237. Mucha, J.A., Flamm, D.L. & Ibbotson, D.E. (1989) *J. Appl. Phys.*, **65**, 3448.

238. Balestrino, G., Marinelli, M., E.Milani, Paoletti, A., Pinter, I. & Tebano, A. (1993) *Appl. Phys. Lett.*, **62**, 879.
239. Gomez-Aleixandre, C., Sanchez, O., Castro, A. & Albella, J.M. (1993) *J. Appl. Phys.*, **74**, 3752.
240. Kobashi, K. (1995) *R&D Report of Kobe Steel Ltd.*, **45**, 65, in Japanese.
241. Wolter, S.D., Glass, J.T. & Stoner, B.R. (1995) *J. Appl. Phys.*, **77**, 5119.
242. Joffreau, P.O., Haubner, R. & Lux, B. (1998) *Int. J. Ref. Hard. Met.*, **7**, 186.
243. Kamo, M., Chawanya, H., Tanaka, T., Sato, Y. & Setaka, N. (1988) *Mater. Sci. Eng.*, **A105/106**, 535.
244. Stoner, B.R., Malta, D.M., Glass, R.C., Sowers, A., and Nemanich, R.J. (1994) *Advanced Materials '93* I/B: Magnetic, Fullerene, Dielectric, Ferroelectric, *Diamond and Related Materials*, Eds. Homma, M. et al., *Trans. Mat. Res. Soc. Jpn.*, **14B**, Elsevier, 1511.
245. He, X.C., Shen, H.S., Zhang, Z.M., Hu, X.J., Wan, Y.Z., Shen, T. (2000) *Diamond and Related Mater.*, **9**, 1626.
246. Carter, C.H., Davis, R.F. & Nutt, S.R. (1986) *J. Mater. Res.*, **1**, 811.
247. Kong, H.S., Wang, Y.C., Glass, J.T. & Davis, R.F. (1988) *J. Mater. Res.*, **3**, 521.
248. Kohl, R., Wild, C., Herres, N., Koidl, P., Stoner, B.R. & Glass, J.T. (1993) *Appl. Phys. Lett.*, **63**, 1792.
249. Kawarada, H., Suenaga, T. & Nagasawa, H. (1995) *Appl. Phys. Lett.*, **66**, 583.
250. Suesada, T., Nakamura, N., Nagasawa, H. & Kawarada, H. (1995) *Jpn. J. Appl. Phys.*, **34**, 4948.
251. Nagasawa, H., Yamaguchi, Y., Izumi, T. & Tonosaki, K. (1993) *Appl. Surf. Sci.*, **70/71**, 542.
252. Mizuochi, Y., Nagasawa, H. & Kawarada, H. (1997) *Diamond and Related Mater.*, **6**, 277.
253. Maréchal, N., Ishikura, T. & Yamashita, S. (1996) *Diamond Films and Technol.*, **6**, 265.
254. Maréchal, N. & Yamashita, S. (1995) *Jpn. J. Appl. Phys.*, **34**, L1671.
255. Maréchal, N. & Yamashita, S. (1996) *Thin Solid Films*, **290–291**, 131.
256. Nagasawa, H. & Yamaguchi, Y. (1993) *Thin Solid Films*, **225**, 230.
257. Gutheit, T., Heinau, M., Füsser, H.-J., Ch. Wild, Koidl, P. & Abstreiter, G. (1995) *J. Crystal Growth*, **157**, 426.
258. Wittorf, D., Jäger, W. Urban, K., Gutheit, T., Güttler, H., Schulz, S. & Zachai, R. (1997) *Diamond and Related Mater.*, **6**, 649.
259. Kono, S., Goto, T., Abukawa, T., Wild, C., Koidl, P. & Kawarada, H. (2000) *Jpn. J. Appl. Phys.*, **39**, Part 1, 4372.
260. Wolter, S.D., Stoner, B.R., Glass, J.T., Ellis, P.J., Buhaenko, D.S., Jenkins, C.E. & Southworth, P. (1993) *Appl. Phys. Lett.*, **62**, 1215.
261. Stoner, B.R., Sahaida, S.R., Bate, J.P., Southworth, P. & Ellis, P.J. (1993) *J. Mater. Res.*, **8**, 1334.
262. *The Experimental Strategies Foundation*, P. O. Box 27254, Seattle, Washington 98125.
263. Jiang, X., Klages, C.-P., Zachai, R., Hartweg, M. & Füsser, H.-J. (1993) *Appl. Phys. Lett.*, **62**, 3438.
264. Kistenmacher, T.J., Ecelberger, S.A. & Stoner, B.R. (1995) *Diamond and Related Mater.*, **4**, 1289.
265. Jiang, X., Schiffmann, K., C.-Klages, P., Wittorf, D., Jia, C.L., Urban, K. & Jäger, W. (1998) *J. Appl. Phys.*, **83**, 2511.

266. Saito, T., Tsuruga, S., Maeda, H., Kusakabe, K. & Morooka, S. (1997) *Diamond and Related Mater.*, **6**, 668.
267. Maeda, H., Irie, M., Hino, T., Kusakabe, K. & Morooka, S. (1995) *J. Mater. Res.*, **10**, 158.
268. Yang, B.X., Zhu, Y., Ahn, J., Tan, H.S. & Lu, D. (1995) *Thin Solid Films*, **270**, 210.
269. Chen, C.J., Chang, L., Lin, T.S. & Chen, F.R. (1996) *J. Mater. Res.*, **11**, 1002.
270. Ownby, P.D., Yang, X. & Liu, L. (1992) *J. Am. Ceram. Soc.*, **75**, 1876.
271. Verwoerd, W.S. & Weimer, K. (1991) *Surf. Coat. Technol.*, **47**, 578.
272. Verwoerd, W.S. (1992) *Diamond and Related Mater.*, **1**, 195.
273. Verwoerd, W.S. (1994) *Surf. Sci.*, **304**, 24.
274. Maillard-Schaller, E., Küttel, O.M., Gröning, P., Gröning, O., Agostino, R.G., Aebi, P., Schlapbach, L., Wurzinger, P. & Pongratz, P. (1997) *Phys. Rev. B*, **55**, 15895.
275. Hessmer, R., Schreck, M., Geier, S., Rauschenbach, B. & Stritzker, B. (1995) *Diamond and Related Mater.*, **4**, 410.
276. Jiang, X. & Klages, C.-P. (1993) *Diamond and Related Mater.*, **2**, 1112.
277. Jiang, X., Schiffmann, K. & Klages, C.-P. (1994-II) *Phys. Rev. B*, **50**, 8402.
278. Jiang, X., Paul, M. & Klages, C.-P. (1996) *Diamond and Related Mater.*, **5**, 251.
279. Jiang, X. & Klages, C.-P. (1996) *phys. stat. sol. (a)*, **154**, 175.
280. Jiang, X. & Jia, C.L. (1996) *Appl. Phys. Lett.*, **69**, 3902.
281. Zhang, W. J., Jiang, X. & Klages, C.-P. (1997) *J. Cryst. Growth*, **171**, 485.
282. Geier, S., Hessmer, R., Schreck, M., Stritzker, B., Rauschenbach, B., Helming, K., Kunze, K. & Erfurth, W. (1996) *Mater. Sci. Forum*, **228–231**, 445.
283. Thürer, K.-H., Schreck, M. & Stritzker, B. (1998-II) *Phys. Rev. B*, **57**, 15454.
284. Schreck, M. & Stritzker, B. (1996) *phys. stat. sol. (a)*, **154**, 197.
285. Schaller, E., Küttel, O.M. Aebi, P. & Schlapbach, L. (1995) *Appl. Phys. Lett.*, **67**, 1533.
286. Küttel, O.M., Agostino, R.G., Fasel, R., Osterwalder, J. & Schlapbach, L. (1994) *Surf. Sci.*, **321**, 131.
287. Maillard-Schaller, E., Küttel, O.M., Gröning, P., Aebi, P. & Schlapbach, L. (1997) *Diamond and Related Mater.*, **6**, 282.
288. Jiang, X., Klages, C.-P., Rösler, M., Zachai, R., Hartweg, M. & Füsser, H.-J. (1993) *Appl. Phys.*, **A57**, 483.
289. Schreck, M., Hessmer, R., Geier, S., Rauschenbach, B. & Stritzker, B. (1994) *Diamond and Related Mater.*, **3**, 510.
290. Helming, K., Geier, S., Schreck, M., Hessmer, R., Stritzker, B. & Rauschenbach, B. (1995) *J. Appl. Phys.*, **77**, 4765.
291. Ellis, P.J., Buhaenko, D.S. & Stoner, B.R. (1995) *Diamond and Related Mater.*, **4**, 406.
292. Schreck, M., Thürer, K.-H., Klarmann, R. & Stritzker, B. (1997) *J. Appl. Phys.*, **81**, 3096.
293. Schreck, M., Thürer, K.-H. & Stritzker, B. (1997) *J. Appl. Phys.*, **81**, 3092.
294. Fox, B.A., Stoner, B.R., Malta, D.M., Ellis, P.J., Glass, R. C. & Sivazlian, F. R. (1994) *Diamond and Related Mater.*, **3**, 382.
295. Jiang, X., Boettger, E., Paul, M. & Klages, C.-P. (1994) *Appl. Phys. Lett.*, **65**, 1519.
296. Jiang, X., Schiffmann, K., Westphal, A. &. Klages, C.-P. (1993) *Appl. Phys. Lett.*, **63**, 1203.
297. Klarmann, R., Schreck, M., Hessmer, R. & Stritzker, B. (1996) *Diamond and Related Mater.*, **5**, 266.
298. Hessmer, R., Schreck, M. & Stritzker, B. (1995) *Diamond and Related Mater.*, **4**, 416.
299. Schiffmann, K. & Jiang, X. (1994) *Appl. Phys. A: Solids Surf.*, **59**, 17.

300. Gerber, J., Robertson, J., Sattel, S. & Ehrhardt, H. (1996) *Diamond and Related Mater.*, **5**, 261.
301. Gerber, J., Weiler, M., Sohr, O., Jung, K. & Ehrhardt, H. (1994) *Diamond and Related Mater.*, **3**, 506.
302. McKenize, D. R., Muller, D., Pailthorpe, B. A., Wang, Z. H., Kravtchinskaia, E., Segal, D., Lukins, P. B., Swift, P. D., Martin, P. J., Amaratunga, G., Gaskell, P. H. & Saeed, A. (1991) *Diamond and Related Mater.*, **1**, 51.
303. Chen, C.J., Chang, L., Lin, T.S. & Chen, F.R. (1995) *J. Mater. Res.*, **10**, 3041.
304. Chang, L., Chen, C.J., Chen, F.R., Hu, S.F. & Lin, T.S. (1996) *Diamond and Related Mater.*, **5**, 326.
305. Chang, L., Chen, F.R., Chen, C.J. & Lin, T.S. (1996) *Diamond and Related Mater.*, **5**, 1282.
306. Zhu, W., Wang, X.H., Stoner, B.R., Ma, G.H.M., Kong, H.S., Braun, M.W.H. & Glass, J.T. (1993-I) *Phys. Rev. B*, **47**, 6529.
307. Stoner, B.R., Ma, G.H., Wolter, S.D., Zhu, W., Wang, Y.-C., Davis, R.F. & Glass, J.T. (1993) *Diamond and Related Mater.*, **2**, 590.
308. Jiang, X. & Jia, C.L. (2000) *Phys. Rev. Lett.*, **84**, 3658.
309. Jiang, X. & Jia, C.L. (1995) *Appl. Phys. Lett.*, **67**, 1197.
310. Jia, C.L., Urban, K. & Jiang, X. (1995) *Phys. Rev. B*, **52**, 5164.
311. Wurzinger, P., Pongratz, P., Gerber, J. & Ehrhardt, H. (1996) *Diamond and Related Mater.*, **5**, 345.
312. Bozzolo, N., Barrat, S., Dieguez, I. & Bauer-Grosse, E. (1997) *Diamond and Related Mater.*, **6**, 41.
313. Plitzko, J., Rösler, M. & Nickel, K.G. (1997) *Diamond and Related Mater.*, **6**, 935.
314. Tzou, Y., Bruley, J., Ernst, F., Rühle, M. & Raj, R. (1994) *J. Mater. Res.*, 9, 1566.
315. Wurzinger, P., Fuchs, N., Pongratz, P., Schreck, M., Heßmer, R. & Stritzker, B. (1997) *Diamond and Related Mater.*, **6**, 752.
316. Jeon, S.-J., Chawla, A.K., Baik, Y.-J. & Sung, C. (1997) *Mat. Res. Soc. Symp.*, **458**, 429.
317. Chang, L., Yang, J.E., Chen, F.R. & Kai, J.J. (2000) *Diamond and Related Mater.*, **9**, 283.
318. Chang, L., Lin, T.-S., Chen, J.-L. & Chen, F.-R. (1993) *Appl. Phys. Lett.*, **62**, 3444.
319. Sivazlian, F.R., Glass, J.T. & Stoner, B.R. (1994) *J. Mater. Res.*, **9**, 2487.
320. Thürer, K.-H., Schreck, M., Stritzker, B., Fuchs, N. & Pongratz, P. (1997) *Diamond and Related Mater.*, **6**, 1010.
321. Lee, J.S., Liu, K.S. & Lin, I-Nan (1995) *Appl. Phys. Lett.*, **67**, 1555.
322. Jiang, X., Zhang, W.J., Paul, M. & Klages, C.-P. (1996) *Appl. Phys. Lett.*, **68**, 1927.
323. Zhang, W.J. & Zhang, X. (1996) *Appl. Phys. Lett.*, **68**, 2195.
324. Jiang, X., Zhang, W.J. & Klages, C.-P. (1998) *Phys. Rev. B*, **58**, 7064.
325. Kobashi, K., Tachibana, T., Yokota, Y., Kawakami, N., Hayashi, K., Yamamoto, K., Koga, Y., Fujiwara, S., Gotoh, Y., Nakahara, H., Tsuji, H., Ishikawa, J., Köck, F. & Nemanich, R.J. (2003) *J. Mater. Res.*, **18**, 305.
326. Barrat, S., Saada, S., Dieguez, I. & Bauer-Grosse, E. (1998) *J. Appl. Phys.*, **84**, 1870.
327. Saada, S., Barrat, S. & Bauer-Grosse, E. (2000) *Diamond and Related Mater.*, **9**, 300.
328. Barrat, S., Saada, S., Thiebaut, J.M. & Bauer-Grosse, E. (2001) *Diamond and Related Mater.*, **10**, 1637.
329. John, P., Graham, C., Milne, D.K., Jubber, M.G. & Wilson, J.I.B. (1996) *Diamond and Related Mater.*, **5**, 256.

330. Stoner, B.R., Williams, B.E., Wolter, S.D., Nishimura, K. & Glass, J.T. (1992) *J. Mater. Res.*, **7**, 257.
331. Wolter, S.D., Glass, J.T. & Stoner, B.R. (1995) *Thin Solid Films*, **261**, 4.
332. Duncan Marshall, R. Whitfield, M.D., Tracey, S., Thompson, D.J., Foord, J.S. & Jackman, R.B. (1997) *Diamond and Related Mater.*, **6**, 676.
333. Guo, X.J., Sung, S.L., Lin, J.C., Chen, F.R. & Shih, H.C. (2000) *Diamond and Related Mater.*, **9**, 1840.
334. Tachibana, T., Hayashi, K. & Kobashi, K. (1996) *Appl. Phys. Lett.*, **68**, 1491.
335. Kim, S.H., Park, Y.S., Han, I.T., Lee, J.-W. & Yun, W.S. (1996) *Appl. Phys. Lett.*, **69**, 2184.
336. Lee, J.-W., Kim, S.H., Park, Y.S. & I.-Han, T. (1997) *Thin Solid Films*, **303**, 264.
337. Kim, Y.K., Lee, K.H., Lee, M.J. & Lee, J.Y. (1999) *Thin Solid Films*, **341**, 211.
338. Jiang, X., Fryda, M. & Jia, C.L. (2000) *Diamond and Related Mater.*, **9**, 1640.
339. Zhang, W.J., Jiang, X. & Xia, Y.B. (1997) *J. Appl. Phys.*, **82**, 1896.
340. Gu, C., Jiang, X. & Jin, Z. (1999) *Diamond and Related Mater.*, **8**, 262.
341. Wolter, S.D., Borst, T.H., Vescan, A. & Kohn, E. (1996) *Appl. Phys. Lett.*, **68**, 3558.
342. Seong, T.-Y., Kim, D.-G., Choi, K.-K. & Baik, Y.-J. (1997) *Appl. Phys. Lett.*, **70**, 3368.
343. Kim, D.-G., Seong, T.-Y. & Baik, Y.-J. (1998) *J. Electrochem. Soc.* **145**, 2095.
344. Flöter, A., Güttler, H., Schulz, G., Steinbach, D., Lutz-Elsner, C., Zachai, R., Bergmaier, A. & Dollinger, G. (1998) *Diamond and Related Mater.*, **7**, 283.
345. Flötter, A., Gütter, H., Schultz, G. & Zachai, R. (1998) *The 5th NIRIM International Symposium on Advanced Materials (ISAM '98)*, Tsukuba, Japan, March 1–5, 85.
346. Asmussen, J., Mossbrucker, J., Khatami, S., Huang, W.S., Wright, B. & Ayres, V. (1999) *Diamond and Related Mater.*, **8**, 220.
347. Schreck, M., Thürner, K.H. Christensen, C., Müller, M. & Stritzker, B. (1999) *Diamond and Related Mater.*, **8**, 160.
348. Choi, I.-H., Weisbecker, P., Barrat, S. & Bauer-Grosse, E. (2004) *Diamond and Related Mater.*, **13**, 574.
349. Suzuki, T., Yagi, M. & Shibuki, K. (1994) *Appl. Phys. Lett.*, **64**, 557.
350. Mizoguchi, K. & Nakashima, S. (1989) *J. Appl. Phys.*, **65**, 2583.
351. John, P., Milne, D.K., Roberts, P.G., Jubber, M.G., Liehr, M. & Wilson, J.I.B. (1994) *J. Mater. Res.*, **9**, 3083.
352. Jubber, M. G. & Milne, D.K. (1996) *phys. stat. sol. (a)*, **154**, 185.
353. von Kaenel, Y., Stiegler, J., Michler, J. & Blank, E. (1997) *J. Appl. Phys.*, **81**, 1726.
354. Michler, J., von Kaenel, Y., Stiegler, J. & Blank, E. (1998) *J. Appl. Phys.*, **83**, 187.
355. Nishitani-Gamo, M., Ando, T., Yamamoto, K., Watanabe, K., Dennig, P.A., Sato, Y. & Sekita, M. (1997) *Appl. Phys. Lett.*, **70**, 1530.
356. Buhaenko, D.S., Southworth, P., Jenkins, C.E., Ellis, P.E. & Stoner, B.R. (1994) *Diamond and Related Mater.*, **3**, 926.
357. Steeds, J.W. (1997) *Inst. Phys. Conf. Ser. No. 153: Section 10, Electron Microscopy and Analysis Group Conf: EMAG97*, Cambridge, 407.
358. Wolter, S.D., Borca-Tasciuc, D.-A., Chen, G., Govindaraju, N., Collazo, R., Okuzumi, F., Prater, J.T. & Sitar, Z. (2003) *Diamond and Related Mater.*, **12**, 61.
359. Inoue, T., Tachibana, H., Kumagai, K., Miyata, K., Nishimura, K., Kobashi, K. & Nakaue, A. (1990) *J. Appl. Phys.*, **67**, 7329.
360. Robertson, J. (1995) *Diamond and Related Mater.*, **4**, 549.

361. Kim, Y.-K., Han, Y.-S. & J.-Lee, Y. (1998) *Diamond and Related Mater.*, **7**, 96.
362. Stöckel, R., Stammler, M., Janischowsky, K., Ley, L., Albrecht, M. & Strunk, H.P. (1998) *J. App. Phys.*, **83**, 531.
363. Stöckel, R., Janischowsky, K., Rohmfeld, S., Ristein, J., Hundhausen, M. & Ley, L. (1996) *Diamond and Related Mater.*, **5**, 321.
364. Stöckel, R., Janischowsky, K., Rohmfeld, S., Ristein, J., Hundhausen, M. & Ley, L. (1996) *J. Appl. Phys.*, **79**, 768.
365. Stammler, M., Stöckel, R., Ley, L., Albrecht, M. & Strunk, H.P. (1997) *Diamond and Related Mater.*, **6**, 747.
366. Yokota, Y., Kawakami, N., Hayashi, K. & Inoue, K. (2001) New diamond, Ohm, Tokyo **56**, 20, in Japanese.
367. Harris, S.J., Belton, D.N., Weiner, A.M. & Schmeig, S.J. (1989) *J. Appl. Phys.*, **66**, 5353.
368. Belton, D.N. & Schmeig, S.J. (1990) *Surf. Sci.* **233**, 131.
369. Belton, D.N. & Schmeig, S.J. (1991) *J. Appl. Phys.*, **69**, 3032.
370. Takaya, M. & Sakamoto, Y. (993) in *First International Conference on Processing Materials for Properties*, Eds. Henein, H. & Oki, T., *The Minerals, Metals & Materials Society*, Warrendale, PA, 1157.
371. Kalss, W., Haubner, R. & Lux, B. (1997) *Diamond and Related Mater.*, **6**, 240.
372. Kalss, W., Haubner, R., Lippold, G. & Lux, B. (1998) *Diamond and Related Mater.*, **7**, 158.
373. Kalss, W., Haubner, R., & Lux, B. (1997) *Diamond and Related Mater.*, **6**, 240.
374. Saito, T., Tsuruga, S., Ohya, N., Kusakabe, K., Morooka, S., Maeda, H., Sawabe, A. & Suzuki, K. (1998) *Diamond and Related Mater.*, **7**, 1381.
375. Sawabe, A., Fukuda, H., Suzuki, T., Ikuhara, Y. & Suzuki, T. (2000) *Surf. Sci.* **467**, L845.
376. Suzuki, K., Fukuda, H., Yamada, T. & Sawabe, A. (2001) *Diamond and Related Mater.*, **10**, 2153.
377. Schreck, M., Roll, H. & Strizker, B. (1999) *Appl. Phys. Lett.*, **74**, 650.
378. Hörmann, F., Schreck, M. & Stritzker, B. (2001) *Diamond and Related Mater.*, **10**, 1617.
379. Bauer, Th., Schreck, M., Hörmann, F., Bergmaier, A., Dollinger, G. & Stritzker, B. (2002) *Diamond and Related Mater.*, **11**, 493.
380. Fujisaki, T., Tachiki, M., Taniyama, N., Kudo, M. & Kawarada, H. (2002) *Diamond and Related Mater.*, **11**, 478.
381. Fujisaki, T., Tachiki, M., Taniuyama, N., Kudo, M. & Kawarada, H.(2003) *Diamond and Related Mater.*, **12**, 246.
382. Ikeda, S. & Nagano, M. (1999) *Jpn. J. Appl. Phys.*, **38**, Part 2, L882.
383. Tachibana, T., Yokota, Y., Miyata, K., Ohnishi, T., Kobashi, K., Tarutani, M., Takai, Y., Shimizu, R. & Shintani, Y. (1997) *Phys. Rev. B*, **56**, 15967.
384. Tachibana, T., Yokota, Y., Miyata, K., Kobashi, K. & Shintani, Y. (1997) *Diamond and Related Mater.*, **6**, 266.
385. Tachibana, T., Yokota, Y., Kobashi, K. & Shintani, Y. (1997) in *Advanced Materials '97, Proc. of the 4th NIRIM Int'l Symp. on Advance Mater.* NIRIM, Japan, 129.
386. Tachibana, T., Yokota, Y., Hayashi, K. & Kobashi, K. (2001) *Diamond and Related Mater.*, **10**, 1633.
387. Tachibana, T., Yokota, Y., Hayashi, K., Miyata, K., Kobashi, K. & Shintani, Y. (2000) *Diamond and Related Mater.*, **9**, 251.

388. Jiang, N., Kujime, S., Yamamoto, H., Inaoka, T., Naoi, Y., Shinatani, Y., Makita, H. & Hiraki, A. (2000) *Jpn. J. Appl. Phys.*, **39**, Part 2, L623.
389. Ishikura, T., Yamashita, S., Ojika, S. & Kawarada, H. (1994) in *Advances in New Diamond Science and Technology*, Eds. Saito, S., Fujimori, N., Fukunaga, O., Kamo, M., Kobashi, K. & Yoshikawa, M. (MYU, Tokyo), p. 283.
390. *Binary Alloy Phase Diagrams*, (1990) 2nd edition Eds. Massalski, T.B., Okamoto, H., Subramanian, P.R. & Kacprzak, L., (ASM International, Materials Park, OH).
391. Roy, R., Dewan, H.S. & Ravindranathan, P. (1993) *Diamond Materials*, Eds. Dismukes, J.P. & Ravi, K.V., *The Electrochemical Society Proceedings*, **93–17**, 160.
392. Roy, R., Dewan, H.S. & Ravindranathan, P. (1993) *Mater. Res. Bull.*, **28**, 861.
393. Uemura, H., Mizoguchi, A., Hanabusa, T., Takeda, K., Inaoka, T., Makita, H., Naoi, Y., Shinatani, Y. & Tachibana, T. (1997) *Diamond Films and Technology*, **7**, 293.
394. Tachibana, T., Yokota, Y., Kobashi, K. & Shintani, Y. (1997) *J. Appl. Phys.*, **82**, 4327.
395. Kong, H.S., Jiang, B.L., Glass, J.T., Rozgoni, G.A. & More, K.L. (1988) *J. Appl. Phys.*, **63**, 2645.
396. Tarutani, M., Zhou, G., Takai, Y., Shimizu, R., Tachibana, T., Kobashi, K. & Shintani, Y. (1997) *Diamond and Related Mater.*, **6**, 272.
397. Zhou, G., Tarutani, M., Takai, Y. & Shimizu, R. (1997) *Jpn. J. Appl. Phys.*, **36**, 2298.
398. Zhou, G., Takai, Y. & Shimizu, R. (1998) *Jpn. J. Appl. Phys.*, **37**, (Part 2, No. 6B) L752.
399. Jiang, N., Kujime, S., Yamamoto, H., Inaoka, T., Naoi, Y., Shinatani, Y., Makita, H., Hatta, A. & Hiraki, A. (2000) *Jpn. J. Appl. Phys.*, **39**, L623.
400. Wang, C. & Ito, T. (2000) *Jpn. J. Appl. Phys.*, **39**, 1286.
401. Nishitani-Gamo, M., Tachibana, T., Kobashi, K., Sakaguchi, I., Loh, K.P., Yamamoto, K. & Ando, T. (1998) *Diamond and Related Mater.*, **7**, 783.
402. Nishitani-Gamo, M., Tachibana, T., Kobashi, K., Sakaguchi, I. & Ando, T. (1998) *J. Mater. Res.*, **13**, 774.
403. Hayashi, Y., Matsushita, Y., Soga, T., Umeno, M. & Jimbo, T. (2002) *Diamond and Related Mater.*, **11**, 499.
404. Tachibana, T., Yokota, Y., Kobashi, K. & Yoshimoto, M. (1999) *J. Cryst. Growth*, **205**, 163.
405. Sawabe, A., Yamada, T., Okamura, H., Katagiri, M. & Suzuki, K. (2002) *New Diamond and Frontier Carbon Technol.*, **12**, 343.
406. Iakoubovskii, K., Stesmans, S., Suzuki, K., Kuwabara, J. & Sawabe, A. (2003) *Diamond and Related Mater.*, **12**, 511.
407. Tsubota, T., Ohta, M., Kusakabe, K., Morooka, S., Watanabe, M. & Maeda, H. (2000) *Diamond and Related Mater.*, **9**, 1380.
408. Schreck, M., Bauer, Th., Gsell, S., Hörmann, F., Bielefeldt, H. & Stritzker, B. (2003) *Diamond and Related Mater.*, **12**, 262.
409. Tachiki, M., Fujisaki, T., Taniyama, N., Kudo, M. & Kawarada, H. (2002) *New Diamond and Frontier Carbon Technol.* **12**, 333.
410. Golding, B., Bednarski-Meinke, C. & Dai, Z. (2004) *Diamond and Related Mater.*, **13**, 545.
411. Dai, Z., Bednarski-Meinke, C. & Golding, B. (2004) *Diamond and Related Mater.*, **13**, 552.
412. Roy, R., Dewan, H.S. & Ravindranathan, P. (1993) *J. Mater. Chem.*, **3**, 685.
413. Roy, R., Dewan, H.S., Cherian, K.A., Cheng, J.P., Badzian, A., Drawl, W. & Langlade, C. (1995) *Mater. Lett.*, **25**, 191.

414. Cherian, K.A., Wolter, S.D. & Roy, R. (1997) *Diamond and Related Mater.*, **6**, 1747.
415. Graham, R.J., Shaapur, F., Kato, Y. & Stoner, B.R. (1994) *Appl. Phys. Lett.*, **65**, 292.
416. Borst, T.H. & Weis, O. (1996) *phys. stat. sol. (a)*, **154**, 423.
417. Malta, D.M., von Windheim, J.A., Wynands, H.A. & Fox, B.A. (1995) *J. Appl. Phys.*, **77**, 1536.
418. Dorsch, O., Holzner, K., Werner, M., Obermaier, E., Harper, R.E., Johnston, C., Chalker, P.R. & Buckley-Golder, I.M. (1993) *Diamond and Related Mater.*, **2**, 1096.
419. Adamschik, M., Müller, R., Gluche, P., Flöter, A., Limmer, W., Sauer, R. & Kohn, E. (2001) *Diamond and Related Mater.*, **10**, 1670.
420. Ager, J.W., III & Drory, M.D. (1993) *Phys. Rev. B*, **48**, 2601.
421. Wang, W.L., Liao, K.J., Feng, B., Sánchez, G., Polo, M.C. & Esteve, J. (1998) *Diamond and Related Mater.*, **7**, 528.
422. Kohn, E., Gluche, P. & Adamschik, M. (1999) *Diamond and Related Mater.*, **8**, 934.
423. Miyata, K., Saito, K., Nishimura, K. & Kobashi, K. (1994) *Rev. Sci. Instrum.*, **65**, 3799.
424. Bate, J.P., Sahaida, S.R., Stoner, B.R., von Windheim, J. A., Glass, J.T., Miyata, K., Nishimura, K. & Kobashi, K. (1993) *Diamond and Related Materials*, **2**, 816.
425. Chalker, P.R., Johnston, C., Crossley, J.A.A., Ambrose, J., Ayres, C.F., Harper, R.E., Buckley-Golder, I.M. & Kobashi, K. (1993) *Diamond and Related Materials*, **2**, 1100–1106.
426. Miyata, K., Saito, K., Nishimura, K. & Kobashi, K. (1994) *Rev. Sci. Instrum.*, **65**, 3799.
427. Adamschik, M., Hinz, M., Maier, C., Schmid, P., Seliger, H., Hofer, E.P. & Kohn, E. (2001) *Diamond and Related Mater.*, **10**, 722.
428. Kohn, E., Adamschik, M., Schmid, P., Ertl, S. & Flöter, A. (2001) *Diamond and Related Mater.*, **10**, 1684.
429. Ertl, S., Adamschik, M., Schmid, P., Gluche, P., Flöter, A. & Kohn, E. (2000) *Diamond and Related Mater.*, **9**, 970.
430. Hayashi, K., Yokota, Y., Tachibana, T., Kobashi, K., Achard, J., Gicquel, A., Olivero, C., Castex, M.-C. & Treshchalov, A. (2001) *Diamond and Related Mater.*, **10**, 1794.
431. Achard, J., Tardieu, A., Kanaev, A., Gicquel, A., Castex, M.-C., Yokota, Y., Hayashi, K., Tachibana, T. & Kobashi, K. (2002) *Diamond and Related Mater.*, **11**, 423.
432. McKeag, R.D., Marshall, R.D., Baral, B., Chan, S.S.M. & Jackman, R.B. (1997) *Diamond and Related Mater.*, **6**, 374.
433. Whitfield, M.D., Lansley, S.P., Gaudin, O., McKeag, R.D., Rizvi, N. & Jackman, R.B. (2001) *Diamond and Related Mater.*, **10**, 715.
434. Lansley, S.P., McKeag, R.D., Whitfield, M.D., Rizvi, R. & Jackman, R.B. (2003) *Diamond and Related Mater.*, **12**, 677.
435. Yokota, Y., Tachibana, T., Miyata, K., Hayashi, K., Kobashi, K., Hatta, A., Ito, T., Hiraki, A. & Shintani, Y. (1999) *Diamond and Related Mater.*, **8**, 1587.
436. Yokota, Y. et al. unpublished data.
437. Yokota, Y., Tachibana, T., Miyata, K., Kobashi, K., Hatta, A., Ito, T., Hiraki, A. & Shintani, Y. (1998) *Appl. Phys. Lett.*, **73**, 1493.
438. Kawarada, H., Wild, C., Herres, N., Koidl, P., Mizouchi, Y., Hokazono, A. & Nagasawa, H. (1998) *Appl. Phys. Lett.*, **72**, 1878.
439. Aleksov, A., Kubovic, M., Kasu, M., Schmid, P., Grobe, D., Ertl, S., Schreck, M., Stritzker, B. & Kohn, E. (2004) *Diamond and Related Mater.*, **13**, 233.
440. Miyata, K., Nishimura, K. & Kobashi, K. (1995) *IEEE Trans. Electron Devices*, **42**, 2010.
441. Yamada, T., Maeda, J. & Sawabe, A. (1999) *Jpn. J. Appl. Phys.*, **38**, Part 2, L902.

442. Yamada, T., Yokoyama, T. & Sawabe, A. (2002) *Diamond and Related Mater.*, **11**, 780.
443. Hirao, T., personal communication.
444. Zaitsev, A.M. (1998) Chap. 7, Ref. [24], p. 227.
445. Takeuchi, D., Yamanaka, S., Watanabe, H., Sawada, S., Ichinose, H., Okushi, H. & Kajimura, K. (1999) *Diamond and Related Mater.*, **8**, 1046.
446. Sawada, H., Ichinose, H., Watanabe, H., Takeuchi, D., & Okushi, H. (2001) *Diamond and Related Mater.*, **10**, 2030.
447. Zaitsev, A. (2001) *Optical Properties of Diamond* (Springer, Berlin).
448. Sawabe, A. & Inuzuka, T. (1986) *Thin Sold Films*, **137**, 89.
449. Suzuki, K., Sawabe, A. & Inuzuka, T. (1987) *Appl. Phys. Lett.*, **50**, 728.
450. Alfonso, D., Drabold, D. & Ulloa, S. (1996) *J. Phys: Condensed Matter*, **8**, 641.
451. Tsuda, M., Oikawa, S., Sekine, C. & Furukawa, S. (1991) *Diamond Materials*, Eds. Purdes, A.J., Angus, J.C., Davis, R.F., Meyerson, B.M., Spear, K.E. & Yoder, M. *The Electrochem. Soc.*, Pennington, NJ, Proc., **91–8**, 154.
452. Hata, M., Murayama, Y., Hoshino, T. & Tsuda, M. (1998) *Appl. Surf. Sci.*, **130–132**, 689.
453. Tsuda, M., Hata, M. & Shin-no, Y. (1997) *J. Cryst. Growth*, **182**, 65.
454. Tuda, M., Hata, M., Sin-no, Y. & Oikawa, S. (1996) *Surf. Sci.*, **357/358**, 844.
455. Tsuda, M., Hata, M. & Oikawa, S. (1996) *Appl. Surf. Sci.* **107**, 116.
456. Tuda, M., Oikawa, S., Fukuhara, S., Sekine, C. & Hata, M. (1992) *J. Electrochem Soc.*, **139**, 1482.
457. Yang, Y.L. & D'Evelyn, M.P. (1992) *J. Am. Chem. Soc.*, **114**, 2796.
458. Yang, S.H., Drabold, D.A. & Adams, J.B. (1993-II) *Phys. Rev. B*, **48**, 5261.
459. Kawarada, H., Aoki, M., Sasaki, H. & Tsugawa, K. (1994) *Diamond and Related Mater.*, **3**, 961.
460. Tsugawa, T. & Kawarada, H. (1994) *Advanced Materials '93 I/B: Magnetic, Dielectric, Ferroeelectric, Diamond and Related Materials*, Eds. Homma, M. et al., *Trans. Mater. Res. Soc. Jpn*, **14B**, 1489.
461. Pate, B.B. (1986) *Surface Sci.*, **165**, 83.
462. Haneman, D. (1982) *Adv. Phys.*, **31**, 165.
463. Pandy, K.C. (1981) *Phys. Rev. Lett.*, **47**, 1913.
464. Chadi, D.J. (1982) *Phys. Rev. B*, **26**, 4762.
465. Hamza, A.V., Kubiak, G.D. & Stulen, R.H. (1988) *Surf. Sci.*, **206**, L833.
466. Yaxin Wang, & John Angus, C. (1995) Highly Oriented Diamond Growth Silicon Substrates, *Carbon*, **33**(3), 328–329.
467. Geier, S., Hessmer, R., Preckwinkel, U., Schweizer, D., Schreck, M. & Rauschenbach, B. (1996) *J. Appl. Phys.*, **79**, 1907.

Index